Haematococcus

Rathinam Raja • Shanmugam Hemaiswarya •
Mathiyazhagan Narayanan •
Sabariswaran Kandasamy • K. R. Jayappriyan
Editors

Haematococcus

Biochemistry, Biotechnology and Biomedical Applications

Springer

Editors
Rathinam Raja
Research and Development Wing
Sree Balaji Medical College and Hospital (SBMCH), BIHER
Chennai, India

Shanmugam Hemaiswarya
Department of Biotechnology
Indian Institute of Technology Madras
Chennai, India

Mathiyazhagan Narayanan
Department of Biotechnology, Saveetha School of Engineering
SIMATS
Chennai, India

Sabariswaran Kandasamy
Department of Biotechnology
PSGR Krishnammal College for Women
Coimbatore, India

K. R. Jayappriyan
Centre for Advanced Studies in Botany
University of Madras
Chennai, India

ISBN 978-981-99-2900-9 ISBN 978-981-99-2901-6 (eBook)
https://doi.org/10.1007/978-981-99-2901-6

This Springer imprint is published by the registered company Springer Nature Singapore Pte Ltd.
The registered company address is: 152 Beach Road, #21-01/04 Gateway East, Singapore 189721, Singapore

Preface

Algae are the oldest living photosynthetic forms in the world. In the marine ecosystem, algae are the primary producers that supply energy required to a diverse group of marine organisms. Particularly macroalgae provide a habitat for invertebrates, fishes, etc. Even though there are numerous scattered documents available, there are some recent fields that are yet to be explored. This book, *Haematococcus: Biochemistry, Biotechnology and Biomedical Applications*, offers a comprehensive analysis of microalgal cultivation methods, the optimization of astaxanthin production and its diverse applications such as clinical uses, algae polymers, proteins and pigments, food applications and packaging, algae forming, cosmetics, and some more. It also sheds light on emerging topics and the industrial exploitation of microalgae and their products, supported by current studies, making it a valuable resource for academia. The book serves as an important resource for scholars and professionals in various industries, offering adequate knowledge on a wide range of topics.

Chennai, India — Rathinam Raja
Chennai, India — Shanmugam Hemaiswarya
Chennai, India — Mathiyazhagan Narayanan
Coimbatore, India — Sabariswaran Kandasamy
Chennai, India — K. R. Jayappriyan

Contents

Contributors

Refad Ahmed Biological Sciences Division, CSIR-North East Institute of Science and Technology (CSIR-NEIST), Naharlagun, Arunachal Pradesh, India
Academy of Scientific and Innovative Research (AcSIR), Ghaziabad, Uttar Pradesh, India

Chalani Akmeemana Department of Food Science and Technology, Faculty of Applied Sciences, University of Sri Jayewardenepura, Nugegoda, Sri Lanka

Arun Alagarsamy Department of Microbiology, Alagappa University, Karaikudi, Tamil Nadu, India

Mohamed Alouani Department of Biology, Faculty of Sciences, Laboratory of Biotechnology and Valorization of Natural Resources, Ibn Zohr University, Agadir, Morocco
Faculty of Applied Sciences, Ibn Zohr University, AitMelloul, Morocco

Nagarajan Arumugam Department of Biotechnology, Applied and Industrial Microbiology Laboratory, Indian Institute of Technology Madras, Chennai, Tamil Nadu, India

K. Arunkumar Phycoscience Laboratory, Department of Plant Science, School of Biological Sciences, Central University of Kerala, Periye, Kerala, India

Rachid Bouharroud National Institute for Agronomic Research, Agadir, Morocco

Vikas Singh Chauhan Plant Cell Biotechnology Department, CSIR- Central Food Technological Research Institute, Mysuru, Karnataka, India
Academy of Scientific and Innovative Research (AcSIR), Ghaziabad, Uttar Pradesh, India

Channakeshavaiah Chikkaputtaiah Academy of Scientific and Innovative Research (AcSIR), Ghaziabad, Uttar Pradesh, India
Biological Sciences and Technology Division, CSIR-North East Institute of Science and Technology (CSIR-NEIST), Jorhat, Assam, India

G. Dharani Marine Biotechnology Division, National Institute of Ocean Technology, Chennai, Tamil Nadu, India

Alves Dutra Bioprocess Intensification Group, Federal University of Santa Maria, UFSM, Santa Maria, RS, Brazil

Raja ElKheloui Department of Biology, Faculty of Sciences, Laboratory of Microbial Biotechnology and Plants Protection, Ibn Zohr University, Agadir, Morocco

Ahmed A. El-Refaey Botany and Microbiology Department, Faculty of Science, Al-Azhar University, Cairo, Egypt

Bahar Aslanbay Guler Department of Bioengineering, Faculty of Engineering, Ege University, Izmir, Turkey

Fatima Hamadi Department of Biology, Faculty of Sciences, Laboratory of Microbial Biotechnology and Plants Protection, Ibn Zohr University, Agadir, Morocco

Mohammed Hassi Faculty of Sciences, Laboratory of Microbial Biotechnology and Plants Protection, Department of Biology, Ibn Zohr University, Agadir, Morocco
Taroudant's Polydisciplinary Faculty, Ibn Zohr University, Taroudant, Morocco

Esra Imamoglu Department of Bioengineering, Faculty of Engineering, Ege University, Izmir, Turkey

Eduardo Jacob-Lopes Bioprocess Intensification Group, Federal University of Santa Maria, UFSM, Santa Maria, RS, Brazil

A. Josephine Central Research Laboratory, Meenakshi Academy of Higher Education and Research (MAHER – Deemed to be University), Chennai, Tamil Nadu, India

M. K. Kalaivani Research and Development Wing, Sree Balaji Medical College and Hospital (SBMCH), BIHER, Chennai, India

Prasad Kaparaju School of Engineering and Built Environment, Griffith University (Nathan Campus), Nathan, QLD, Australia

Sankha Karunarathna Department of Food Science and Technology, Faculty of Applied Sciences, University of Sri Jayewardenepura, Nugegoda, Sri Lanka

T. S. Kumar Marine Biotechnology, National Institute of Ocean Technology, Ministry of Earth Sciences, Government of India, Chennai, Tamil Nadu, India

Chegu Krishnamurthi Madhubalaji Plant Cell Biotechnology Department, CSIR- Central Food Technological Research Institute, Mysuru, Karnataka, India
Academy of Scientific and Innovative Research (AcSIR), Ghaziabad, Uttar Pradesh, India

Yogeswar Mohanasundaram Center for Biotechnology, Alagappa College of Technology, Anna University, Chennai, Tamil Nadu, India

Tirthankar Mukherjee School of Engineering and Built Environment, Griffith University (Nathan Campus), Nathan, QLD, Australia

Sanjivkumar Muthusamy Department of Microbiology, K.R. College of Arts & Science, Kovilpatti, Tamil Nadu, India

Thangavel Muthusamy Research and Development Wing, Sree Balaji Medical College and Hospital (SBMCH), Bharath Institute of Higher Education and Research (BIHER), Chennai, Tamil Nadu, India

Suphi Ş. Öncel Department of Bioengineering, Faculty of Engineering, University of Ege, Izmir, Turkey

Oumaima Ouaddi Faculty of Sciences, Laboratory of Microbial Biotechnology and Plants Protection, Ibn Zohr University, Agadir, Morocco

Dhamodharan Prabhu Centre for Drug Discovery, Department of Biotechnology, Karpagam Academy of Higher Education, Coimbatore, India

S. Pujithaa Research and Development Wing, Sree Balaji Medical College and Hospital (SBMCH), BIHER, Chennai, India

Redouan Qessaoui National Institute for Agronomic Research, Agadir, Morocco

Rathinam Raja Research and Development Wing, Sree Balaji Medical College and Hospital (SBMCH), BIHER, Chennai, India

Sundarraj Rajamanikandan Centre for Drug Discovery, Department of Biochemistry, Karpagam Academy of Higher Education, Coimbatore, India

M. Rajamiriyam Research and Development Wing, Sree Balaji Medical College and Hospital (SBMCH), BIHER, Chennai, India

Sarada Ravi Plant Cell Biotechnology Department, CSIR- Central Food Technological Research Institute, Mysuru, Karnataka, India
Academy of Scientific and Innovative Research (AcSIR), Ghaziabad, Uttar Pradesh, India

Diganta Kumar Saikia Biological Sciences Division, CSIR-North East Institute of Science and Technology (CSIR-NEIST), Naharlagun, Arunachal Pradesh, India

Salem S. Salem Botany and Microbiology Department, Faculty of Science, Al-Azhar University, Cairo, Egypt

Babu Sarangam Mahathi Biotech, Ramapuram, Chennai, India

Nilay Kumar Sarker School of Engineering and Built Environment, Griffith University (Nathan Campus), Nathan, QLD, Australia

Silambarasan Tamil Selvan Department of Microbiology, School of Allied Health Sciences, Vinayaka Missions Research Foundation (DU), Salem, Tamil Nadu, India

M. P. Sudhakar Marine Biotechnology Division, National Institute of Ocean Technology, Chennai, Tamil Nadu, India

Sathyaramya Thangavel Department of Biological Chemistry, University of Louisville, Louisville, KY, USA

Gamze Turan Aquaculture Department, Fisheries Faculty, Ege University, Izmir, Turkey

Natarajan Velmurugan Biological Sciences Division, CSIR-North East Institute of Science and Technology (CSIR-NEIST), Naharlagun, Arunachal Pradesh, India
Academy of Scientific and Innovative Research (AcSIR), Ghaziabad, Uttar Pradesh, India

D. Vidya Phycoscience Laboratory, Department of Plant Science, School of Biological Sciences, Central University of Kerala, Periye, Kerala, India

Indira Wickramasinghe Department of Food Science and Technology, Faculty of Applied Sciences, University of Sri Jayewardenepura, Nugegoda, Sri Lanka

Çağla Yarkent Department of Bioengineering, Faculty of Engineering, University of Ege, Izmir, Turkey

Leila Queiroz Zepka Bioprocess Intensification Group, Federal University of Santa Maria, UFSM, Santa Maria, RS, Brazil

Chapter 1
An Introduction to *Haematococcus*

Yogeswar Mohanasundaram, Nagarajan Arumugam, Babu Sarangam, Arun Alagarsamy, and Rathinam Raja

The production of microalgae on an industrial scale began in the 1950s, as an alternative protein source for animals and plants proposed by Burlew (Mularczyk et al. 2020). The cultivation of microalgae nowadays is beyond animal feed as it gained more popularity in the food industry due to its high nutrient content and healthy diet option (Ruiz-Domínguez et al. 2019). Microalgae are cultivated for various purposes such as the production of active biomolecules in pharmaceutical and cosmetic industries, and the production of biodegradable plastics and biofuels (Ariede et al. 2017; Koller et al. 2014). Many bioactive compounds, polysaccharides such as starch, cellulose, and β-1,3 glucan, and lipids like polyunsaturated fatty acids, and hydrocarbons have found application in pharmaceutical and cosmetic industries (Mularczyk et al. 2020). Pharmaceutical industries also utilize enzymes, amino acids, polypeptides, and pigments like carotenoids or chlorophylls produced from microalgae, these were also employed in food industries (Harasym and Bogacz-Radomska 2016).

There are more than 7000 species that come under the green algae classification. The genus *Haematococcus* under the family Haematococcaceae belongs to the class

Y. Mohanasundaram
Center for Biotechnology, Alagappa College of Technology, Anna University, Chennai, India

N. Arumugam (✉)
Applied and Industrial Microbiology Laboratory, Department of Biotechnology, Indian Institute of Technology Madras, Chennai, India

B. Sarangam
Mahathi Biotech, Ramapuram, Chennai, India

A. Alagarsamy
Department of Microbiology, Alagappa University, Karaikudi, India

R. Raja
Research and Development Wing, Sree Balaji Medical College and Hospital (SBMCH), BIHER, Chennai, India

R. Raja et al. (eds.), *Haematococcus*, https://doi.org/10.1007/978-981-99-2901-6_1

Chlorophyceae and the order Volvocales. J. Von Flotow in 1844 described the Genus *Haematococcus* with the unicellular freshwater biflagellate green microalgae *Haematococcus pluvialis*. *Haematococcus* has been found in North America, Europe, and Africa, and in different climatic zones such as temperate, continental, alpine, and tropical (Turan 2019). *Haematococcus* genus comprises as many as six species, namely *H. lacustris* (*H. pluvialis*), *H. capensis*, *H. droebakensis*, *H. rubens*, *H. rubicundus*, and *H. zimbabwiensis*, however, the most known and researched species is the *H. pluvialis* also known as *Haematococcus lacustris* or *Sphaerella lacustris*. The beneficial effects of microalgae components and their biological properties such are antioxidant and anti-inflammatory, etc. are backed up by researchers (Nawrocka et al. 2017). The biology and biochemical composition of the species *Haematococcus pluvialis* will be discussed in this chapter.

1.1 Biology of *Haematococcus lacustris*

1.1.1 Cell Cycle and Morphology

H. lacustris shows a similar cellular structure to that of the other members of volvocalean unicellular green algae. The cellular morphology of *H. lacustris* was found to vary depending on its life cycle stages. There are four distinguishable cell types are observed in its life cycle: macrozooids (zoospores), microzooids, palmella, and hematocysts (aplanospores) (Elliot 1934). The aplanospores are usually called the "red nonmotile astaxanthin accumulated encysted phase," whereas the rest is termed as the "green vegetative phase" of the *H. lacustris* life cycle. Zoospores are ellipsoidal, spherical, or pear-shaped cells, biflagellate emerging from the anterior end with equal lengths. The zoospores are 8–20 μm long with variable thickness of distinct gelatinous extracellular matrix and have a cup-shaped chloroplast with numerous scattered pyrenoids. The protoplast surface of the cell houses irregularly distributed contractile vacuoles. These type of cells grows predominately in favorable culture conditions and can divide into 2–32 daughter cells by mitosis through sporangium formation (Hagen et al. 2002; Wayama et al. 2013).

The zoospores lose their flagella and increase in cell size during unfavorable environment or culture conditions and become amorphous, multi-layered inside the extracellular matrix, nonmotile resting vegetative cells called "palmella" within 1–2 days (Fig. 1.1). The cessation of cell division caused by continued stress conditions (nutrient deprivation, high salinity, high light irradiance) leads to transformation into the asexual aplanospores (Hagen et al. 2002; Kim et al. 2015). A thick and rigid trilaminar sheath and secondary cell wall of acetolysis-resistant material (algaenan, an aliphatic biopolymer with great resistance to chemicals) protect the cell against acetolysis, high salinity, and high light irradiance (María Cecilia Damiani et al. 2006). Secondary carotenoids, especially astaxanthin are stored in the cytoplasm inside lipid droplets. This large accumulation of astaxanthin in mature aplanospores causes the characteristic bright red color (Hagen et al. 2002) and is induced by high

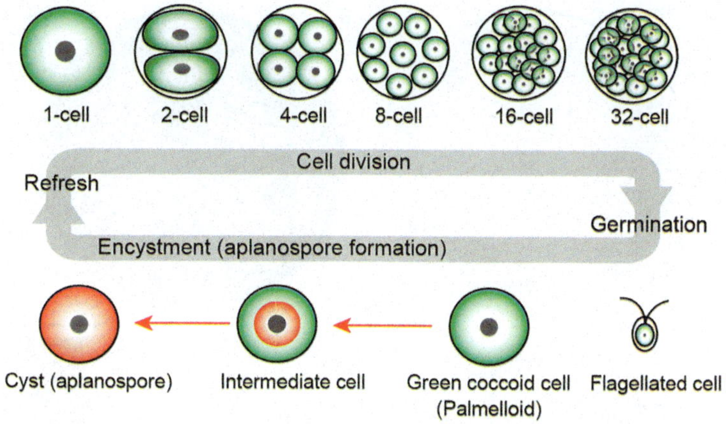

Fig. 1.1 Life cycle of *H. lacustris*. Figure reproduced from (Wayama et al. 2013) distributed under the terms of the Creative Common Attribution License

light exposure or nutrient deprivation (Butler et al. 2018). The red aplanospores initiate a new vegetative growth cycle by germinating to form zoospores when subject to optimal growth conditions. Aplanospores in rare cases undergo sexual reproduction when returned to optimal condition after exposure to extremely adverse conditions (desiccation, freezing, or starvation). Gametogenesis occurs in aplanospores and produces up to 64 gametes called microzooids, though conjugates of the gametes are rare and largely replaced by vegetative reproduction. After being released from gametocysts, the microzooids show high-speed motility and are usually smaller in size (<10 μm) compared to the zoospores (20–50 μm) (Triki et al. 1997).

1.1.2 *Ultrastructural Changes of* H. lacustris *During Carotenogenesis*

Lang, and Santos and Mesquita, both reported that the ultrastructural changes such as a large number of cytosolic lipid body formations were involved in the astaxanthin accumulation in *H. lacustris* which appeared as a small red spot in the center of the cell when observed under a light microscope. These small lipid bodies contain carotenoids in the perinuclear or hyaloplasm region and were present adjacent to the rough endoplasmic reticulum (ER) cisternae and appeared as electron-dense droplets under electron microscopic (EM) observations (Lang 1968; Santos and Mesquita 1984). The lipid bodies prevent the chloroplast from high exposure to light by covering the entire chloroplast beneath the plasma membranes as they are coalesced and gradually moved away from the center upon an increment of pigment quantities. The secondary cell wall development is usually

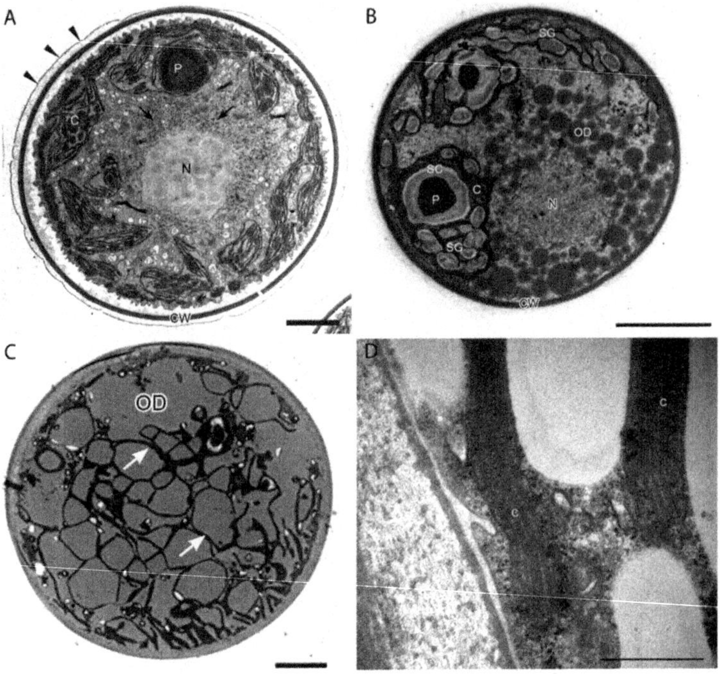

Fig. 1.2 Transmission electron micrographs of *H. lacustris* during different stages of life cycle. (**a**) General ultrastructure of green coccoid cells. The cell wall is surrounded by extracellular matrix (arrowheads). Arrows indicate astaxanthin granules. (**b**) General ultrastructure of intermediate cells. (**c**) General ultrastructure of cyst cells, showing small granules that contain astaxanthin. (**d**) High-magnification view of chloroplasts. C, chloroplast; CW, cell wall; N, nucleus; P, pyrenoid; OD, oil droplets; SC, starch capsule; SG, starch grain. Scale bars in a–c and d: 5 μm and 0.5 μm, respectively. Figure reproduced from (Wayama et al. 2013) distributed under the terms of the Creative Common Attribution License

synonymous with lipid bodies formation and the aplanospores cell wall consists of 70% carbohydrate, 6% protein, 3% cellulose, and 3% acetolysis-resistant materials (Hagen et al. 2002).

The ultrastructural changes happening in *H. lacustris* at different stages of the cell cycle were elucidated using transmission electron microscopy (TEM) (Fig. 1.2) (Wayama et al. 2013). During the green vegetative stage, two layers of the extracellular matrix are found near the thick cell wall, and the chloroplasts and pyrenoids with the electron-dense matrix are located in the periphery and stroma, respectively. Astaxanthin granules are found around the centrally located nucleus. The intermediate stage or encystment process in *Haematococcus* is observed in middle-aged culture (1–2 weeks) indicated by the greenish-orange cells. The pyrenoids in the stroma were encompassed by thick starch capsules and the nucleus is surrounded by lipid bodies of various sizes containing astaxanthin. At this stage, the chloroplast volume reduces due to the accumulation of astaxanthin but still maintains the photosynthetic activity and the partial degradation of thylakoids also occurs. Accelerated accumulation of astaxanthin by the strong and/or continuous light exposure leads to cyst cell formation.

The lipid bodies and astaxanthin accumulation pattern varies among the cyst cells with high astaxanthin granules in some cells while others have large lipid bodies with less electron-dense astaxanthin (fusion of lipid bodies occurs occasionally). Degenerated chloroplasts were localized in the interspace between lipid bodies having a net-like appearance. The association of cellulose synthesis with primary cell wall formation was revealed by the characterization of protein composition in the cell wall of *H. lacustris* at different developmental stages (Wang et al. 2004). Although the cell wall modification and cellulose synthesis are unknown in *H. lacustris*, a framework for future characterization of biogenesis and function of cell walls in this organism is provided by the reference proteome.

1.2 Biochemical Composition of *Haematococcus lacustris*

The cellular composition of *H. lacustris* varies tremendously between its "green" and "red" stages of cultivation due to its unique life cycle. The composition during both stages is shown in Table 1.1.

Table 1.1 Typical composition of *H. lacustris* in both green and red stages

Composition (% of D.C.W.)	Green stage	Red stage
Proteins	29–45	17–25
Carbohydrates	15–17	36–40
Lipids (Total %)	20–25	32–37
Neutral lipids	59	51.9–53.5
Phospholipids	23.7	20.6–21.1
Glycolipids	11.5	25.7–26.5
Carotenoids (Total %)	0.5	2.0–5.0
Lutein	56.3	0.5
β-carotene	16.7	1
Violaxanthin	12.5	n.d
Neoxanthin	8.3	n.d
Zeaxathin	6.3	n.d
Astaxanthin (all forms)	n.d	81.2
Canthaxanthin	n.d	5.1
Adonirubin	n.d	0.6
Adonixanthin	n.d	0.4
Echinenone	n.d	0.2
Chlorophylls	1.5–2	0

1.2.1 Protein

During the green vegetative stage under favorable growth conditions, most *H. lacustris* strains are rich in protein (29–45%) and around 17–25% in the red stage. Gacheva et al. reported the lowest protein content of 23.6% during green stage cultivation in a Bulgarian strain *Haematococcus* cf. pluvialis Rozhen-12 (Gacheva et al. 2015). The protein content reduces to 36% in palmella cells, while a much lower 21 (Kim et al. 2015) and 23% (Lorenz 1999) during the red stage cultivation. Proteins in the red stage were mainly composed of glutamic acid, aspartic acids, leucine, and alanine with 10.02/100 mg of total amino acid content and about 46% belongs to essential amino acids (Kim et al. 2015; Lorenz 1999).

1.2.2 Carbohydrates

The cells survive the prolonged stress conditions by consuming starch as carbohydrates. The carbohydrate content is around 15–17% in the green stage and increases rapidly to double-triple the content in the red stage under stress (e.g., light stress, nutrient starvation, temperature variations, high acidity, etc.), for example, 38 (Lorenz 1999), 60 (Recht et al. 2012), and 74% (Recht et al. 2012) starch content. The carbohydrate content during the first day of stress increased rapidly up to 63% and reduced to 41 % on the next day and remained the same till cultivation (Recht et al. 2012).

1.2.3 Carotenoids

The carotenoid content in *H. pluvialis* increases from 0.5% d.c.w. in the green stage to 2–5% d.c.w. in the red stage. The highest fraction of carotenoids is lutein (75–80%) followed by β-carotene (16.5%) and others such as primary carotenoids: violaxanthin (12.5%), neoxanthin (8.3%), lactucaxanthin, and zeaxanthin in trace in the green stage along with 1.5–2% d.c.w of chlorophyll (a and b) (Harker et al. 1996; Shah et al. 2016). In the red stage, the characteristic primary carotenoids are replaced by secondary carotenoids, especially astaxanthin (80–99% of the total carotenoids) the most important carotenoid obtained *H. lacustris* which is accumulated only in the red stage (Dragoş et al. 2010; Harker et al. 1996). The carotenoids to chlorophylls ratio increase drastically from 0.2 in the green stage by an order of magnitude to about 2–9 in the red stage. Fatty acid esters of astaxanthin, usually mono- or diesters of palmitic (16:0), oleic (18:1), or linoleic (18:2) acids are the form of the majority of astaxanthin that exists in the cell and not in its free form. The non-polar matrix of lipid bodies makes these modifications necessary for the highly polar nature of astaxanthin. The mature red cells contain about 70% monoesters, 25% diesters, and only 5% of free ketocartenoids (Solovchenko 2015). Astaxanthin accumulation of up to 3–5% d.c.w. was reported under certain conditions of stress (Chekanov et al. 2014; Han et al. 2013).

1.2.4 Lipids

Astaxanthin start to accumulate as fatty acid mono- or diesters in cytoplasmic lipid droplets (LD) after exposure to stress condition during the transition from green stage to red stage (Aflalo et al. 2007), and both chlorophyll and protein content drops. The total lipid content during the green stage varies from 20% to 25% with predominantly 10% of lipids composed of short polyunsaturated fatty acids (C16, C18) deposited in the chloroplast. Both green and red cells have neutral lipids as major lipid class (Table 1.1). The synthesis of neutral lipids-triacylglycerols (TAG) is increased greatly due to the prolonged stress conditions during red stage. About 40% of the cell weight of red cells consist of cytoplasmic lipid droplets, with upto 4% of ketocarotenoid astaxanthin (Sammy Boussiba et al. 1999; Saha et al. 2013). The glycolipid content nearly doubles, while the phospholipid fraction remains the same in red stage when compared to the green stage (Cecilia Damiani et al. 2010). The total fatty acid profile of *H. pluvialis* is relatively complex. The fatty acid composition varies between the strains (Table 1.2) with the predominant components such as the palmitic (16:0), oleic (18:1), linoleic (18:2), and linolenic (18:3) acids (Kim et al. 2015; Lei et al. 2012; Lorenz 1999; Scodelaro Bilbao et al. 2016). Several factors such as culture environment, culture parameters, stress conditions, strain origin, nutrient, etc. are associated with the variation in fatty acid composition. The possibility of biodiesel production from *H. lacustris* well indicated by its high lipid content and fatty acids profile under nutrient starvation (Cecilia Damiani et al. 2010; Saha et al. 2013). The enhanced biosynthesis of triacylglycerols (TAG) increasing the accumulation of astaxanthin greatly along with the cellular response to stress conditions (Zhekisheva et al. 2005), and the loss of cytochrome f, reduction in photosynthetic activity of PSII, and subsequent reduction in electron transport, and increased respiration rate (Boussiba 2000).

Table 1.2 Total fatty acids composition of different *H. lacustris* strains in green vegetative stage

Fatty acids	*H. pluvialis*	*H. pluvialis*	*Haematococcus* sp. KORDI03	*H. pluvialis* CCALA 1081
C6:0 caproic	0.1	n.d	n.d	1.99
C8:0 caprylic	0.1	n.d	n.d	0.87
C12:0 lauric	0.1	0.48	n.d	n.d
C14:0 myristic	0.5	1.11	0.1	0.61
C15: 0 pentadecanoic	n.d	0.43	0.1	n.d
C15:1 pentadecenoic	n.d	1.03	n.d	n.d
C16:0 palmitic	29	21.76	13.7	22.94
C16:1 palmitoleic	0.6	1.20	0.5	0.35
C16:2	n.d	n.d	0.4	n.d
C16:3	n.d	n.d	3.5	n.d
C16:4	n.d	n.d	3.3	n.d
C17:0 margaric	0.2	0.39	n.d	n.d
C17:1 margaroleic	1.3	n.d	n.d	n.d
C18:0 stearic	2.1	8.20	0.7	1.15
C18:1 elaidic	n.d	n.d	n.d	5.46
C18:1 oleic	25.9	19.14	4.9	16.28
C18:2 linolelaidic	n.d	n.d	n.d	6.64
C18:2 linoleic	20.8	22.26	24.9	23.9
C18:3 linolenic (γ)	n.d	3.13	n.d	2.19
C18:3 linolenic (α)	12.8	4.86	39.7	12.5
C18:4 octadecatetraenoic	1.4	n.d	5.8	n.d
C20:0 arachidic	0.6	0.60	n.d	n.d
C20:1 gadoleic	0.3	2.23	0.5	n.d
C20:2 eicosadenoic	1.2	1.49	n.d	2.21
C20:3 eicosatrienoic gamma	0.5	0.31	n.d	n.d
C20:4 arachidonic	1.4	3.03	0.9	1.92
C20:5 eicosapentaenoic	0.6	1.69	0.6	0.66
C22:0 behenic	0.4	0.27	n.d	n.d
C22:1 erucic	n.d	5.46	n.d	n.d
C24:0 lignoceric	0.2	0.68	0.3	0.33
C24:1 nervonic acid	0.1	0.24	0.1	n.d
∑SFAs	33.2	37.77	15	25.3
∑MUFAs	28.1	24.74	6	16.6
∑PUFAs	38.7	37.49	79.1	41.2
Total fatty acid	100.00	100.00	100.00	100.00

References

Aflalo C, Meshulam Y, Zarka A, Boussiba S (2007) On the relative efficiency of two- vs. one-stage production of astaxanthin by the green alga *Haematococcus pluvialis*. Biotechnol Bioeng 98(1): 300–305. https://doi.org/10.1002/bit.21391

Ariede MB, Candido TM, Jacome ALM, Velasco MVR, de Carvalho JCM, Baby AR (2017) Cosmetic attributes of algae - a review. Algal Res 25:483–487. https://doi.org/10.1016/j.algal.2017.05.019

Boussiba S (2000) Carotenogenesis in the green alga *Haematococcus pluvialis*: cellular physiology and stress response. In: Physiologia Plantarum, vol 108, no 2. Wiley, Hoboken, pp 111–117. https://doi.org/10.1034/j.1399-3054.2000.108002111.x

Boussiba S, Bing W, Yuan JP, Zarka A, Chen F (1999) Changes in pigments profile in the green alga *Haematococcus pluvialis* exposed to environmental stresses. Biotechnol Lett 21(7): 601–604. https://doi.org/10.1023/A:1005507514694

Butler TO, McDougall GJ, Campbell R, Stanley MS, Day JG (2018) Media screening for obtaining *Haematococcus pluvialis* red motile macrozooids rich in astaxanthin and fatty acids. Biology 7(1):2. https://doi.org/10.3390/biology7010002

Chekanov K, Lobakova E, Selyakh I, Semenova L, Sidorov R, Solovchenko A (2014) Accumulation of astaxanthin by a new *Haematococcus pluvialis* strain BM1 from the white sea coastal rocks (Russia). Mar Drugs 12(8):4504–4520. https://doi.org/10.3390/md12084504

Damiani MC, Leonardi PI, Pieroni OI, Cáceres EJ (2006) Ultrastructure of the cyst wall of *Haematococcus pluvialis* (Chlorophyceae): Wall development and behaviour during cyst germination. Phycologia 45(6):616–623. https://doi.org/10.2216/05-27.1

Damiani MC, Popovich CA, Constenla D, Leonardi PI (2010) Lipid analysis in *Haematococcus pluvialis* to assess its potential use as a biodiesel feedstock. Bioresour Technol 101(11): 3801–3807. https://doi.org/10.1016/j.biortech.2009.12.136

Dragoş N, Bercea V, Bica A, Drugă B, Nicoară A, Coman C (2010) Astaxanthin production from a new strain of *Haematococcus pluvialis* grown in batch culture. Ann Romanian Soc Cell Biol 15(2):353–361

Elliot AM (1934) Morphology and life history of *Haematococcus pluvialis*. Archiv Für Protistenkunde 82(Usnm 186797):250–274. https://eurekamag.com/research/025/037/025037549.php

Gacheva G, Dimitrova PH, Pilarski P (2015) New strain *Haematococcus* Cf. *pluvialis* Rozhen-12 – growth, biochemical characteristics and future perspectives. Genet Plant Physiol 5(1):29–38. https://www.researchgate.net/publication/275849029%

Hagen C, Siegmund S, Braune W (2002) Ultrastructural and chemical changes in the cell wall of *Haematococcus pluvialis* (Volvocales, Chlorophyta) during aplanospore formation. Eur J Phycol 37(2):217–226. https://doi.org/10.1017/S0967026202003669

Han D, Li Y, Hu Q (2013) Biology and commercial aspects of *Haematococcus pluvialis*. In: Handbook of microalgal culture: applied phycology and biotechnology, 2nd edn. Wiley, Hoboken, pp 388–405. https://doi.org/10.1002/9781118567166.ch20

Harasym J, Bogacz-Radomska L (2016) Colorants in foods - from past to present. Eng Sci Technol 3(22):21–35. https://doi.org/10.15611/NIT.2016.3.02

Harker M, Tsavalos AJ, Young AJ (1996) Factors responsible for astaxanthin formation in the Chlorophyte *Haematococcus pluvialis*. Bioresour Technol 55(3):207–214. https://doi.org/10.1016/0960-8524(95)00002-X

Kim JH, Affan MA, Jang J, Kang MH, Ko AR, Jeon SM, Oh C, Heo SJ, Lee YH, Ju SJ, Kang DH (2015) Morphological, molecular, and biochemical characterization of astaxanthin-producing green microalga *Haematococcus* sp. KORDI03 (haematococcaceae, chlorophyta) isolated from Korea. J Microbiol Biotechnol 25(2):238–246. https://doi.org/10.4014/jmb.1410.10032

Koller M, Muhr A, Braunegg G (2014) Microalgae as versatile cellular factories for valued products. Algal Res 6(PA):52–63. https://doi.org/10.1016/j.algal.2014.09.002

Lang NJ (1968) Electron microscopic studies of extraplastidic astaxanthin in *Haematococcus*. J Phycol 4(1):12–19. https://doi.org/10.1111/j.1529-8817.1968.tb04670.x

Lei A, Chen H, Shen G, Hu Z, Chen L, Wang J (2012) Expression of fatty acid synthesis genes and fatty acid accumulation in *Haematococcus pluvialis* under different stressors. Biotechnol Biofuels 5(1):1–11. https://doi.org/10.1186/1754-6834-5-18

Lorenz R (1999) A technical review of *Haematococcus* algae. Naturose Technical Bulletin, pp 1–15. http://www.cyanotech.com/pdfs/bioastin/axbul60.pdf

Mularczyk M, Michalak I, Marycz K (2020) Astaxanthin and other nutrients from *Haematococcus pluvialis*-multifunctional applications. Mar Drugs 18(9):459. https://doi.org/10.3390/md18090459

Nawrocka D, Kornicka K, Śmieszek A, Marycz K (2017) *Spirulina platensis* improves mitochondrial function impaired by elevated oxidative stress in adipose-derived mesenchymal stromal cells (ASCs) and intestinal epithelial cells (IECs), and enhances insulin sensitivity in equine metabolic syndrome (EMS) horses. Mar Drugs 15(8). https://doi.org/10.3390/MD15080237

Recht L, Zarka A, Boussiba S (2012) Patterns of carbohydrate and fatty acid changes under nitrogen starvation in the microalgae *Haematococcus pluvialis* and *Nannochloropsis* sp. Appl Microbiol Biotechnol 94(6):1495–1503. https://doi.org/10.1007/S00253-012-3940-4/FIGURES/6

Ruiz-Domínguez MC, Espinosa C, Paredes A, Palma J, Jaime C, Vílchez C, Cerezal P (2019) Determining the potential of *Haematococcus pluvialis* oleoresin as a rich source of antioxidants. Molecules 24(22):4073. https://doi.org/10.3390/MOLECULES24224073

Saha SK, McHugh E, Hayes J, Moane S, Walsh D, Murray P (2013) Effect of various stress-regulatory factors on biomass and lipid production in microalga *Haematococcus pluvialis*. Bioresour Technol 128:118–124. https://doi.org/10.1016/j.biortech.2012.10.049

Santos MF, Mesquita JF (1984) Ultrastructural study of *Haematococcus lacustris* (Girod.) Rostafinski (Volvocales) I. Some aspects of carotenogenesis. Cytologia 49(1):215–228. https://doi.org/10.1508/cytologia.49.215

Scodelaro Bilbao PG, Damiani C, Salvador GA, Leonardi P (2016) *Haematococcus pluvialis* as a source of fatty acids and phytosterols: potential nutritional and biological implications. J Appl Phycol 28(6):3283–3294. https://doi.org/10.1007/s10811-016-0899-z

Shah MMR, Liang Y, Cheng JJ, Daroch M (2016) Astaxanthin-producing green microalga *Haematococcus pluvialis*: from single cell to high value commercial products. Front Plant Sci 7:531. https://doi.org/10.3389/fpls.2016.00531

Solovchenko AE (2015) Recent breakthroughs in the biology of astaxanthin accumulation by microalgal cell. Photosynth Res 125(3):437–449. https://doi.org/10.1007/s11120-015-0156-3

Triki A, Maillard P, Gudin C (1997) Gametogenesis in *Haematococcus pluvialis* flotow (Volvocales, Chlorophyta). Phycologia 36(3):190–194. https://doi.org/10.2216/i0031-8884-36-3-190.1

Turan G (2019) Isolation and cultivation of a newly-discovered astaxanthin-rich green microalga-*Haematococcus* sp. Flotow strain from homeros valley (bornova creek, izmir, turkey). Turkish J Fish Aquat Sci 19(9):785–791. https://doi.org/10.4194/1303-2712-v19_9_07

Wang SB, Hu Q, Sommerfeld M, Chen F (2004) Cell wall proteomics of the green alga *Haematococcus pluvialis* (Chloroplyceae). Proteomics 4(3):692–708. https://doi.org/10.1002/pmic.200300634

Wayama M, Ota S, Matsuura H, Nango N, Hirata A, Kawano S (2013) Three-dimensional ultrastructural study of oil and astaxanthin accumulation during encystment in the green alga *Haematococcus pluvialis*. PloS One 8(1):e53618. https://doi.org/10.1371/journal.pone.0053618

Zhekisheva M, Zarka A, Khozin-Goldberg I, Cohen Z, Boussiba S (2005) Inhibition of astaxanthin synthesis under high irradiance does not abolish triacylglycerol accumulation in the green alga *Haematococcus pluvialis* (Chlorophyceae). J Phycol 41(4):819–826. https://doi.org/10.1111/j.0022-3646.2005.05015.x

Part I
Cultivation and Astaxanthin Optimization

Chapter 2
Nutritional Studies and Optimization of Biomass from Unicellular Microalgae *Haematococcus* sp.

Silambarasan Tamil Selvan and Sanjivkumar Muthusamy

2.1 Introduction

A unicellular fresh water microalga in the *Haematococcus pluvialis* can be found in many habitats all over the world. Green alga (Chlorophyta) *Haematococcus pluvialis* flourishes in shallow freshwater bodies such as small ponds, generally after desiccation, when it forms red layers attached to the pond walls (Bouyahya et al. 2021; Allewaert et al. 2015). A green alga's life cycle is divided into two distinct stages. Its cells display a pear-shaped morphology under favorable conditions, are motile because of two flagella and are green in color when it grows under favorable conditions. The flagella and oval shape of the cells are lost later in the cell cycle when the cellular division is restricted or restricted completely. The aplanospores develop into rounded cells of larger size. During the development of aplanospores, their cell walls thicken and their pigmentation becomes deep red. In lipid vesicles, astaxanthin accumulates in large amounts during enantiomeric synthesis, mainly as esterified C18 and C16 fatty acids. The color change occurs because chlorophyll degrades concomitantly with astaxanthin synthesis (Butnariu et al. 2014; Kitaya et al. 2005). At some point, the cytoplasm will be occupied by these vesicles. In nutrient-deficient environments, high irradiance, or extreme temperatures, or when a carotenoid accumulates, a cell is protected against conditions affecting cellular homeostasis, which result in decreased oxygen production (Mao et al. 2020; Lin et al. 2016).

The metabolism of microalgae results in a variety of metabolites being accumulated. In microalgae with diverse cell compositions (autotrophic, heterotrophic, and

S. Tamil Selvan (✉)
Department of Microbiology, School of Allied Health Sciences, Vinayaka Missions Research Foundation (DU), Salem, Tamil Nadu, India

S. Muthusamy
Department of Microbiology, K.R. College of Arts & Science, Kovilpatti, Tamil Nadu, India

R. Raja et al. (eds.), *Haematococcus*, https://doi.org/10.1007/978-981-99-2901-6_2

mixotrophic), these metabolites are involved in carbon partitioning, which is influenced by factors such as light intensity, pH, and dissolved oxygen as well as nutrient limited conditions (Park et al. 2014; Kitaya et al. 2005). A sufficient amount of nutrients and light are needed for microalgae to produce primary metabolites, and adverse conditions such as nutrient deficit, high light intensity, and salt stress activate the defense systems of microalgae, leading to secondary metabolites accumulation (Daliry et al. 2017; Liang et al. 2015). As an alternative to traditional plant sources, algae-derived lutein has been considered a good source of free, non-esterified lutein (Qiang et al. 2019; Abalde et al. 1991). In terms of replacing marigold plants for secondary metabolites production, microalgae have shown promise in terms of harvesting costs, extraction of crude lutein, purification, and bioavailability analyses. In recent years, synthesis of nutritional secondary metabolites has been studied in a wide range of microalgal genera, including *Scenedesmus, Chlorella, Coccomyxa, Parachlorella,* and *Tetraselmis* (Daliry et al. 2017; Domınguez-Bocanegra et al. 2004; Khan et al. 2016).

The exact mechanism of action of this response has not yet been clarified, as some authors have proposed that the massive synthesis of astaxanthin merely acts as a physical screen against light that prevents cellular damage, whereas others believe that astaxanthin acts as a protective mechanism by reacting with ROS, specifically singlet oxygen, through its antioxidant capacity. The astaxanthin is not the protective agent itself, but merely a collateral effect resulting from other protective compounds working within the chloroplast, close to the photosynthetic membranes. While it is evident that there is a strong physiological reaction, more effort needs to be put into characterizing it. However, astaxanthin is a valuable microalgae-derived product regardless of how carotenogenesis occurs in *H. pluvialis* (Fábregas et al. 2001; Liang et al. 2015).

Astaxanthin is derived from this organism and considered to be the best natural source. There are many individuals whose bodies produce this red secondary carotenoid, but algae, plants, yeast, and bacteria are the ones that produce this pigment naturally (Lin et al. 2015). The chiral centers of astaxanthin allow it to exist in three different stereoisomer shapes. Biosynthesis takes place primarily in *H. pluvialis* and it produces the stereoisomer, which is the most valuable (Hirt 2000; Po-Fung and Chen 2005; Harker et al. 1996). There is an inverse correlation between astaxanthin synthesis in *H. pluvialis* and lipid droplet deposition under conditions of cellular stress. A carotenoid pathway synthesizes astaxanthin from glyceraldehyde-3-phosphate and pyruvate, which are both produced by a carotenoid pathway. Photosynthesis and/or glycolysis are both involved in the creation of these compounds. Astaxanthin is made from these two key metabolic intermediates by entering the non-mevalonate (MEP) pathway, which generates IPP—the key intermediate in the synthesis of all carotenoids, including mevalonate (MEP) (Gimpel et al. 2013; Mao et al. 2020). The free radical scavenging ability of astaxanthin has enabled it to be used in many different industries, including food, feed, cosmetics, aquaculture, nutraceuticals, and pharmaceuticals. The antioxidant activity of astaxanthin is 65 times greater than vitamin C, 54 times greater than *s*-carotene, 10 times greater than canthaxanthin, zeaxanthin, and lutein; and 100 times greater than α-tocopherol.

By keeping this view, this chapter was heighted the nutritional impact of Haematococcus spp., with the incorporation of new technologies for their mass production,harvesting, and higher biomass production (Oncel et al. 2011; Mata et al. 2010).

2.2 Nutritional Compounds of *Haematococcus*

2.2.1 Astaxanthin

Astaxanthin is a natural pigment found in many types of microalgae. The antioxidant power of this compound is 65 times that of vitamin C and ten times greater than that of other carotenoids. Its outstanding properties in free radical scavenging, anti-inflammatory, anti-hypertensive, and anticancer therapies have made it widely used in functional foods, pharmaceuticals, and cosmetics. It is biologically inefficient due to its structural heterogeneity. In the current market for astaxanthin, microalgae-derived astaxanthin represents only 1% (Ren et al. 2021; Oslan et al. 2021). Astaxanthin produced naturally is also much more expensive than that produced chemically. Commercialization of astaxanthin requires improving its productivity by microalgae. In 2008, the FDA approved astaxanthin derived from microalgae for direct consumption by humans (García-Malea et al. 2009). The secondary carotenoid astaxanthin is an accumulation of carotenoid under adverse conditions in microalgae (García-Malea et al. 2009). Microalgae can survive oxidative stress by mediating cellular redox imbalance. Currently, *Haematococcus pluvialis*, with attached cultivation, accumulates up to 5% of astaxanthin, and industry to produce astaxanthin with an annual yield of 300 tons. As the growth conditions become adverse, *H. pluvialis* transforms from green cells to red cysts containing high astaxanthin levels. A number of studies have explored the effects of nutrient deficiency, high light, adding chemical compounds, and combining these strategies on astaxanthin accumulation (Fernández-Sevilla et al. 2010; Cuellar-Bermudez et al. 2015). Mixotrophy allows *H. pluvialis* to take up glucose as a carbon source and achieve high biomass concentrations. The ability of this organism to produce carotenoids, lipids, and exopolysaccharides under different conditions has drawn a lot of attention. There is evidence that *H. pluvialis* contains up to 6.5 mg of astaxanthin per gram dry weight, and its robust biotechnological characteristics may make it a better organism for producing astaxanthin at a large scale (Ceron et al. 2007).

Astaxanthin also have various bioactivity like antioxidant, anti-inflammatory and anticancer properties. It is often used as an aquaculture and food colorant and also acted as an important food supplement to improve the growth hormones and performances of fishes. Moreover, dietary astaxanthin provision at increasing doses markedly reduced the circulating levels of serum cholesterol and triglycerides in fish (Allewaert et al. 2015). Astaxanthin can also be synthesized from different green algae such as *H. pluvialis, Chlorococcum* sp., *C. zofingiensis, Chlorella vulgaris* or the heterobasidiomycetes yeasts *Xanthophyllomyces dendrorhous,* etc. (Soletto et al. 2005; Wayama et al. 2013). Other than pigments, microalgae are also rich sources of protein for nourishment. They can synthesize different polysaccharides (starch, cellulose, and hemicelluloses) too and the algal cells make them an important food source (Soletto et al. 2005). The selected candidate microalgae synthesize and store huge amounts of lipids in the form of glycerol, esterified sugars with various accounts of fatty acids. This may differ from species to species and are altered by various factors. The algal derived fatty acids have both nutritional and medicinal applications (Cuellar-Bermudez et al. 2015; Soletto et al. 2005).

2.2.2 *β-Carotenoid*

The β-Carotenoids are hydrocarbon compounds that are commonly found in plants and algae. As a precursor to Vitamin A, it benefits the treatment of night blindness since it is lipid-soluble (Liu et al. 2021; Lin et al. 2016). β-Carotene is a powerful antioxidant, anticancer agent, and immune booster, and is widely used in food and medicine today. Consumers prefer natural β-carotene over chemically synthesized B-carotene, although chemically synthesized β-carotene is less expensive. β-carotene acts as a primary carotenoid in microalgae by transferring light energy to chlorophyll for greater light absorption. In addition, it is the first product from microalgae ever to be commercialized with a high value (Sun and Li 2020). It is approved for human consumption to consume β-carotene from microalgae. Upon exposure to nutrient deprivation, high light, high temperature, and high salinity, *Haematococcus* sp. accumulates β-carotene to a maximum of 13% DW, which is its main source of natural β-carotene. Considering that *Haematococcus* sp. can grow in high salinity, its cultivation in seawater may be profitable and prevent contamination by microorganisms (Del Campo et al. 2000; Campenni et al. 2013). Furthermore, microalgae are capable of synthesizing β-carotene. The antioxidant properties of *Haematococcus* sp., *Chlorella zofingiensis, Spirulina platensis,* and *Caulerpa taxifolia* have been documented at 0.1–2% dry weight on average. There are many studies, which demonstrated that the accumulation total amount of beta carotene from Haematococcus spp., is ten times higher than that of carrot. In high light and salt conditions, *Haematococcus* sp. accumulates high levels of lutein, β-carotene, and zeaxanthin. β-carotene derived from microalgae has a different profile than that derived from chemically synthesized sources. Microalgae can produce natural β-carotene, which plays a key role in antioxidation, at low cost and with improved bioactivity (Kobayashi et al. 1993; Yuan et al. 2011).

2.2.3 *Lutein*

A light protector, lutein protects eye tissues from damage caused by sunlight by protecting them from damage. During the development of cataracts, oxidation of the lens (clouding of the lens) is a major factor contributing to light collecting on the retina. Lutein is neutralized by free radicals to prevent oxidative stress and retinal damage (Roberts et al. 2009). In order to prevent age-related macular degeneration (AMD), lutein has gained popularity as a functional food ingredient. Additionally, lutein inhibits inflammation, atherogenic growth, hypertension, diabetes, ulcers, and cancer. A carotenoids called lutein forms 66–77% of the total amount of carotenoids found in a newborn's brain. Lutein has been studied as a nutraceutical functional food ingredient due to its health-promoting properties and potential to be a natural food colorant (Del Campo et al. 2000).

2.2.4 *Zeaxanthin*

Zeaxanthin is a xanthophyll pigment. In cornea, zeaxanthin combines with lutein to create a macular pigment that protects the retinal membrane from blue light and enhances vision. The antioxidant and anti-inflammatory properties of zeaxanthin also give it potential for preventing neurological diseases. Zeaxanthin is in high demand because of the worldwide increase in age-related macular degeneration (Bouyahya et al. 2021; Cerón et al. 2008) and the algal derived Zeaxanthin pigment is nine-fold higher than the red pepper. Also, microalgae-derived zeaxanthin is partly free rather than mono- and di-esters found in flowers and fruits. Hence, microalgae have higher productivity and bioavailability for processing zeaxanthin (Butnariu et al. 2014). A variety of microalgae species and cultivation factors contribute to the production of zeaxanthin from microalgae. Depending on the species of microalgae, zeaxanthin requires different light conditions. However, high light is not always preferred for microalgae to accumulate zeaxanthin. Specifically, a mutant *Haematococcus* sp. can accumulate 5.9 mg/g DW of zeaxanthin under continuous low light (100 moles photons), an increase compared to the 4.18 mg/g DW acquired under continuous high light (100 moles photons). The cultivation of *Haematococcus* sp. in seawater may be profitable since this strain grows well in hypersaline medium. Separation and extraction of zeaxanthin downstream are also necessary. Zeaxanthin extraction from microalgae can be impacted by solvent system efficiencies, with chloroform: methanol (1:2) having the highest zeaxanthin extraction efficiency (30.2 mg/g DW). Zeaxanthin was also extracted from *C. ellipsoidea* using a pressurized liquid extraction method. The chemical structure of zeaxanthin is similar to that of lutein, which is even more important. The ultra-high performance liquid chromatography (UHPLC) equipped with a Waters YMC Carotenoid C30 column was found to be a promising method for separation of zeaxanthin and lutein (Barceló-Villalobos et al. 2019).

2.2.5 *Canthaxanthin*

Canthaxanthin is a carotenoid with nine conjugated double bonds, terminated by two oxo substituents at positions and of the β-ionone backbone. A red-orange color is defined by these structural features (Venugopalan et al. 2013). Astaxanthin is a commercially important ketocarotenoid, as well as an additional high-value product, canthaxanthin. It is formed as a result of an enzyme carotene hydroxylase add hydroxyl ions to the canthaxanthin rings. Canthaxanthin is one among a small number of carotenoids that is marketed commercially for a variety of applications. Foods that are "colorful" and "healthier" are being offered to consumers (Elia et al. 2019). Animals' reddish-orange color can be attributed to their diet, either directly or indirectly, through ingestion of organisms containing carotenoid or through consumption of feed supplements containing carotenoids. Ketocarotenes, such as canthaxanthin, are more antioxidants and free radical scavengers than other carotenes. In studies, canthaxanthin was found to be capable of quenching singlet oxygen and scavenging reactive oxygen species (Sujak 2009). The supplementation of canthaxanthin also reduced lipid peroxidation in rats to prevent induced liver DNA damage and enhanced antioxidant defenses. As a protective agent against infectious disease and chronic inflammation, canthaxanthin may reduce the damaging effects of lipopolysaccharides. The effects and benefits of canthaxanthin in humans. A natural skin-tanner, canthaxanthin is used in cosmetics and can produce an orange brown color if used in large quantities. In addition to its many benefits for human and animal health, canthaxanthin is used as a feed additive for livestock and as a food ingredient in animal feeds (Grama et al. 2014).

2.3 Optimization of Physical and Chemical Parameters for the Production of Biomass

The production of algal biomass with huge amount of metabolically active components were attained under optimized conditions using different parameters like pH, temperature, and nutrient levels (Gimpel et al. 2013). To improve the growth and productivity, the production medium can be supplemented by the addition of acetate (Oncel et al. 2011; Gimpel et al. 2013). Based on the requirement of nutrients for the growth and productivity, the microalgae species may be different. Basically, the macronutrients such as carbon, nitrogen, and phosphorus are acted as backbone for their growth and also absorb O_2 and H_2 from water Fan et al. (1994). The required volume of macronutrients for their growth may vary for species to species. Apart from macro-elements, the growth and productivity of microalgae also needed the micronutrients such as Mo, K, Co, Fe, Mg, Mn, B, and Zn, etc. The maximum ratios of carbon to nitrogen have been recommended for the production of bioactive metabolites especially astaxanthin in mixotrophic cultures (Devi and Mohan 2012; Lin et al. 2016). The deficiency of C/N ratio were greatly affects the growth rate of microalgae and results in low biomass. Subsequently, the low range of C/N ration diminish the synthesis and accumulation of major bioactive components carbohydrates and lipids in microalgae (Devi and Mohan 2012). The growth parameters influence the growth and productivity of microalga are as follows.

2.3.1 *Salinity*

The salinity of the environment has a complex influence on microalgae nutritional production. Additionally, NaCl salt damages the oxygen-evolving complex (OEC) and the photosystem II reaction center (PS II), inhibiting electron transport on both sides of the light energy transfer, potentially inhibiting and killing algae growth. Increasing the extracellular concentration of inorganic ions alters the osmotic balance of the extracellular and intracellular compartments, leading to exosmosis or cellular water loss (Minhas et al. 2016). In addition, excessive NaCl can disrupt equilibrium between reactive oxygen species consumption and production, thereby leading to cell death. In order to improve biomass, lipid, and carotenoid content of the microalgal variety, a comparatively cheap and readily available NaCl salt would be applied. Additionally, adding NaCl to the media can induce astaxanthin and carotenoid formation (Gao et al. 2015). Additionally, astaxanthin and carotenoid accumulation may be increased when NaCl and sodium acetate are added together. Accordingly, the effects of salinity stress on cellular growth can be suppressed by adding NaCl, which increases astaxanthin synthesis. Gao et al. (2015) concluded that different species of *H. pluvialis* have distinct salinity tolerance patterns, as shown by the growth of three isolates under 1% NaCl salinity. A study by Tam et al. (2012) compared the growth and accumulated astaxanthin and carotenoid of *H. pluvialis* at different NaCl concentrations. Salinity stress reduced cell growth and increased carotenoid content per cell (Nahidian et al. 2018).

2.3.2 *Temperature*

Temperature can also be employed as a stress inducer for the enhancement of biomass as well as the production of valuable bioactive metabolites. Fan et al. (1994) examined that the maximum biomass and productivity of a microalga *H. pluvialis* at the temperature range of 30–35 °C, respectively. Similarly, the growth rate of the candidate microalga *H. pluvialis* was maximally attained at 30 °C by Allewaert et al. (2015). Converti et al. (2009) described that the selected algal strain *Chlorella vulgaris* synthesized the maximum amount of carbohydrates and lipids at the temperature 25 °C. Kitaya et al. (2005) examined that the influences of different temperature ranges on microalgae growth. Who were also stated that the temperature ranges between 27–31 °C was suitable forthe betterment of several microalgae species. Oncel et al. (2011) performed an experiment with the influences of glucose

(5 g L^{-1}) on the growth of *C. zofingiensis* and *H. pluvialis* for 14 days. After incubation, the strain *H. pluvialis* exhibited the highest biomass rate of 0.56 g L^{-1} day^{-1} than *C. zofingiensis* (0.49 g L^{-1} day^{-1}), subsequently the production rate of carotenoid astaxanthin 40 and 10 mg g^{-1} were also attained at 28 °C.

The temperature change sensitivity of *Haematococcus* cultures has been demonstrated. When *H. pluvialis* is exposed to optimum temperatures and maximum intracellular levels of oxidative stress, astaxanthin can accumulate. Two culture conditions were used for *H. pluvialis* such as 20 and 25 °C (Thana et al. 2008). Astaxanthin accumulated 37% more in a nitrogen-starved high-temperature culture compared to a control (156 mg vs. 115 mg), showing a very positive relationship between temperature and astaxanthin accumulation (Wan et al. 2014). Astaxanthin production was increased when the increased temperature was combined with nitrogen starvation stress. In addition, Hong et al. (2015) found that cells grown at moderate temperatures (25–28 °C) produced more astaxanthin than cells grown at low temperatures (23 °C). In 18 days, photoautotrophic induction reduced the biomass concentration in cells cultured at 30 and 36 °C by 20% compared to 23 °C (1.21 g L^{-1}). A significant decline in astaxanthin content was also observed in the cells cultured at 30 and 36 °C, as compared to 23 °C (31.8 mg per g dry weight, but significantly higher at 36 °C). Astaxanthin levels in *H. pluvialis* are inversely related to cellular photosynthetic activity under heat stress from 30 to 36 °C (Wayama et al. 2013).

Sarada et al. (2002a, b) described the production of bioactive molecule like astaxanthin from *H. pluvialis* was maximally at pH 7.0 with the temperature range of below 30 °C. Harker et al. (1996) performed the growth of microalgae *H. pluvialis* using a closed photobioreactor with the flow rate (1.5–3.0 L/min) of 1.5% of CO_2 at 28 °C. In an another study, Thana et al. (2008) documented that the optimal condition for astaxanthin yield (83.78 wt%) from *H. pluvialis* was found to be at the temperature 70 °C for 4 h with the maximum antioxidant property (IC_{50} 1.84 mg L^{-1}). Ceron et al. (2007) studied the influences of nitrogen on the growth of microalgae for the production of fatty acids. Who were also observed that the medium containing lower amount of nitrogen (1.7 Mm) yielded 7.60% of fatty acids in dry weight, while increasing the concentration of nitrogen (4.7 Mm) exhibited least amount 2.1% d.w. of fatty acids. Liang et al. (2015) reported that the growth of microalgae expressed the maximum (56.92% d.w.) amount of lipid was attained when the medium exposure to the high light intensity of 350μmoL photons m^{-2} s^{-1}.

2.3.3 Light Intensity

The use of illumination enhances the biomass productivity of *Haematococcus pluvialis* in mixotrophic culture as well. A photosynthetic electron transport chain converts photosynthesis energy into ATP and NADPH through interference with radiation in the life cycle of *Haematococcus pluvialis*. In the Calvin cycle, microalgae fix CO_2 in starch to store chemical energy. *Haematococcus pluvialis* growth rate was assessed by examining the effects of illumination and culture media. While increasing the light intensity (50–100 μmol photons m^{-2} s^{-1}), the biomass production rate was also increased up to 14.7% and decreasing the light intensity makes the reduction of growth rate to 11% (Imamoglu et al. 2009). As the light intensity increased from 50 to 400 μmol photons m^{-2} s^{-1}, the dry cells weight of *Haematococcus pluvialis* dramatically increased, coupled with nitrogen depletion. It increased by 1.39 times the amount of dry weight when the light intensity was 400 μmol photons m^{-2} s^{-1} compared to 50 μmol photons m^{-2} s^{-1}. The significant variation in light intensity was found to be associated with high net photosynthetic production (Zhang et al. 2018). Cultivating microorganisms in a vertical twin-layer photobioreactor at high light intensities of up to 1000 μmol photons m^{-2} s^{-1} *H. pluvialis* produced 19.4 g of biomass productivities and 213 g of final biomass dry weight m growth area after using CO_2 supplementation of 1–10% (Kiperstok et al. 2017).

In an another report, Boussiba and Vonshak (1991) documented that the maximum production of bioactive metabolites especially astaxanthin from *H. pluvialis* was attained under some modified environmental conditions such as the intensity of light intensity 170 Em^{-2} s^{-1}, phosphate starvation and 0.8% of NaCl concentration. Similarly, 2–3 days of after stressed condition the candidate strain *H. pluvialis* showed highest production of red aplanospores (hematocysts) with huge amount (1.5–3.0%) of astaxanthin was achieved by Olaizola and Huntley (2003). Mata et al. (2010) observed that the aerated culture of microalgae *H. pluvialis* under the light intensities 12,000 lx for 12 h of daylight enhance the growth yield, whereas the reduction of light intensity reveal declined the yield of biomass. In an another report, Khan et al. (2016) documented that the maximum growth and carbohydrates productivity of *Microcystis aeruginosa* achieved at the intensity of red LED light 5000 lx. Ip and Chen denoted that the maximum production of bio-metabolites as well as the carotenoids from *C. zofingiensis* expressed the medium supplemented with 1% of glucose with the biomass productivity of 0.151 g L^{-1} day^{-1} over 6 days. Daliry et al. (2017) achieved that the maximum production of lipid as well as the biomass rate of the selected algal strain *Chlorella vulgaris* at the light intensities of 5000–7000 lx.

2.3.4 Carbon Sources

Mixotrophic cultivation of *Haematococcus* sp. using acetic acid or acetate can yield high biomass levels. A carbon source supplemented with potassium acetate and an organic carbon source were used to induce astaxanthin accumulation in *Haematococcus* sp. *Haematococcus* sp. cell productivity increased to 0.243 g after treatment with sodium acetate at 30 mM. *Haematococcus* sp. was grown with sodium acetate additions (1 g) at the beginning or end of the log phase cultivation, and the number of cells added after the log phase was completed increased to almost twice that number, from 21.7 to 42.9 cells (Goksan and Gokpinar 2010). In comparison with adding sodium acetate at the beginning of the cultivation (which resulted in a 1.2-fold increase in biomass), the biomass grew faster. Sodium acetate can provide an organic carbon source that is effective and capable of increasing growth rate. Carbon dioxide (CO_2) influenced C/N balance to promote cyst formation and astaxanthin accumulation, resulting in highly productive biomass and astaxanthin. A specific concentration of gaseous carbon could be added to the media to replace nitrogen deficiency in *Haematococcus* sp. to increase biomass. As an alternative to replacing nitrogen-deficient media with carbon-deficient ones, one method of achieving nitrogen starvation is to alter the carbon/nitrogen ratio (C/N).

Astaxanthin induction and biomass production were highest with 6% CO_2. The addition of additional carbon into the culture system was hypothesized to shift the C/N balance and induce a relative nutrient deficiency that would promote astaxanthin accumulation. The growth of cells would also be diminished if CO_2 was increased further to 20% (Cheng et al. 2016). If CO_2 levels increase by 20%, chloroplasts will be inhibited, resulting in a decrease in cell growth, resulting in high cell mortality. After sodium gluconate at 2 g and 105 μmol photons m^{-2} $s^{-1}$of illumination intensity, recent studies reported this maximum cell number to be 2.43×10^6 cells mL^{-1} in *H. pluvialis*. The gradient fed batch strategy in which they gradually increased the medium C/N ratio from 10 to 50 through gradient feeding. Assimilation of carbon and accumulation of carotenoids would co-regulate the metabolism of *H. pluvialis* cells. At the end of the green stage, the SGF strategy contributed to the hyperdensity production of immotile cyst cells with a total cell yield of 9.18 g that were derived from CH_3COONa and $NaNO_3$. In the second stage, high light intensity treatment applied to SGF-grown *Haematococcus pluvialis* cyst cells led to 15.45 mg of astaxanthin productivity (Lu et al. 2019).

2.3.5 Nitrogen Sources

Nitrogen plays an essential nutrient for the cultivation of *H. pluvialis* and it can boost up the production of astaxanthin as well as cell growth. The size of *H pluvialis* cells has increased as a result of the development of cell growth. Consequently, when the cells lack nutrients during cultivation, biomass accumulation might also be inhibited (Zhang et al. 2018). The microalgae are well known for their ability to metabolize nitrogen sources such as ammonia, urea, and nitrate. In urea, nitrogen accounts for 46% and carbon for 20% of the elements. Microalgae cells increase as a result of the availability of nitrogen and carbon together (Minyuk et al. 2020). A similar inorganic nitrogen source commonly used in aquaculture media preparation, namely nitrate (sodium or potassium salt), provides only 16.5% of nitrogen. Microalgae metabolism pathways could be altered because of different nitrogen sources, especially regarding nitrogen uptake. The mechanisms for nitrogen uptake from ammonium have a fairly straight forward metabolic pathway for direct assimilation by microalgae, but the processes for co-transport, which is supplied exogenously, are more complex (Sarada et al. 2002a, b). In addition, nitrate reductase and urease must first reduce it into ammonium ions. Inorganic nitrogen, phosphorus, and essential trace minerals are contained in the formulation, which is suitable for general-purpose culture medium. The components of the medium initially contain ammonium ions, which favor autotrophic growth of *H pluvialis*. In the previous studies, however, ammonium hydrogen carbonate (NH_4HCO_3) was found to acidify medium and cause cell death when used as a nitrogen source (Wijanarko 2011). The pH of the culture medium become into alkaline condition due to the enzymatic action on urea into ammonia. Using ammonia as a nitrogen source in microalgal cultivation, 55–60% of secondary metabolites could be produced from total cellular biomass. A possible explanation is that trace ammonia is converted into nitrate and ammonium ions, which are then reabsorbed (Pang et al. 2019). In the urea-based medium under light and dark conditions, nitrate and ammonium ions were assumed to be present due to the nitrogen source's ability to increase biomass accumulation (Dhup et al. 2016).

2.4 Microalgal Cultivation Methods for the Production of Biomass

2.4.1 Cultivation of Open Ponds

Microalgae are typically grown in open tanks using the round or raceway method (Fig. 2.1). An open tank has many advantages over a closed one, including ease of construction, complexity, and low cost (Sarada et al. 2002a, b). Despite these limitations, a lack of sunlight and insufficient combining procedures played a part. Even though algae grow quickly and have adequate nutrition, open ponds cannot support organisms that cannot tolerate high salinities and pH levels above 7.0. Open ponds are suitable for growing algae in certain environments, such as lakes, ponds,

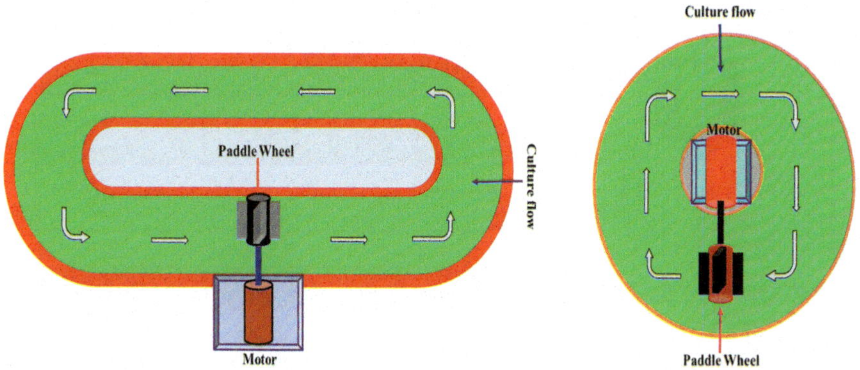

Fig. 2.1 Cultivation method for biomass production using Open pond

lagoons, and artificial ponds. The cultivation of microalgae is commonly done in open reservoirs, using paddle wheels, a closed loop system, and a recirculation system to provide a better ratio of algae, water, and nutrients. Algae can be grown over a large area of water in algae cultivation. Cement or polycarbonate is commonly used for the excavation surfaces of open ponds. The paddle wheels have been installed to support the growth of algae biomass by mixing it sufficiently to connect many linked ponds using a single paddlewheel (Del Campo et al. 2007).

2.4.2 Closed Cultivation Method: Photobioreactors

The substantial improvements in PBR design and maintenance, closed system cultivation models can minimize harvest costs by producing nearly three times the biomass of open systems (Figs. 2.2 and 2.3). It is estimated that about 90% of current biomass production is influenced by environmental factors, even though PBRs have the potential to optimize productivity, environmental effects, and strain selection. However, many closed PBR prototypes developed in laboratories, pilot scale settings, and elsewhere have successfully scaled up to an industrial scale. A tube model of closed loop photoreactors is a common structure used for the cultivation of algae for biomass conversion (Liyanaarachchi et al. 2021). A tube model plays a vital role in the cultivation of biomass that is involved in photosynthesis and respiration. The thin walls of PBR tubes make them ideal for growing algae more quickly than steel, plastic, or fibrous materials. The result is an entire tube of algae in an instant instead of the slower growth associated with steel, plastic, or fibrous materials. Although microalgae cultivation could have been improved to reduce or even eliminate

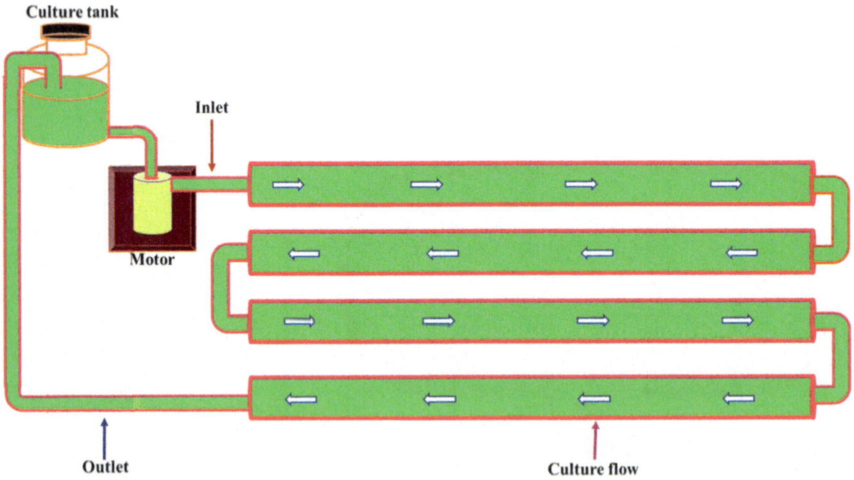

Fig. 2.2 Cultivation method for biomass production using Vertical Tubular Photobioreactors

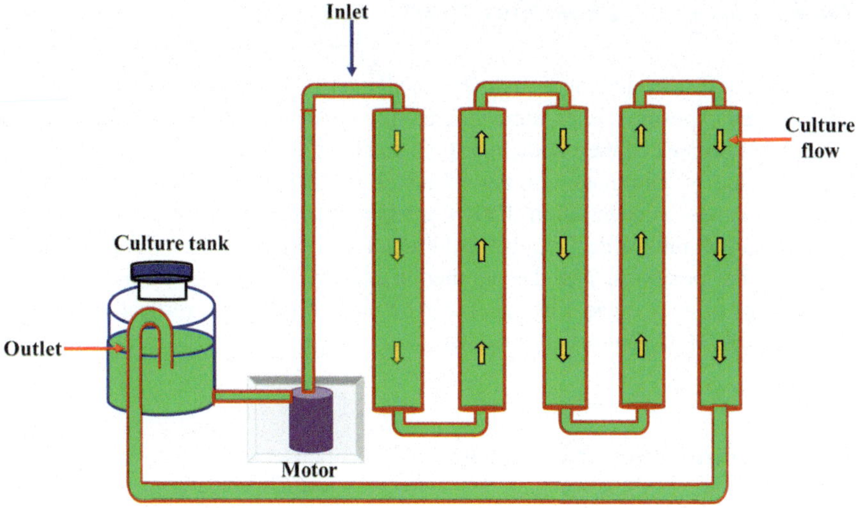

Fig. 2.3 Cultivation method for biomass production using Horizontal Tubular Photobioreactors

external maintenance costs, the performance of the sophisticated system could have been improved (Mao et al. 2020).

2.4.3 Vertical and Horizontal Tubular Photobioreactors

The tubes of a photobioreactor are made from synthetic polymers, such as polyvinyl chloride (PVC), polycarbonate, or porous plastics (Figs. 2.2 and 2.3). In addition to containing gasification equipment that reduces the oxygen produced during photosynthesis, these polymers can be shaped horizontally, vertically, cylindrically, or angled (Endres et al. 2018). In a vertical PBR, photosynthetic species are dispersed in columns rather than in culture areas. The vertical reactor can enhance light, gas, and liquid interaction, thereby increasing the efficiency of the culture (Weise et al. 2020). Throughout the process, algae are grown in vertical or horizontal tubes immersed in liquid. The growth of microalgae can be optimized by using a number of large, symmetrical pipes (helic, flat). In essence, the tube contains gas bubbles and air which circulate a liquid microalgal growth medium, which is subsequently recirculated. Photosynthesis is then encouraged by the use of a certain amount of oxygen, as indicated by the design, regulation, mass exchange, and scaling up issues (Benner et al. 2022).

2.4.4 Flat Panel Photobioreactors

The cultivation of microalgae biomass in flat panel bioreactors is light-efficient. Optimal cell exposure and extreme temperature concentration can be achieved with these PBRs because of their wide light field and high volume-to-area ratio (Fig. 2.4). It is possible to eliminate limiting variables in both types while attempting to increase biomass concentration. PBR implementation tasks are necessary because the volume of microalgae requires additional components. Therefore, flat panel angles, panel positions, and the number of panels per unit of ground affect the productivity of microalgae. Flat panels can be seen horizontally, vertically, or angled (Benner et al. 2022; Endres et al. 2018).

2.5 Column Photobioreactors

2.5.1 Stirred Photobioreactors

In photobioreactors driven by structural blurring, O_2 is absorbed more efficiently by the growing media (Fig. 2.5), but high fluid friction forces adversely affect the algal membrane, thus making light more available for photosynthetic processes and helping O_2 absorb nutrients more rapidly (Soletto et al. 2005).

2.5.2 Airlift Photobioreactors

This type of bioreactors is widely used for the treatment of pollution and cultivation of plants, was not recognized until recently as an alternative to traditional photobioreactors. In this case, liquids are left behind after O_2 bubbles pass through only one section of the column (Fig. 2.5). A steering tube and its flow rate are taken into account in the current analysis. Higher transport phenomena, effective mixing,

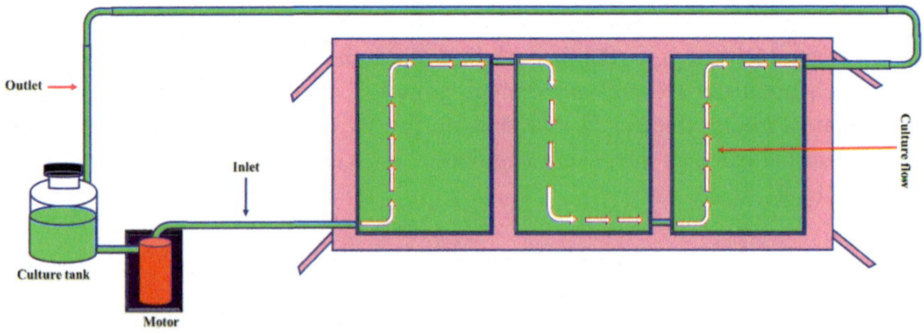

Fig. 2.4 Cultivation method for biomass production using Flat Panel Tubular Photobioreactors

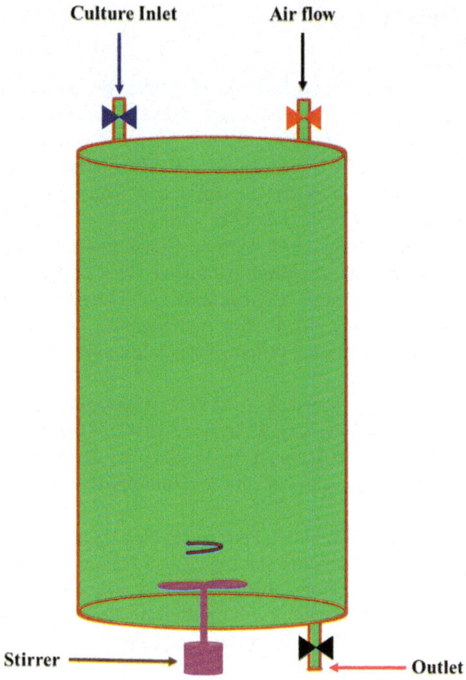

Fig. 2.5 Cultivation method for biomass production using column photobioreactors

low power consumption, high scaling efficiency, relaxed, perfect for algae adsorption, decreased photo impairment make it suitable. Consequently, a smaller surface area with limited development and lighting are potential advantages (Harker et al. 1996).

2.5.3 *Bubble Column Photobioreactors*

The air bubbles in O_2 bubble column bioreactors travel through a sparred base structure and are contained inside circular or rectangular columns. Models using these parameters can obtain the largest gas volumes, which implies that the largest amount of mass can be transported (Fig. 2.5). This modified version of a split-spark column bioreactor would provide effective airflow, minimal fluctuations, and dimensional stability through pore membranes. In addition, gas flow rates can be increased and energy costs can be reduced (Devi and Mohan 2012).

References

Abalde J, Fabregas J, Herrero C (1991) β-Carotene, vitamin C and vitamin E content of the marine microalga *Dunaliella tertiolecta* cultured with different nitrogen sources. Bioresour Technol 38(2–3):121–125

Allewaert CC, Vanormelingen P, Pröschold T, Gómez PI, González MA, Vyverman W (2015) Species diversity in European *Haematococcus pluvialis* (Chlorophyceae, Volvocales). Phycologia 54:583–598

Barceló-Villalobos M, Serrano CG, Zurano AS, García LA, Maldonado SE, Peña J et al (2019) Variations of culture parameters in a pilot-scale thin layer reactor and their influence on the performance of *Scenedesmus almeriensis* culture. Bioresour Technol Rep 6:190–197

Benner P, Meier L, Pfeffer A, Krüger K, Oropeza Vargas JE, Weuster-Botz D (2022) Lab-scale photobioreactor systems: principles, applications, and scalability. Bioprocess Biosyst Eng 45:1–23

Boussiba S, Vonshak A (1991) Astaxanthin accumulation in the green alga *Haematococcus pluvialis*. Plant Cell Physiol 32:1077–1082

Bouyahya A, El Omari N, Hakkur M, El Hachlafi N, Charfi S, Balahbib A, Bakrim S (2021) Sources, health benefits, and biological properties of zeaxanthin. Trends Food Sci Technol 118: 519–538

Butnariu M, Rodino S, Petrache P, Negoescu C, Butu M (2014) Determination and quantification of maize zeaxanthin stability. Dig J Nanomater Biostruct 9(2):745–755

Campenni L, Nobre BP, Santos CA, Oliveira AC, Aires-Barros MR, Palavra AMF, Gouveia L (2013) Carotenoid and lipid production by the autotrophic microalga *Chlorella protothecoides* under nutritional, salinity, and luminosity stress conditions. Appl Microbiol Biotechnol 97(3): 1383–1393

Ceron MC, García-Malea MC, Rivas J, Acien FG, Fernandez JM, Del Río E, Guerrero MG, Molina E (2007) Antioxidant activity of *Haematococcus pluvialis* cells grown in continuous culture as a function of their carotenoid and fatty acid content. Appl Microbiol Biotechnol 74:1112–1119

Cerón MC, Campos I, Sánchez JF, Acién FG, Molina E, Fernández-Sevilla JM (2008) Recovery of lutein from microalgae biomass: development of a process for *Scenedesmus almeriensis* biomass. J Agric Food Chem 56:11761–11766

Cheng J, Li K, Yang Z, Zhou J, Cen K (2016) Enhancing the growth rate and astaxanthin yield of *Haematococcus pluvialis* by nuclear irradiation and high concentration of carbon dioxide stress. Bioresour Technol 204:49–54

Converti A, Casazza AA, Ortiz EY, Perego P, Del Borghi M (2009) Effect of temperature and nitrogen concentration on the growth and lipid content of *Nannochloropsis oculata* and *Chlorella vulgaris* for biodiesel production. Chem Eng Process 48:1146–1151

Cuellar-Bermudez SP, Aguilar-Hernandez I, Cardenas-Chavez DL, Ornelas-Soto N, Romero-Ogawa MA, Parra-Saldivar R (2015) Extraction and purification of high-value metabolites from microalgae: essential lipids, astaxanthin and phycobiliproteins. Microb Biotechnol 8: 190–209

Daliry S, Hallajisani A, Mohammadi Roshandeh J, Nouri H, Golzary A (2017) Investigation of optimal condition for *Chlorella vulgaris* microalgae growth. Global J Environ Sci Manage 3(2): 217–230

Del Campo JA, Moreno J, Rodríguez H, Vargas MA, Rivas J, Guerrero MG (2000) Carotenoid content of chlorophycean microalgae: factors determining lutein accumulation in *Muriellopsis* sp. (Chlorophyta). J Biotechnol 76(1):51–59

Del Campo JA, García-González M, Guerrero MG (2007) Outdoor cultivation of microalgae for carotenoid production: current state and perspectives. Appl Microbiol Biotechnol 74:1163–1174

Devi MP, Mohan SV (2012) CO_2 supplementation to domestic wastewater enhances microalgae lipid accumulation under mixotrophic microenvironment: effect of sparging period and interval. Bioresour Technol 112:116–123

Dhup S, Kannan DC, Dhawan V (2016) Understanding urea assimilation and its effect on lipid production and fatty acid composition of *Scenedesmus* sp. SOJ Biochem 2(7):1

Domınguez-Bocanegra AR, Legarreta IG, Jeronimo FM, Campocosio AT (2004) Influence of environmental and nutritional factors in the production of astaxanthin from *Haematococcus pluvialis*. Bioresour Technol 92(2):209–214

Elia AC, Prearo M, Dörr AJM, Pacini N, Magara G, Brizio P, Abete MC (2019) Effects of astaxanthin and canthaxanthin on oxidative stress biomarkers in rainbow trout. J Toxic Environ Health A 82(13):760–768

Endres CH, Roth A, Brück TB (2018) Modeling microalgae productivity in industrial-scale vertical flat panel photobioreactors. Environ Sci Technol 52(9):5490–5498

Fábregas J, Otero A, Maseda A, Domínguez A (2001) Two-stage cultures for the production of astaxanthin from *Haematococcus pluvialis*. J Biotechnol 89(1):65–71

Fan L, Vonshak A, Boussiba S (1994) Effect of temperature and irradiance on growth of *Haematococcus pluvialis* (Chlorophyceae). J Phycol 30:829–833

Fernández-Sevilla JM, Acién Fernández FG, Molina Grima E (2010) Biotechnological production of lutein and its applications. Appl Microbiol Biotechnol 86(1):27–40

Gao Z, Meng C, Chen YC, Ahmed F, Mangott A, Schenk PM, Li Y (2015) Comparison of astaxanthin accumulation and biosynthesis gene expression of three *Haematococcus pluvialis* strains upon salinity stress. J Appl Phycol 27(5):1853–1860

García-Malea MC, Acién FG, Del Río E, Fernández JM, Cerón MC, Guerrero MG, Molina-Grima E (2009) Production of astaxanthin by *Haematococcus pluvialis*: taking the one-step system outdoors. Biotechnol Bioeng 102(2):651–657

Gimpel JA, Specht EA, Georgianna DR, Mayfield SP (2013) Advances in microalgae engineering and synthetic biology applications for biofuel production. Curr Opin Chem Biol 17:489–495

Goksan T, Gokpinar S (2010) An alternative approach to the traditional mixotrophic cultures of *Haematococcus pluvialis* Flotow (Chlorophyceae). J Microbiol Biotechnol 20(9):1276–1282

Grama BS, Chader S, Khelifi D, Agathos SN, Jeffryes C (2014) Induction of canthaxanthin production in a *Dactylococcus* microalga isolated from the Algerian Sahara. Bioresour Technol 151:297–305

Harker M, Tsavalos AJ, Young AJ (1996) Autotrophic growth and carotenoid production of *Haematococcus pluvialis* in a 30 L air-lift photobioreactor. Ferment Bioeng 82:113–118

Hirt H (2000) Connecting oxidative stress, auxin, and cell cycle regulation through a plant mitogen-activated protein kinase pathway. Proc Natl Acad Sci 97(6):2405–2407

Hong ME, Hwang SK, Chang WS, Kim BW, Lee J, Sim SJ (2015) Enhanced autotrophic astaxanthin production from *Haematococcus pluvialis* under high temperature *via* heat stress-driven Haber–Weiss reaction. Appl Microbiol Biotechnol 99(12):5203–5215

Imamoglu E, Dalay MC, Sukan FV (2009) Influences of different stress media and high light intensities on accumulation of astaxanthin in the green alga *Haematococcus pluvialis*. New Biotechnol 26(3–4):199–204

Khan MI, Lee MG, Seo HJ, Shin JH, Shin TS, Kim MY, Choi JI, Kim JD (2016) Enhancing the feasibility *of Microcystis aeruginosa* as a feedstock for bioethanol production under the influence of various factors. Biomed Res Int 2016:4540826

Kiperstok AC, Sebestyén P, Podola B, Melkonian M (2017) Biofilm cultivation of *Haematococcus pluvialis* enables a highly productive one-phase process for astaxanthin production using high light intensities. Algal Res 21:213–222

Kitaya Y, Azuma H, Kiyota M (2005) Effects of temperature, CO_2/O_2 concentrations and light intensity on cellular multiplication of microalgae, *Euglena gracilis*. Adv Space Res 35(9): 1584–1588

Kobayashi M, Kakizono T, Nagai S (1993) Enhanced carotenoid biosynthesis by oxidative stress in acetate-induced cyst cells of a green unicellular alga, Haematococcus pluvialis. Appl Environ Microbiol 59(3):867–873

Liang C, Zhai Y, Xu D, Ye N, Zhang X, Wang Y, Zhang W, Yu J (2015) Correlation between lipid and carotenoid synthesis and photosynthetic capacity in *Haematococcus pluvialis* grown under high light and nitrogen deprivation stress. Grasas Aceites 66:77

Lin JH, Lee DJ, Chang JS (2015) Lutein production from biomass: marigold flowers versus microalgae. Bioresour Technol 184:421–428

Lin J, Huang L, Yu J, Xiang S, Wang J, Zhang J, Yan X, He S, Wang Q (2016) Fucoxanthin, a marine carotenoid, reverses scopolamine-induced cognitive impairments in mice and inhibits acetylcholinesterase *in vitro*. Mar Drugs 14:67

Liu C, Hu B, Cheng Y, Guo Y, Yao W, Qian H (2021) Carotenoids from fungi and microalgae: a review on their recent production, extraction, and developments. Bioresour Technol 337: 125398

Liyanaarachchi VC, Premaratne M, Ariyadasa TU, Nimarshana PHV, Malik A (2021) Two-stage cultivation of microalgae for production of high-value compounds and biofuels: a review. Algal Res 57:102353

Lu Z, Zheng L, Liu J, Dai J, Song L (2019) A novel fed-batch strategy to boost cyst cells production based on the understanding of intracellular carbon and nitrogen metabolism in *Haematococcus pluvialis*. Bioresour Technol 289:121744

Mao X, Zhang Y, Wang X, Liu J (2020) Novel insights into salinity-induced lipogenesis and carotenogenesis in the oleaginous astaxanthin-producing alga *Chromochloris zofingiensis*: a multi-omics study. Biotechnol Biofuels 13(1):1–24

Mata TM, Martins AA, Caetano NS (2010) Microalgae for biodiesel production and other applications: a review. Renew Sust Energy Rev 14:217–232

Minhas AK, Hodgson P, Barrow CJ, Adholeya A (2016) A review on the assessment of stress conditions for simultaneous production of microalgal lipids and carotenoids. Front Microbiol 7: 546

Minyuk G, Sidorov R, Solovchenko A (2020) Effect of nitrogen source on the growth, lipid, and valuable carotenoid production in the green microalga *Chromochloris zofingiensis*. J Appl Phycol 32(2):923–935

Nahidian B, Ghanati F, Shahbazi M, Soltani N (2018) Effect of nutrients on the growth and physiological features of newly isolated *Haematococcus pluvialis* TMU1. Bioresour Technol 255:229–237

Olaizola M, Huntley ME (2003) Recent advances in commercial production of astaxanthin from microalgae. In: Fingerman M, Nagabhushanam R (eds) Biomaterials and bioprocessing. Science Publishers, Enfield, pp 143–164

Oncel SS, Imamoglu E, Gunerken E, Sukan FV (2011) Comparison of different cultivation modes and light intensities using mono-cultures and co-cultures of *Haematococcus pluvialis* and *Chlorella zofingiensis*. J Chem Technol Biotechnol 86:414–420

Oslan SNH, Shoparwe NF, Yusoff AH, Rahim AA et al (2021) A review on *Haematococcus pluvialis* bioprocess, optimization of green and red stage culture conditions for the production of natural astaxanthin. Biomol Ther 11:1–15

Pang N, Gu X, Fu X, Chen S (2019) Effects of gluconate on biomass improvement and light stress tolerance of *Haematococcus pluvialis* in mixotrophic culture. Algal Res 43:101647

Park JC, Choi SP, Hong ME, Sim SJ (2014) Enhanced astaxanthin production from microalga, *Haematococcus pluvialis* by two-stage perfusion culture with stepwise light irradiation. Bioprocess Biosyst Eng 37(10):2039–2047

Po-Fung IP, Chen F (2005) Production of astaxanthin by the green microalga *Chlorella zofingiensis* in the dark. Process Biochem 40:733–738

Qiang S, Su AP, Li Y, Chen Z, Hu CY, Meng YH (2019) Elevated β-carotene synthesis by the engineered *Rhodobacter sphaeroides* with enhanced CrtY expression. J Agric Food Chem 67(34):9560–9568

Ren Y, Deng J, Huang J, Wu Z, Yi L, Bi Y, Chen F (2021) Using green alga *Haematococcus pluvialis* for astaxanthin and lipid co-production: advances and outlook. Bioresour Technol 340: 125736

Roberts RL, Green J, Lewis B (2009) Lutein and zeaxanthin in eye and skin health. Clin Dermatol 27(2):195–201

Sarada R, Tripathi U, Ravishankar GA (2002a) Influence of stress on astaxanthin production in *Haematococcus pluvialis* grown under different culture conditions. Process Biochem 37(6): 623–627

Sarada R, Usha T, Ravishankar GA (2002b) Influence of stress on astaxanthin production in *Haematococcus pluvialis* grown under different culture conditions. Process Biochem 37:623–627

Soletto D, Binaghi L, Lodi A, Carvalho JCM, Converti A (2005) Batch and fedbatch cultivations of *Spirulina platensis* using ammonium sulphate and urea as nitrogen sources. Aquaculture 243: 217–224

Sujak A (2009) Interactions between canthaxanthin and lipid membranes-possible mechanisms of canthaxanthin toxicity. Cell Mol Biol Lett 14(3):395–410

Sun T, Li L (2020) Toward the 'golden' era: the status in uncovering the regulatory control of carotenoid accumulation in plants. Plant Sci 290:110331

Tam LT, Hoang DD, Mai DTN, Thu NTH, Anh HTL, Hong DD (2012) Study on the effect of salt concentration on growth and Astaxanthin accumulation of microalgae *Haematococcus pluvialis* as the initial basis for two phase culture of astaxanthin production. Acad J Biol 34(2):213–223

Thana P, Machmudah S, Goto M, Sasaki M, Pavasant P, Shotipruk A (2008) Response surface methodology to supercritical carbon dioxide extraction of astaxanthin from *Haematococcus pluvialis*. Bioresour Technol 99:3110–3115

Venugopalan V, Tripathi SK, Nahar P, Saradhi PP, Das RH, Gautam HK (2013) Characterization of canthaxanthin isomers isolated from a new soil *Dietzia* sp. and their antioxidant activities. J Microbiol Biotechnol 23(2):237–245

Wan M, Zhang J, Hou D, Fan J, Li Y, Huang J, Wang J (2014) The effect of temperature on cell growth and astaxanthin accumulation of *Haematococcus pluvialis* during a light–dark cyclic cultivation. Bioresour Technol 167:276–283

Wayama M, Ota S, Matsuura H, Nango N, Hirata A, Kawano S (2013) Three-dimensional ultrastructural study of oil and astaxanthin accumulation during encystment in the green alga *Haematococcus pluvialis*. PLoS One 8:e53618

Weise T, Grewe C, Pfaff M (2020) Experimental and model-based analysis to optimize microalgal biomass productivity in a pilot-scale tubular photobioreactor. Front Bioeng Biotechnol 8:453

Wijanarko A (2011) Effect of the presence of substituted urea and also ammonia as nitrogen source in cultivated medium on chlorella's lipid content. Prog Biomass Bioenergy Prod 14:1–6

Yuan JP, Peng J, Yin K, Wang JH (2011) Potential health-promoting effects of astaxanthin: a high-value carotenoid mostly from microalgae. Mol Nutr Food Res 55(1):150–165

Zhang WW, Zhou XF, Zhang YL, Cheng PF, Ma R, Cheng WL, Chu HQ (2018) Enhancing astaxanthin accumulation in *Haematococcus pluvialis* by coupled light intensity and nitrogen starvation in column photobioreactors. J Microbiol Biotechnol 28(12):2019–2028

Chapter 3
Abiotic Stress Factors and High Astaxanthin Accumulation in *Haematococcus pluvialis*

Mohamed Alouani, Mohammed Hassi, Oumaima Ouaddi, Redouan Qessaoui, and Rachid Bouharroud

3.1 Introduction

The microalgae *Haematococcus pluvialis*, recently called *H. lacustris*, is a freshwater chlorophyceae highly sought after in the pharmaceutical, nutraceutical, aquaculture, and food industries thanks to its ability to synthesize and accumulate a molecule of high added value "astaxanthin" (Kim et al. 2006; Shah et al. 2016; Karuppan et al. 2022). The astaxanthin production in *H. pluvialis* is associated to changes in the morphology, physiology, and photosynthetic characteristics of the microalgae. Under favorable conditions, *Haematococcus* cells are green, biflagellate oval and motile and grow rapidly by direct cell division (Zhang et al. 2017). Under conditions of abiotic stress, they lose their flagella and transform in spherical red cells (a sign of

M. Alouani
Laboratory of Biotechnologies and Valorization of Natural Resources, Faculty of Sciences - Agadir, Ibn Zohr University, Agadir, Morocco

Faculty of Applied Sciences - Ait Melloul, Ibn Zohr University, Agadir, Morocco
e-mail: m.alouani@uiz.ac.ma

M. Hassi
Faculty of Sciences, Laboratory of Microbial Biotechnology and Plants Protection, Ibn Zohr University, Agadir, Morocco

Taroudant's Polydisciplinary Faculty, Ibn Zohr University, Taroudant, Morocco

O. Ouaddi
Faculty of Sciences, Laboratory of Microbial Biotechnology and Plants Protection, Ibn Zohr University, Agadir, Morocco

R. Qessaoui · R. Bouharroud (✉)
Regional Center of Agricultural Research of Agadir, National Institute for Agricultural Research, Rabat, Morocco
e-mail: rachid.bouharroud@inra.ma

R. Raja et al. (eds.), *Haematococcus*, https://doi.org/10.1007/978-981-99-2901-6_3

astaxanthin accumulation) called hematocysts or aplanospores (Poonkum et al. 2014; Shah et al. 2016; Azizi et al. 2020). This microalga is one of the most abundant reservoirs of astaxanthin and synthesizes up to 4% of its weight under stress conditions (Christian et al. 2018; Wan et al. 2021).

This pigment is used as a food color additive and has antioxidant properties giving it several applications in the field of human health and nutrition (Guerin et al. 2003). Several studies have shown that astaxanthin is a more powerful antioxidant than vitamin E and other carotenoids (Igielska-Kalwat et al. 2015; Kim et al. 2015). It is able to neutralize lib radicals 500 times more than vitamin E and 10 times more than β-carotene (López et al. 2004). This is why astaxanthin is considered as "super vitamin E" (Shimidzu et al. 1996). This particular antioxidant activity also contributes in other biological activities, such as anti-inflammatory, antitumor, preventive effects of diabetes, protection against heart liver and ocular diseases, protection against UV rays, lipid peroxidation, oxidative damage to LDL cholesterol (Guerin et al. 2003; Chang et al. 2010; Lu et al. 2010; Park et al. 2011; Ambati et al. 2014; Ranga et al. 2013a, b; Wibrand et al. 2013; Dong et al. 2013; Turkez et al. 2012; Chan et al. 2012; Iizuka et al. 2012; Yoshida et al. 2010; Bhuvaneswari et al. 2014). For this reason, astaxanthin produced by *H. pluvialis* covers more than 50% of the natural astaxanthin market in the global food industry (Shah et al. 2016; Khoo et al. 2019). These benefits have also increased the demand and size of the astaxanthin market, thus the global market for astaxanthin is expected to reach 3.5 billion $USD by 2026, from 1.37 billion $USD in 2020 (Grand View Research Inc. 2021). However, the high cost of astaxanthin production from *H. pluvialis* cultures limits its commercial viability.

Given its unique characteristics, much research has been focused on improving techniques and strategies to increase astaxanthin synthesis and accumulation from *H. pluvialis* cultures while meeting these commercial demands. In order to improve and accelerate the production of astaxanthin in *H. pluvialis*, the culture is carried out under different conditions of nutritional and environmental stress such as high light intensity, nitrogen and phosphorus deprivation, high temperature and salinity, pH of the culture medium and the combination between these different stress factors (Wayama et al. 2013; Wang et al. 2019). Pereira and Otero (2020) showed that these different stress conditions led to the disruption of homeostasis thus inducing cells to self-protect by synthesizing regulators and antioxidants including astaxanthin against these stresses. Several works have been carried out with the aim to monitor the culture system of *H. pluvialis* to produce sufficient biomass quantity during the green stage where the conditions are favorable, to induce the synthesis and accumulation of astaxanthin during the red stage under various exogenous stress (Wang et al. 2019; Zhu et al. 2021; Oslan et al. 2021). This chapter summarizes standard knowledge about *H. pluvialis* and their responses to different abiotic stress, including nutritional deficiency, alone or in combination and their effects on improving astaxanthin production.

3.2 Generalities About *Haematococcus pluvialis*

3.2.1 Taxonomy and Distribution

H. pluvialis belongs to the class Chlorophyceae, order Volvocales, and family Haematococcaseae (Eom et al. 2006). This microalga is an unicellular biflagellate green microalga generally from fresh water, known as *H. lacustris* or *Sphaerella lacustris* and it is widely distributed worldwide in many habitats (Shah et al. 2016). *H. pluvialis* has been found in various environmental and climatic conditions (Gacheva et al. 2015; Kim et al. 2015).

3.2.2 Cell Morphology and Life Cycle of H. pluvialis

The model life cycle of *H. pluvialis* proceeds through four cellular stages (I: vegetative cell growth; II: encystment; III: maturation; and IV: germination) with four types of cells: flagellated macrozooids, palmella, green and red immature aplanospores, and red mature aplanospores (Fig. 3.1). Under favorable culture conditions, flagellated macrozooids (vegetative cells) are oval and motile via two flagella with a single chloroplast containing pyrenoids and grow rapidly due to high levels of chlorophyll

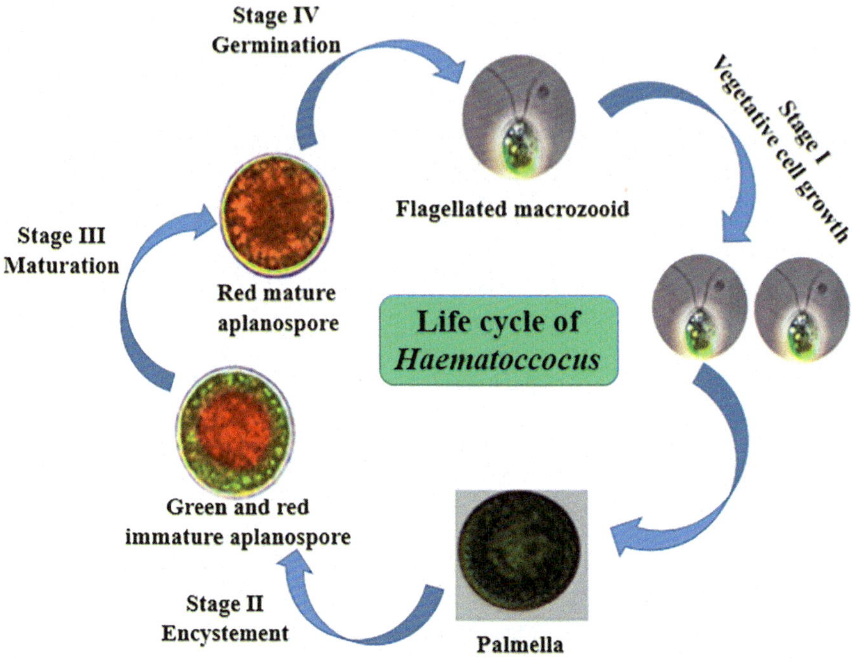

Fig. 3.1 Model life cycle of *H. pluvialis*

and protein and they are characterized by very low carotenoid contents (Kobayashi et al. 1997). Gradually after a few days in culture and especially at rest under unfavorable conditions, these cells lose their flagella and become immobile spherical, surrounded by a rigid cell wall, sometimes forming cells called palmella. During encystment, under continued stress conditions, cells develop into immature cysts (immature green and red aplanospores) marked by the degradation of chlorophyll and proteins and the onset of astaxanthin biosynthesis (Shah et al. 2016). A thick algean-containing cell wall is formed that protects aplanospores from acetolysis through nutrient deprivation or high exposure to stress conditions (Butler et al. 2017). During the process of maturation of cystic cells, the biosynthesis and accumulation of astaxanthin became enhanced and eventually filled all the cytoplasm forming completely red mature aplanospores. When mature aplanospores have been subjected to optimal growth conditions, they germinate to form zoospores, thereby initiating a new cycle of vegetative growth. The different stages of the life cycle of *H. pluvialis* are generally asexual (Shah et al. 2016).

3.2.3 Biochemical Composition of H. pluvialis

Due to its complex life cycle, the biochemical composition of *H. pluvialis* varies according to the stage of its life cycle and especially the macrozoid stage and the aplanospore stage (Table 3.1). At the green stage and under favorable growing

Table 3.1 Biochemical composition of *H. pluvialis* in green and red cultivation stages (Shah et al. 2016)

Composition content (% of DW)	Green stage	Red stage
Proteins	29–45	17–25
Lipids (% of total):	20–25	32–37
Neutral lipids	59	51.9–53.5
Phospholipids	23.7	20.6–21.1
Glycolipids	11.5	25.7–26.5
Carbohydrates	15–17	36–40
Carotenoids (% of total):	0.5	2–5
Neoxanthin	8.3	n.d.
Violaxanthin	12.5	n.d.
β-carotene	16.7	1.0
Lutein	56.3	0.5
Zeaxanthin	6.3	n.d.
Astaxanthin (including esters)	n.d.	81.2
Adonixanthin	n.d.	0.4
Adonirubin	n.d.	0.6
Canthaxanthin	n.d.	5.1
Echinenone	n.d.	0.2
Chlorophylls	1.5–2	0

Adapted from Grewe and Griehl (2012). *n.d.* no data, *% of DW* percentage of dry weight conditions, most strains of *H. pluvialis* are rich in proteins (29–45%), in carbohydrates (15–17%), in lipids vary (20–25%) particularly polyunsaturated fatty acids short (C16, C18) and carotenoids especially lutein (75–80%), β-carotene (10–20%), chlorophyll a and b and primary carotenoids; violaxanthin, neoxanthin, lactucaxanthin, and zeaxanthin (Table 3.1). In the red stage, *H. pluvialis* is rich in amino acids (aspartic acid, glutamic acid, alanine, and leucine). It accumulates a greater quantity of starch which will be consumed in the cell during prolonged stress conditions. Under these latter conditions, the cells of *H. pluvialis* synthesize high amounts of neutral lipids, triacylglycerols (TAG), and secondary metabolites (Saha et al. 2013). The synthesis of secondary carotenoids, mainly astaxanthin (80–99% of total carotenoids) is markedly improved during this stage (Dragos et al. 2010).

3.3 *Haematococcus* Astaxanthin

3.3.1 Structure and Biological Functions of Astaxanthin

The astaxanthin molecule (3,3′-dihydroxy-β,β-carotene-4,4′-dione) is a tertraterpene ($C_{40}H_{52}O_4$) which has a structure similar to that of β-carotene and other carotenoids such as zeaxanthin, lutein, and canthaxanthin (Yang et al. 2013; Basiony et al. 2022; Fig. 3.2a). It shares with the other carotenoids a hydrocarbon chain linked by conjugated double bonds and two terminal carbon rings (β-ionone). However, the structure of astaxanthin differs from β-carotene and canthaxanthin by the presence of

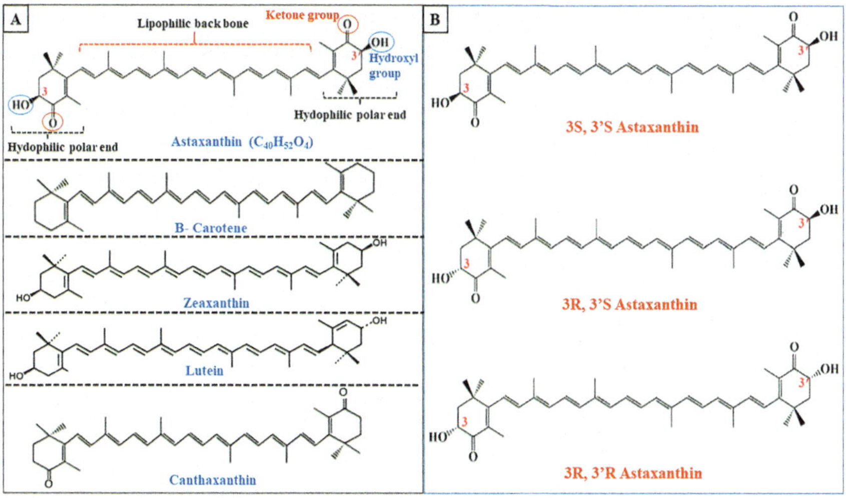

Fig. 3.2 Astaxanthin chemical structure (**a**) and astaxanthin configurational isomers (**b**). (Adapted from Urich 1994; Basiony et al. 2022; Østerlie et al. 1999)

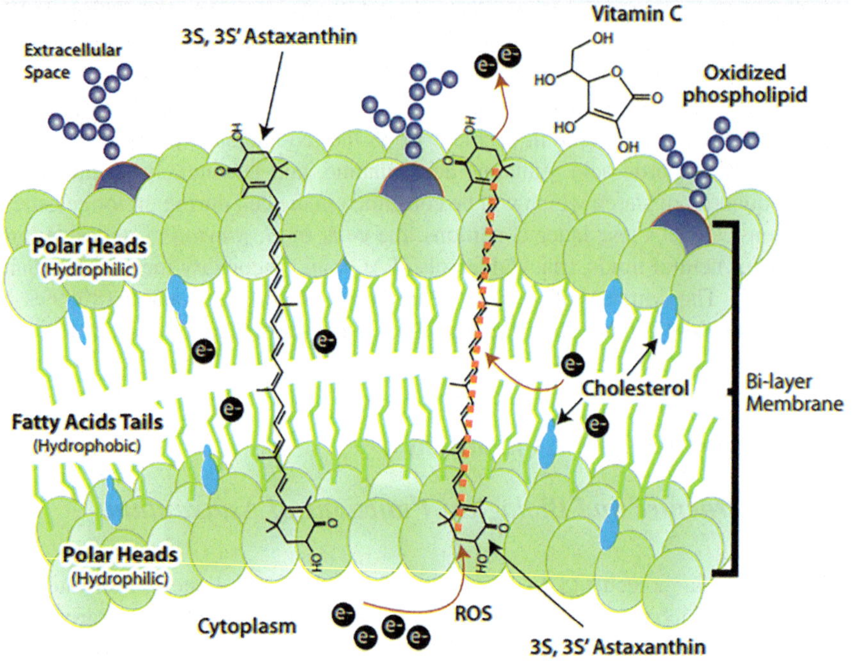

Fig. 3.3 Transmembrane orientation of polar carotenoids facilitates electron shuttle (Pashkow et al. 2008)

a hydroxyl group (OH) and from β-carotene, zeaxanthin and lutein by the presence of a 4-keto group (C=O). The astaxanthin has three optically different isomers including two enantiomers (3S, 3′S), (3R, 3′R) and a mesomeric structure (3R, 3′ S) because of the 3 and 3′ hydroxyl groups which exist in S or R form (Fig. 3.2b).

The structure of astaxanthin and their transmembrane alignment across the phospholipid bilayer of the cell membrane allows the hydrophilic ends to be exposed to the cytoplasm and to the aqueous medium external to the cell (Fig. 3.3). These structural properties either supply electrons or attract unpaired electrons from free radicals which block the chain reaction of free radicals or bond with free radicals to form an unreactive product, thereby achieving anti-oxidation (Kidd 2011; Rao et al. 2013). The 13 conjugated double bonds also give the astaxanthin molecule the ability to transport free radicals from inside the cell to the outside where they will be neutralized by other antioxidants such as vitamin C (Kidd 2011; Fig. 3.3). These double bonds also trap reactive oxygen species (ROS) (Han et al. 2013). Due to this particular structure, astaxanthin is a pigment soluble in water, oil, and the majority of organic solvents at different concentrations and it is found bound to fatty acids or proteins (Peng et al. 2017) which makes astaxanthin applicable in the different fields (Ambati et al. 2019).

The structural arrangements of astaxanthin, especially the presence of both hydroxyl and carbonyl functions in each ionone ring, explain very well their antioxidant properties and their biological activities. This structure also explains

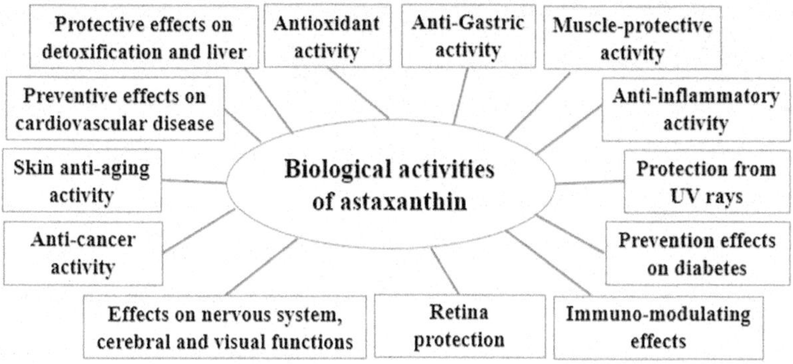

Fig. 3.4 Astaxanthin biological activities in vitro and in vivo experiments generated from several studies (Bennedsen et al. 2000; Kamath et al. 2008; Chang et al. 2010; Lu et al. 2010; Ranga Rao et al. 2010; Yoshida et al. 2010; Choi et al. 2011; Chan et al. 2012; Iizuka et al. 2012; Turkez et al. 2012; Bhuvaneswari et al. 2014; Chew et al. 2013; Dong et al. 2013; Huangfu et al. 2013; Ranga Rao et al. 2013a, b, 2014; Ishiki et al. 2013; Galasso et al. 2018; Lin et al. 2020)

other characteristics, namely the ability to be esterified and polar. In nature, axtaxanthin also exists in other derivatives such as astaxanthin mono- and di-ester, astaxanthin β-D-glucoside, astaxanthin β-D-diglucoside, astaxanthin dirhamnoside, and astaxanthin diglucoside di-ester (Asker et al. 2009, 2018). These derivatives influence the polarity, solubility, and biological activity of astaxanthin (Yokoyama et al. 1995; Řezanka et al. 2008; Ambati et al. 2014; Asker et al. 2018). The hydrocarbon chain in conjugated double bonds, in the form of *cis* or *trans* geometric isomers, gives astaxanthin a distinctive molecular structure, chemical properties, and light absorption characteristics. Indeed, the *trans* isomers are the most thermodynamically stable and the most present in nature (Higuera-Ciapara et al. 2006). According to Lorenz and Cysewski (2000), the astaxanthin enantiomers (3S, 3′S) showed a strong antioxidant activity than the other isomers ((3R, 3′R) and (3R, 3′ S)), which therefore constitutes the most sought form on the nutraceutical market. Astaxanthin act with a wide variety of biological activities that have been demonstrated in several studies and the most important are illustrated in Fig. 3.4 (Ishiki et al. 2013; Galasso et al. 2018; Lin et al. 2020).

3.3.2 *Astaxanthin Biosynthetic Pathway in* H. pluvialis

The natural synthesis of astaxanthin in *H. pluvialis* is a complex process that is strongly induced by stress conditions. It is synthesized from the hydroxylation and ketolation of β-carotene which is in turn synthesized from isopentene pyrophosphate (IPP) and dimethylallyl pyrophosphate (DMAPP). The essential precursors, IPP and DMAPP, play a crucial role in the biosynthesis of astaxanthin (Basiony et al. 2022). The IPP produced from two main pathways: the mevalonic acid (MVA) and the

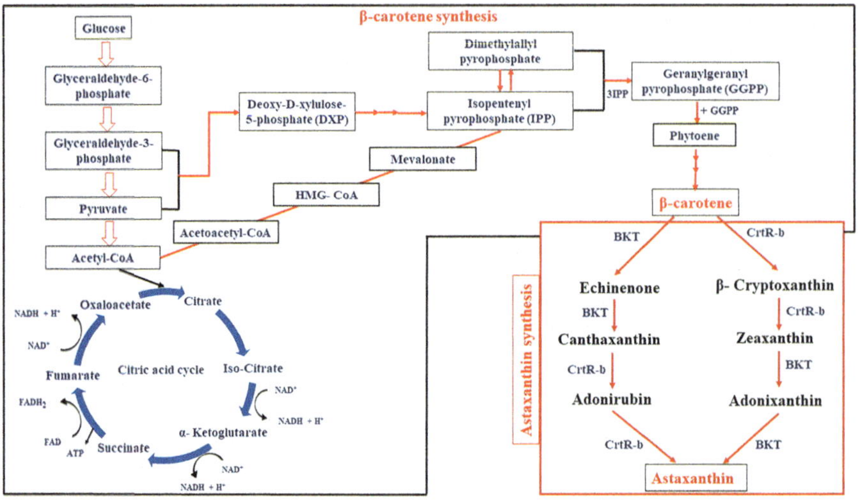

Fig. 3.5 Astaxanthin biosynthetic pathway of microalgae *H. pluvialis*

methyl erythritol phosphate (MEP) (Shah et al. 2016). The synthesis of IPP starts from glycolysis pathway, in which glucose is converted to Glyceraldehyde-6-phosphate, then into Glyceraldehyde-3-phosphate, and into pyruvate that will be converted into acetyl-CoA (Fig. 3.5). The synthesis of β-carotene from IPP and DMAPP from the MVA pathway goes through several steps, the first one is the condensation of two molecules of acetyl-CoA to form acetoacetyl-CoA which condenses with an acetyl-CoA molecule to form 3-hydroxy-3-methylglutaryl-CoA (HMG-CoA) (Fig. 3.5; Wang et al. 2019).

The reduction of HMG-CoA results in the production of mevalonate that is then converted to IPP through three different steps. The isomerization of IPP leads to the formation of dimethylallyl pyrophosphate (DMAPP). The synthesis of β-carotene by the MEP pathway begins with the conversion of pyruvate and glyceraldehyde-3-phosphate to 1-Deoxy-D-xylulose-5-phosphate (DXP) and then to IPP and DMAPP (Fig. 3.5; Gwak et al. 2014). The elongation of the isoprenoid chain begins with the addition of one molecule of DMAPP and three molecules of IPP to form geranylgeranyl pyrophosphate (GGPP) (Takaichi 2011). The condensation of two molecules of GGPP resulting in the formation of phytoene, a first precursor of the compound C40, which transforms into β-carotene after a series of reactions catalyzed by three enzymes (Li et al. 2020).

The synthesis of astaxanthin is initiated by the ketolation or hydroxylation of β-carotene catalyzed by the activity of two enzymes: β-carotene ketolase (BKT) and β-carotene hydroxylase (CrtR-b), respectively. It arises from two distinct pathways of which the first pathway is the conversion of β-carotene to β-cryptoxanthin and then to zeaxanthin by CrtR-b catalyzed hydroxylation and further ketolation to astaxanthin *via* adonixanthin by the action of BKT (Lemoine and Schoefs 2010). The other pathway begins with the ketolation of β-carotene by BKT action to form echinenone and then canthaxanthin and ends with two CrtR-b catalyzed hydroxylation reactions to form astaxanthin via adonirubin (Wan et al. 2021).

3.3.3 *Influence of Stress Factors on Improvement of Astaxanthin Accumulation*

The enhancement of astaxanthin accumulation in *H. pluvialis* under extreme nutritional and environmental conditions (nutrient deficit, light intensity, salinity, temperature, pH) is well studied and demonstrated (Del Campo et al. 2007; Han et al. 2013). In this chapter, we elucidate and discuss the main factors and their possible interactions on the induction and accumulation of this pigment in the microalgae *H. pluvialis*.

3.3.4 *Effect of Light Intensity*

Light plays an important role for both cell growth and astaxanthin synthesis and accumulation in *Haematococcus* (Huang et al. 2019; Oslan et al. 2021). The cell culture of *H. pluvialis* for the production of astaxanthin goes through two stages (green and red: Fig. 3.6), the first of which is characterized by green mobile cells and significant growth, under favorable conditions, in order to obtain a high biomass

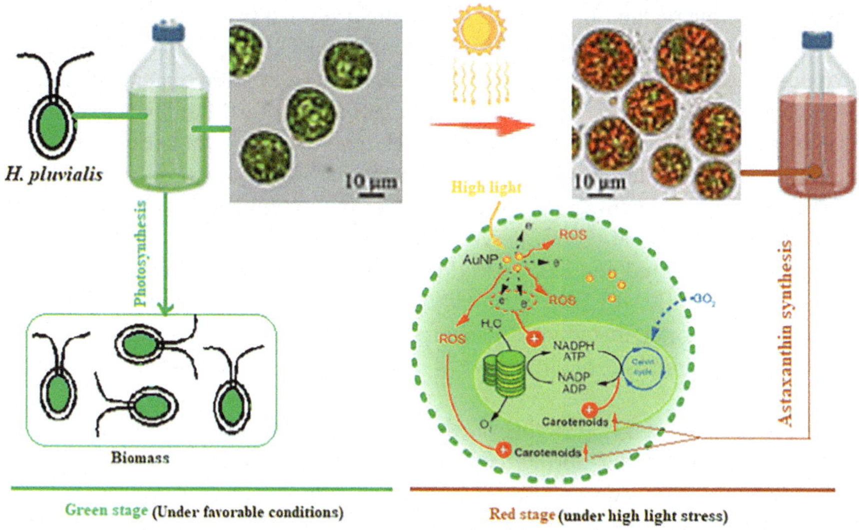

Fig. 3.6 The colors and phenotype of *H. pluvialis* at different conditions. Green stage: Green cells under favorable conditions and Red stage: red cells under high light stress

(Hu et al. 2020; Oslan et al. 2021). While the second stage is dedicated to the synthesis and accumulation of astaxanthin in response to stress inducers and characterized by non-mobile red cysts. Hu et al. (2020) reported that the longer the duration of the light supply, the greater the accumulation of astaxanthin in *H. pluvialis* and especially at high light intensities and under culture conditions in outdoors (García-Malea et al. 2009; Zhang et al. 2009). However, all cells turned into non-motile red cysts after 14 days under high light stress. In the life cycle of *H. pluvialis*, light contributes to the conversion of antennae pigments into chemical energy in the form of ATP and NADPH *via* photosynthetics (Pang et al. 2019). The microalgae will store chemical energy in the starch by fixing CO_2 *via* the Calvin cycle by producing precursors necessary for other biosynthetic pathways. High light intensities enhance astaxanthin production and rapidly induce astaxanthin accumulation in cells but inhibit cell growth and even cause cell death (Zhang et al. 2018; Sadoud et al. 2019).

Karuppan et al. (2022) showed that the increase in light intensity induces an increase in the accumulation of total carotenoids, in particular astaxanthin, inhibits cell growth and decreases the amount of chlorophyll a and especially after prolonged exposure (more than 20 days) at a light intensity of 40.0 μmol/m^2s (Karuppan et al. 2022). Li et al. (2019) reported higher astaxanthin accumulation and 50% and 74% loss of motile cells at light intensities of 50 and 300 μmol/m ^{2}s photons, respectively. The quality and wavelength of light significantly influence the growth and accumulation of this pigment in *Haematococcus* (Lababpour et al. 2004; Katsuda et al. 2006; Kim et al. 2006; Vidhyavathi et al. 2007; Ma et al. 2018). *Haematococcus* growth is better under red light, while induction of astaxanthin accumulation is improved by blue light (Lee et al. 2018; Ma et al. 2018). Kobayashi et al. (1992) demonstrated that stimulation of astaxanthin formation and accumulation is significantly enhanced by exposing the microalgae *H. pluvialis* to blue light instead of white or red light.

The exposure of *H. pluvialis* to blue light induces ROS suggesting their exposure to oxidative stress that leads to upregulate the level of expression of genes coding for β-carotene ketolase and hydroxylase thus promoting astaxanthin biosynthesis (Fu et al. 2012; Lee et al. 2018). They also reported that continuous lighting is more favorable to the formation of astaxanthin. Astaxanthin accumulation in *Haematococcus* is significantly stimulated by the combination of blue light and flash (Vidhyavathi et al. 2007; Praveenkumar et al. 2015) and is optimal (25 mg/L day) with mixed red and blue light (Pereira and Otero 2020). It has proven to be enhanced by the use of LEDs which adjust the composition of the light. Telli and Şahin (2020) reported that gradually increasing light input is the best strategy to keep the maximum number of living cells accumulating a significant amount of astaxanthin.

The production and scavenging of reactive oxygen species (ROS) in *H. pluvialis* cells is in balance under normal culture conditions (Dring 2005). However, high light intensities can cause oxidative stress and promote excess reactive oxygen species (ROS) in *H. pluvialis* that disrupts cellular homeostasis and thus induces the biosynthesis and accumulation of astaxanthin (Han et al. 2012; Pereira and Otero 2020). In this case, *H. pluvialis* accumulates a significant amount of astaxanthin to respond to the oxidative stress caused by the high accumulation of ROS under conditions of high light stress (Hong et al. 2015; Baxter et al. 2013). This microalga adjusts its metabolic pathways to prevent stress by synthesizing the pigment astaxanthin with antioxidant activity (Gwak et al. 2014; Recht et al. 2014; Miao et al. 2008). The accumulation of astaxanthin and the biosynthesis of ROS during high light stress is accompanied by the increase of many metabolites, in particular those with antioxidant properties, such as the oligosaccharides of the family raffinose (RFOs), glutamate and GSH (reduced glutathione (GSH)), and by decreasing levels of purines, pyrimidines, tricarboxylic acid cycle (TCA) intermediates and lysophospholipids (Hu et al. 2020). Lv et al. (2016) also demonstrated that the increase in light intensity leads to the production, by the Calvin cycle and the TCA cycle, of more precursors for the other biosynthetic pathways, including that of astaxanthin.

In practice, high light intensities are often associated with other stress, such as nutritional deficiency, high salinity, heavy metals, acetic acid, and carbon dioxide (Christian et al. 2018; Orosa et al. 2001a, b; Zhang et al. 2018; Pan-utai et al. 2017; Wei et al. 2018). Astaxanthin accumulation in *H. pluvialis* is better when combining two or more stressors and especially when combining high light intensities and nitrogen deprivation (Zhang et al. 2018). In addition, the combination of exogenous supplements such as Fe^{2+}, to the high light stress factor, increase oxidative stress by influencing photosynthesis then by positive regulation of photosynthesis antenna genes that promotes the induction of astaxanthin synthesis (He et al. 2018). Li et al. (2020) found that the addition of γ-aminobutyric acid leads to increased resistance to stress by microalgal cells under the high light conditions by regulating the increased expression of stress-related genes, which resulted in a significant improvement in cell growth and astaxanthin accumulation. They detected also an upregulation of Lhca2 and PTOX gene expressions during this combination, which resulted in a marked improvement of protection against high light stress and an improvement of the photosynthetic activity of the cells. Niizawa et al. (2018) showed that prolonged nitrogen deprivation followed by moderate exposure to light intensity resulted in a 25% increase of astaxanthin yield. The nitrogen deprivation enhances the resistance of *H. pluvialis* cells to photo-oxidation by enhancing the stress response inducing cells to rapidly counter photo-oxidative stress induced by high irradiation (Scibilia et al. 2015).

3.3.5 *Effects of Nutrient Concentration*

In order to minimize the high costs of continuous light input to ensure the culture of *H. pluvialis* cells and the production of astaxanthin especially on a large scale, nutrient deprivation remains the most suitable strategy than high light stress (Han et al. 2013; Solovchenko 2015). Boussiba et al. (1999) and Scibilia et al. (2015) reported that deprivation of all nutrients except CO_2 may be the least expensive and most effective strategy to enhance astaxanthin biosynthesis and accumulation in

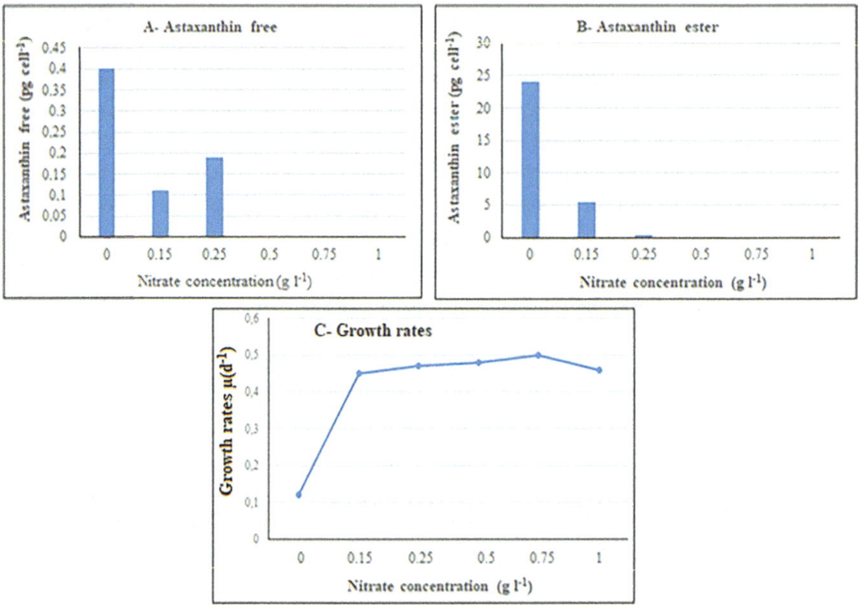

Fig. 3.7 (**a**) Free and (**b**) esterified astaxanthin (in pg/cell) and (**c**) growth rates µ (day) of *H. pluvialis* cells in response to nitrate concentrations. (Generated from the results of Orosa et al. 2005)

Haematococcus. Droop (1954) pointed out that carotenogenesis in *H. pluvialis* may be nitrogen independent. Other studies have shown that nitrogen deficiency markedly improves astaxanthin accumulation in *H. pluvialis* (Fábregas et al. 1998; Orosa et al. 2005). Decreasing the nitrogen concentration is therefore an effective way to improve the accumulation of free and esterified astaxanthin in *Haematococcus* (Fig. 3.7a, b). However, nitrogen deprivation negatively influences cell growth (Fig. 3.7c) which remains weak and even null because of the cessation of cell division that requires nitrogen. Nitrogen is one of the crucial and essential factors that affect cell growth and enzymatic activity of *H. pluvialis* (Zhang et al. 2018) and therefore its deficiency blocks cell division and stimulates the accumulation of astaxanthin (Boussiba and Vonshak 1991; Hagen et al. 2000). Scibilia et al. (2015) demonstrated that nitrogen deficiency induces the synthesis and accumulation of astaxanthin but it promotes chlororespiration and cyclic electron transport, deteriorates chlorophylls and destabilizes photosystems. Orosa et al. (2005) pointed out that a minimum nitrate concentration of around 0.15 g/L of $NaNO_3$ is necessary, especially at the start of growth, to avoid this problem.

The accumulation of astaxanthin is also influenced by the carbon source (Orosa et al. 2005). Acetate is an important carbon source that markedly enhances growth and carotenogenesis and the formation of free or esterified astaxanthin (Fig. 3.8a, b).

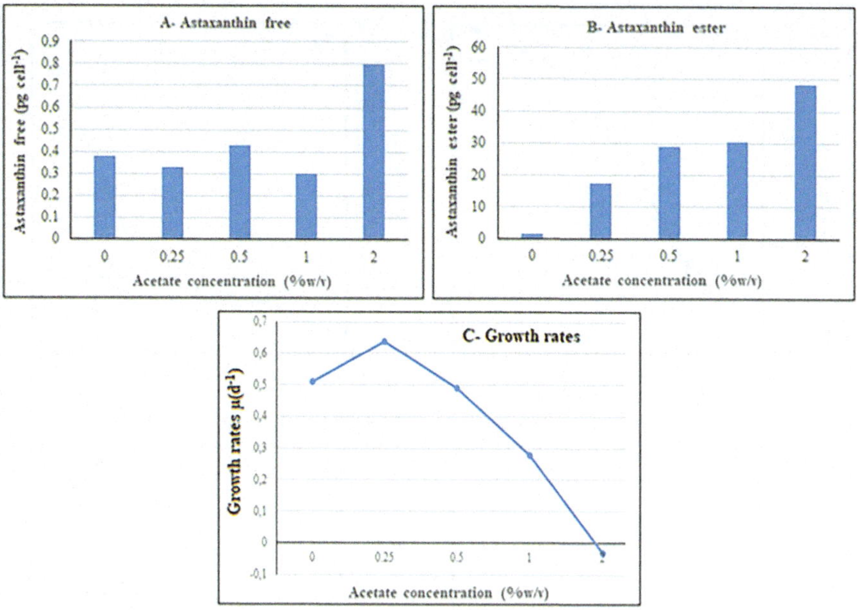

Fig. 3.8 Free (**a**) and esterified astaxanthin (in pg/cell) (**b**) and growth rates μ (per day) (**c**) of *H. pluvialis* cells according to acetate concentrations. (Generated from the results of Orosa et al. 2005)

However, increasing the acetate concentration inhibits growth and enhances astaxanthin accumulation (Fig. 3.8c).

Phosphorus is a nutrient that plays a crucial role in the growth of microalgae by contributing to the synthesis of nucleic acids, the formation of cell membranes and the transfer of cellular energy (Karuppan et al. 2022). It is also involved in the activation of metabolic intermediates, enzymatic regulation, and signal transduction cascades (Sadoud et al. 2019). For this reason, phosphorus deficiency in cultured cells of *H. pluvialis* induces their transformation into resistant forms and also triggers the process of astaxanthin accumulation in these cells (Sadoud et al. 2019; Fig. 3.9). Kang et al. (2006) reported that phosphorus promotes cell growth at low to moderate concentrations. Harker et al. (1996) demonstrated that the accumulation of carotenoids decreases with a phosphate intake greater than 0.85 mM. It should be noted that the reduction of phosphate concentration in the cultures is similar to the effects of nitrogen deficiency but it did not block their growth as severely as nitrogen deprivation. The accumulation of astaxanthin is also enhanced by the presence of relatively high levels of iron (Harker et al. 1996; Fig. 3.9). However, the growth of *H. pluvialis* is inhibited by iron concentrations above 72 μM.

The decrease in the concentration of nitrogen or phosphorus induces the rapid accumulation of astaxanthin and the transformation of mobile cells into red cysts more resistant to stress, however, nitrogen deprivation remains more effective than phosphorus deficit (Boussiba et al. 1999; Scibilia et al. 2015; Sadoud et al. 2019).

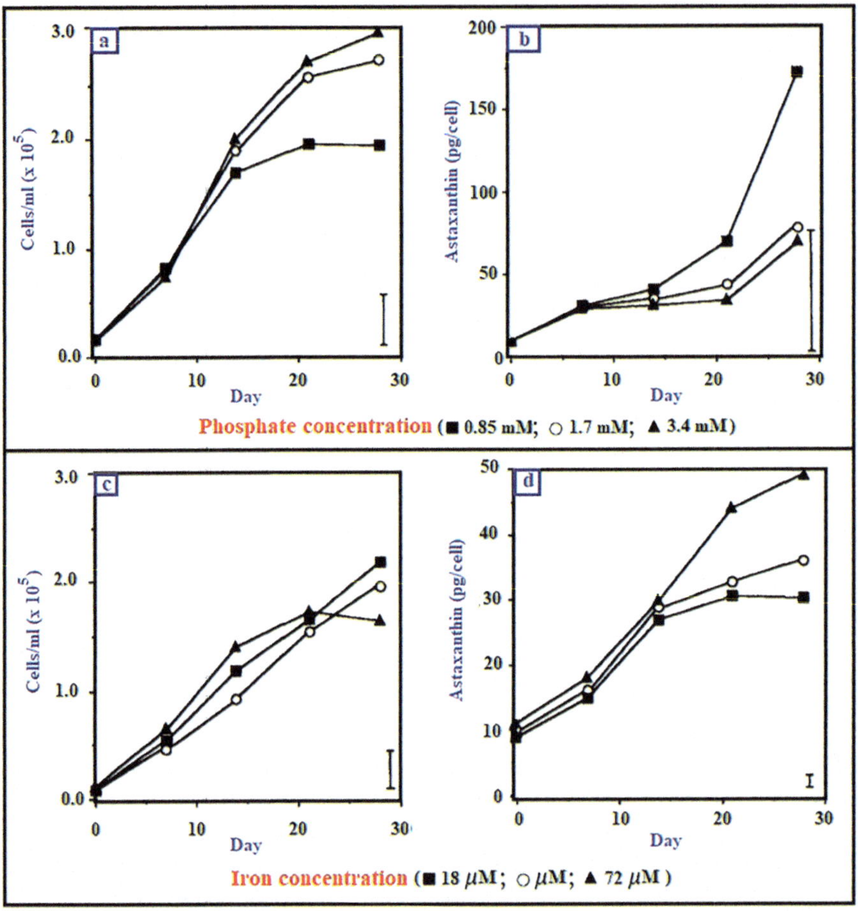

Fig. 3.9 Effect of phosphate and iron concentrations on (**a**; **c**) cell number and (**b**; **d**) astaxanthin formation in cells of *H. pluvialis*. (Generated from the results of Harker et al. 1996)

Low-intensity light supply can support astaxanthin accumulation in the case of nutrient deprivation (Scibilia et al. 2015). The combination of nitrogen deficiency and high light intensity greatly influences the synthesis of astaxanthin pigment in the motile cells of *H. pluvialis*; in this case, the yield of astaxanthin increases up to 49.52 ± 5.9 μg/mL against 37.5 ± 6.8 μg/mL recorded in cultures exposed to high light stress (Fábregas et al. 1998). A similar study (Sadoud et al. 2019) showed that this combination significantly improves astaxanthin accumulation, approximately 67.25 ± 2.28 mg/L, compared to other treatments. The Mg^{+2} deficiency also stimulated astaxanthin synthesis but at lower concentrations than in the case of nitrogen and phosphorus deprivation (Fábregas et al. 1998). Nagaraj et al. (2012) reported that astaxanthin yield in *H. pluvialis* is enhanced by nitrogen starvation, phosphorus, or any other nutrient that delays cell multiplication to protect the photosynthetic apparatus.

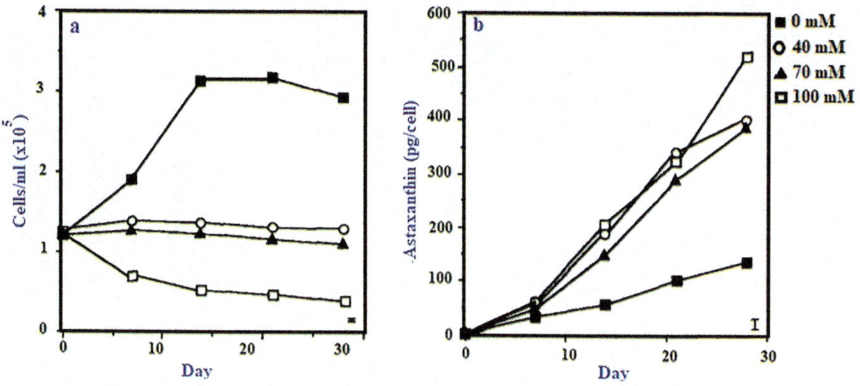

Fig. 3.10 Effect of NaCl concentration on (**a**) cell number and (**b**) astaxanthin formation in cells of *H. pluvialis* (Harker et al. 1996)

3.3.6 Effect of Salinity

Using high light intensities to enhance astaxanthin production requires a large amount of energy, while applying an appropriate concentration of NaCl is more practical in terms of cost and implementation for production of astaxanthin at an industrial production (Ji et al. 2018). Addition of NaCl to the culture medium of *H. pluvialis* can induce astaxanthin formation and reduce cell growth (Tam et al. 2012; Ji et al. 2018; Fig. 3.10). Boussiba and Vonshak (1991) reported that the increase in salt concentration induces the accumulation of ROS which inhibits cell growth and triggers the biosynthesis of antioxidants namely astaxanthin. Ji et al. (2018) found that salt stress affects photosystem II (PS II) and decreases electron transport by affecting photosynthesis, which can lead to microalgae growth inhibition and cell death. Cell death may also be related to disruption of ROS balance and increased extracellular osmotic pressure by excess NaCl (Guilian et al. 2004). The increase in salinity triggers the formation and accumulation of astaxanthin in *H. pluvialis* with a maximum at 100 mM NaCl and leads to a high rate of cell death, while cells that remained alive have the increased ability to produce high levels of astaxanthin (Harker et al. 1996). A 45% mortality of *H. pluvialis* cells was recorded at 0.8% NaCl (138 mM) (Sarada et al. 2002). The transformation of green and motile cells of *H. pluvialis* into non-motile red cells is an adaptation strategy to exogenous stress including salt stress. Li et al. (2019) claimed that non-motile cells have significantly higher tolerance, astaxanthin accumulation, and lower mortality than motile green cells after stress induction. Red cells develop several strategies to neutralize harmful ROS released in excess during high stress and avoid photo-oxidative damage by regulating excess electrons produced by the PS-II system decreasing the level of cytochrome. The yield of astaxanthin in cultures induced by salt stress also depends on the age of the cultures; the older the culture, the more the number of non-motile cells increases, leading to a higher survival rate and an

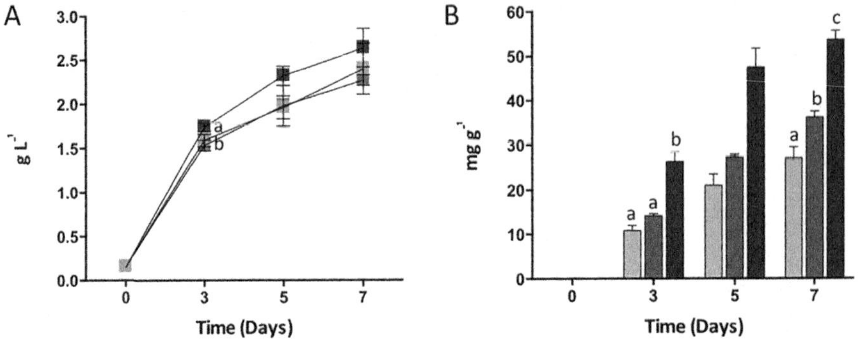

Fig. 3.11 Effect of temperature on biomass concentration (g/L) (**a**), astaxanthin content (mg/g) (**b**). The temperatures tested were: 20 °C (light gray), 25 °C (gray), 30 °C (dark gray) (Pereira and Otero 2020)

accumulation of astaxanthin (Sarada et al. 2002; Telli and Şahin 2020). Shah et al. (2016) reported that adding 0.25–0.5% NaCl to culture media of *H. pluvialis* gives better production of astaxanthin.

3.3.7 Effect of Temperature

Several studies have investigated the effect of temperature on cell growth (green stage) and astaxanthin production and accumulation (red stage) in *H. pluvialis*. The culture of *H. pluvialis* at 30 °C improves the production of astaxanthin threefold more than that obtained at 20 °C. Pereira and Otero (2020) showed that cell growth in the green stage is optimal at 20 °C under monochromatic red light and the synthesis of astaxanthin increases significantly with increasing temperature and this from the third day coinciding with the onset of the stationary phase. This phase is triggered by the depletion of nitrate which contributes to the increase of ROS in the cells and consequently the production of astaxanthin (Domínguez et al. 2019). At the end of the culture period, the astaxanthin content reaches 50 mg/g at 30 °C and around 28 mg/g at 20 °C and 38 mg/g at 25 °C (Fig. 3.11). However, the rate of astaxanthin accumulation is better and reaches a maximum of 25 mg/L/day at 30 °C under a mixture of red and blue light. Wan et al. (2014) found that astaxanthin productivity reach a higher value (120 mg/L/day) at 28 °C. Other study also showed a 1.4-fold improvement in astaxanthin synthesis and accumulation in *H. pluvialis* under a temperature of 27 °C (Giannelli et al. 2015).

High temperatures can generate maximum intracellular levels of oxidative stress that cells of *H. pluvialis* can cope by inducing biosynthesis and increasing astaxanthin accumulation. Tjahjono et al. (1994) reported that cells of *H. pluvialis* withstand and accumulate astaxanthin at high temperatures of 35 °C but for a short time. However, Fan et al. (1994) and Wan et al. (2014) reported that *H. pluvialis* green cell viability decreases at temperatures above 30 °C. Other studies have reported that any temperature above 35–40 °C will induce cell lysis in *H. pluvialis* (Shah et al. 2016; Pereira and Otero 2020). Similarly, biomass productivity in *H. pluvialis* is reduced by 20 and 48% at 30 and 36 °C, respectively (Hong et al. 2015).

Wan et al. (2014) reported that increasing daytime or night time temperature up to 28 °C stimulates astaxanthin accumulation. They also showed that the astaxanthin content, in outdoor cultures, under controlled temperature is five times higher than that obtained at natural temperature. The combination of high temperature and other stressors leads to an improvement in astaxanthin buildup. High temperature has a positive effect on the final astaxanthin content when combined with the stress of nitrogen deprivation (Giannelli et al. 2015). The production of astaxanthin is twice as high if the culture is carried out at 30 °C with the addition of acetate (Tjahjono et al. 1994). Addition of ferrous ions to the culture medium of cells cultured at high temperatures significantly increases astaxanthin content by 75% (30 °C) and 133% (36 °C) compared to that of cells exposed to stress thermal alone (Hong et al. 2015).

3.3.8 *Effect of pH*

The maximum growth and accumulation of astaxanthin (21.85 mg/L) in *H. pluvialis* is obtained at pH 7.5 (Karuppan et al. 2022). While Nagaraj et al. (2012) and Sarada et al. (2006) reported that the highest cell density value is obtained at an initial pH of 7.0 and the total amount of astaxanthin is highest at pH 7.5. They also showed that growth is inhibited at an acidic pH (pH 5.0) and this is the result of the inhibition of chlorophyll formation by the activity of H^+ ions in the media. Synthesis and accumulation of astaxanthin is in significant at acidic pH. Similar study showed also that cell growth of *H. pluvialis* grown in cultures at different pH is maximal at pH 7.0, absent at pH 5.0 and lowest at pH 9.0, while astaxanthin content increases significantly in media at pH 6–8 with a maximal at pH 7.0 (Sarada et al. 2002). However, the use of pH shock (pH = 4.5, 60s) imposes stress in cells and induces astaxanthin synthesis and accumulation in *H. pluvialis* (Han et al. 2020). The combination of an acid-base pH shock (H_2SO_4–KOH) induces an increase in astaxanthin content per cell to about 39% compared to the control. Han et al. (2020) reported also that the enhancement of astaxanthin production is increased by 65% with the application of pH shock and high irradiance and by almost 105% with the application of pH shock and the addition of carbon source compared to those of the control.

3.4 Conclusion

Haematococcus pluvialis is an excellent source of astaxanthin named "king of antioxidants" which represents a great commercial value in various fields of applications. This microalga accumulates less than 1% of commercial astaxanthin production under normal growth conditions. This chapter discusses some recent strategies for improving and optimizing astaxanthin synthesis and accumulation. These strategies are generally based on a two-step process that is used in culture systems: first of all, the optimization of culture conditions favorable to the production of biomass at the green stage and then the induction of astaxanthin production at red stage by application of various stress factors, such as nutrient deficiency or amendment of culture medium, high light intensity, pH, high temperature, etc. Most of these factors, alone or in combination, have been optimized to obtain a better yield from this pigment.

The growth of cells of *H. pluvialis* in the green stage is optimal at a temperature of 20 °C, a pH varies from 7 to 7.5 under low to medium light intensity and in the presence of a source of nitrogen and carbon. The induction of astaxanthin production and accumulation at the red stage is ensured by the culture of *H. pluvialis* cells under stress conditions of which the most effective are the deprivation of nitrate and phosphate and the addition of NaCl and of acetate to the culture medium and the use of high light intensities and pH close to 7. Selective combinations between these culture and induction conditions have mounted much higher yields of astaxanthin. The combination between these strategies and the selection of strains resistant to stress factors and doubted of a high growth potential during the green phase and the use of new generation photobioreactor systems will be necessary to improve the massive production of natural astaxanthin with relatively lower cost. Implementing these advances in the research and development of enhanced commercial production of astaxanthin in *H. pluvialis* on a large scale will still require continued effort and careful research.

References

Ambati R, Phang S-M, Ravi S, Aswathanarayana R (2014) Astaxanthin: sources, extraction, stability, biological activities and its commercial applications-a review. Mar Drugs 12(1): 128–152. https://doi.org/10.3390/md12010128

Ambati RR, Gogisetty D, Aswathanarayana RG, Ravi S, Bikkina PN, Bo L, Yuepeng S (2019) Industrial potential of carotenoid pigments from microalgae: current trends and future prospects. Crit Rev Food Sci Nutr 59(12):1880–1902. https://doi.org/10.1080/10408398.2018.1432561

Asker D, Amano S, Morita K, Tamura K, Sakuda S, Kikuchi N, Furihata K, Matsufuji H, Beppu T, Ueda K (2009) Astaxanthin dirhamnoside, a new astaxanthin derivative produced by a radio-tolerant bacterium, *Sphingomonasas taxanthinifaciens*. J Antibiotics 62(7):397–399. https://doi.org/10.1038/ja.2009.50

Asker D, Awad TS, Beppu T, Ueda K (2018) Purification and identification of astaxanthin and its novel derivative produced by radio-tolerant Sphingomonas astaxanthinifaciens. In: Barreiro, C., Barredo, JL. (eds) Microbial Carotenoids. Methods in Molecular Biology, vol 1852. Humana Press, New York, NY. https://doi.org/10.1007/978-1-4939-8742-9_10

Azizi M, Moteshafi H, Hashemi M (2020) Distinctive nutrient designs using statistical approach coupled with light feeding strategy to improve the *Haematococcus pluvialis* growth performance and astaxanthin accumulation. Bioresour Technol 300:122594. https://doi.org/10.1016/j.biortech.2019.122594

Basiony M, Ouyang L, Wang D, Yu J, Zhou L, Zhu M, Wang X, Feng J, Dai J, Shen Y, Zhang C, Hua Q, Yang X, Zhang L (2022) Optimization of microbial cell factories for astaxanthin production: biosynthesis and regulations, engineering strategies and fermentation optimization strategies. Synth Syst Biotechnol 7(2):689–704. https://doi.org/10.1016/j.synbio.2022.01.002

Baxter A, Mittler R, Suzuki N (2013) ROS as key players in plant stress signalling. J Exp Bot 65(5): 1229–1240. https://doi.org/10.1093/jxb/ert375

Bennedsen M, Wang X, Willén R, Wadström T, Andersen LP (2000) Treatment of H. pylori infected mice with antioxidant astaxanthin reduces gastric inflammation, bacterial load and modulates cytokine release by splenocytes. Immunol Lett 70(3):185–189. https://doi.org/10.1016/s0165-2478(99)00145-5

Bhuvaneswari S, Yogalakshmi B, Sreeja S, Anuradha CV (2014) Astaxanthin reduces hepatic endoplasmic reticulum stress and nuclear factor-κB-mediated inflammation in high fructose and high fat diet-fed mice. Cell Stress Chaperones 19(2):183–191. https://doi.org/10.1007/s12192-013-0443-x

Boussiba S, Vonshak A (1991) Astaxanthin accumulation in the green alga *Haematococcus pluvialis*. Plant Cell Physiol 32:1077–1082

Boussiba S, Bing W, Yuan J-P, Zarka A, Chen F (1999) Changes in pigments profie in the green alga *Haeamtococcus pluvialis* exposed to environmental stresses. Biotechnol Lett 21:601–604. https://doi.org/10.1023/A:1005507514694

Butler T, McDougall G, Campbell R, Stanley M, Day J (2017) Media screening for obtaining *Haematococcus pluvialis* red motile macrozooids rich in astaxanthin and fatty acids. Biology 7(1):2. https://doi.org/10.3390/biology7010002

Chan K, Pen P-J, Yin M (2012) Anticoagulatory and antiinflammatory effects of astaxanthin in diabetic rats. J Food Sci 77(2):H76–H80. https://doi.org/10.1111/j.1750-3841.2011.02558.x

Chang C-H, Chen C-Y, Chiou J-Y, Peng RY, Peng C-H (2010) Astaxanthine secured apoptotic death of PC12 cells induced by β-amyloid peptide 25–35: its molecular action targets. J Med Food 13(3):548–556. https://doi.org/10.1089/jmf.2009.1291

Chew W, Mathison B, Kimble L, Mixter P, Chew B (2013) Astaxanthin decreases inflammatory biomarkers associated with cardiovascular disease in human umbilical vein endothelial cells. Am J Adv Food Sci Technol. https://doi.org/10.7726/ajafst.2013.1001

Choi YE, Yun YS, Park JM, Yang JW (2011) Determination of the time transferring cells for astaxanthin production considering twostage process of *Haematococcus pluvialis* cultivation. Bioresour Technol 102:11249–11253. https://doi.org/10.1016/j.biortech.2011.09.092

Christian D, Zhang J, Sawdon AJ, Peng C-A (2018) Enhanced astaxanthin accumulation in *Haematococcus pluvialis* using high carbon dioxide concentration and light illumination. Bioresour Technol 256:548–551. https://doi.org/10.1016/j.biortech.2018.02.074

Del Campo JA, García-González M, Guerrero MG (2007) Outdoor cultivation of microalgae for carotenoid production: current state and perspectives. Appl Microbiol Biotechnol 74(6): 1163–1174. https://doi.org/10.1007/s00253-007-0844-9

Domínguez A, Pereira S, Otero A (2019) Does *Haematococcus pluvialis* need to sleep? Algal Res 44:101722. https://doi.org/10.1016/j.algal.2019.101722

Dong L-Y, Jin J, Lu G, Kang X-L (2013) Astaxanthin attenuates the apoptosis of retinal ganglion cells in db/db mice by inhibition of oxidative stress. Mar Drugs 11(12):960–974. https://doi.org/10.3390/md11030960

Dragos N, Bercea V, Bica A, Druga B, Nicoara A, Coman C (2010) Astaxanthin production from a new strain of *Haematococcus pluvialis* grown in batch culture. Ann Roman Soc Cell Biol 15: 353–361

Dring MJ (2005) Stress resistance and disease resistance in seaweeds: the role of reactive oxygen metabolism. Incorporating Adv Plant Pathol 43:175–207. https://doi.org/10.1016/s0065-2296(05)43004-9

Droop MR (1954) Conditions governing haematochrome formation and loss in the alga *Haematococcus pluvialis* flotow. Arch Mikrobiol 20(4):391–397. https://doi.org/10.1007/bf00690882

Eom H, Lee CG, Jin E (2006) Gene expression profile analysis in astaxanthin-induced *Haematococcus pluvialis* using a cDNA microarray. Planta 223:1231–1242. https://doi.org/10.1007/s00425-005-0171-2

Fábregas J, Domínguez A, Álvarez DG, Lamela T, Otero A (1998) Induction of astaxanthin accumulation by nitrogen and magnesium deficiencies in *Haematococcus pluvialis*. Biotechnol Lett 20(6):623–626. https://doi.org/10.1023/a:1005322416796

Fan L, Vonshak A, Boussiba S (1994) Effect of temperature and irradiance on growth of *Haematococcus pluvialis* (chlorophyceae). J Phycol 30(5):829–833. https://doi.org/10.1111/j.0022-3646.1994.00829.x

Fu W, Guðmundsson Ó, Paglia G, Herjólfsson G, Andrésson ÓS, Palsson BØ, Brynjólfsson S (2012) Enhancement of carotenoid biosynthesis in the green microalga *Dunaliella salina* with light-emitting diodes and adaptive laboratory evolution. Appl Microbiol Biotechnol 97(6): 2395–2403. https://doi.org/10.1007/s00253-012-4502-5

Gacheva G, Dimitrova P, Pilarski P (2015) New strain *Haematococcus cf. pluvialis* Rozhen-12 - growth, biochemical characteristics and future perspectives. Genet Plant Physiol 5:29–38

Galasso C, Orefice I, Pellone P, Cirino P, Miele R, Ianora A, Brunet C, Sansone C (2018) On the neuroprotective role of astaxanthin: new perspectives? Mar Drugs 16(8):247. https://doi.org/10.3390/md16080247

García-Malea MC, Acién FG, Del Río E, Fernández JM, Cerón MC, Guerrero MG et al (2009) Production of astaxanthin by *Haematococcus pluvialis*: taking the one-step system outdoors. Biotechnol Bioeng 102:651–657. https://doi.org/10.1002/bit.22076

Giannelli L, Yamada H, Katsuda T, Yamaji H (2015) Effects of temperature on the astaxanthin productivity and light harvesting characteristics of the green alga *Haematococcus pluvialis*. J Biosci Bioeng 119(3):345–350. https://doi.org/10.1016/j.jbiosc.2014.09.002

Grand View Research Inc (2021) Astaxanthin market size, share & trends analysis report by product (oil, softgel, liquid), by source (natural, synthetic), by application (aquaculture & animal feed, nutraceuticals), by region, and segment forecasts, 2021-2028. Grand View Research

Grewe CB, Griehl C (2012) The carotenoid astaxanthin from *Haematococcus pluvialis*. In: Posten C, Walter C (eds) Microalgal biotechnology: integration and economy. De Gruyter, Berlin, pp 129–144. https://doi.org/10.1515/9783110298321.129

Guerin M, Huntley ME, Olaizola M (2003) *Haematococcus astaxanthin*: applications for human health and nutrition. Trends Biotechnol 21(5):210–216. https://doi.org/10.1016/s0167-7799(03)00078

Guilian M, Xing X, Zhaozhen X (2004) Advances in physiological and biochemical research of salt tolerance in plant. Chin J Eco-Agric 2004(12):43–46

Gwak Y, Hwang YS, Wang B, Kim M, Jeong J, Lee CG et al (2014) Comparative analyses of lipidomes and transcriptomes reveal a concerted action of multiple defensive systems against photooxidative stress in *Haematococcus pluvialis*. J Exp Bot 65:4317–4334. https://doi.org/10.1093/jxb/eru206

Hagen C, Grunewald K, Schmidt S, Muller J (2000) Accumulation of secondary carotenoids in flagellates of *Haematococcus pluvialis* (Chlorophyta) is accompanied by an increase in per unit chlorophyll productivity of photosynthesis. Eur J Phycol 35:75–82

Han D, Wang J, Sommerfeld M, Hu Q (2012) Susceptibility and protective mechanisms of motile and non-motile cells of *Haematococcus pluvialis* (Chlorophyceae) to photooxidative stress. J Phycol 48(3):693–705. https://doi.org/10.1111/j.1529-8817.2012.01147.x

Han D, Li Y, Hu Q (2013) Biology and commercial aspects of *Haematococcus pluvialis*. In: Richmond A, Hu Q (eds) Handbook of microalgal culture: applied phycology and biotechnology, 2nd edn. Blackwell, Hoboken, pp 388–405. https://doi.org/10.1002/9781118567166.ch20

Han S-I, Chang SH, Lee C, Jeon MS, Heo YM, Kim S, Choi Y-E (2020) Astaxanthin biosynthesis promotion with pH shock in the green microalga, *Haematococcus lacustris*. Bioresour Technol 314:123725. https://doi.org/10.1016/j.biortech.2020.123725

Harker M, Tsavalos AJ, Young AJ (1996) Factors responsible for astaxanthin formation in the chlorophyte *Haematococcus pluvialis*. Bioresour Technol 55:207–214. https://doi.org/10.1016/0960-8524(95)00002-X

He B, Hou L, Dong M, Shi J, Huang X, Ding Y, Cong X, Zhang F, Zhang X, Zang X (2018) Transcriptome analysis in *Haematococcus pluvialis*: astaxanthin induction by high light with acetate and Fe2+. Int J Mol Sci 19(1):175. https://doi.org/10.3390/ijms19010175

Higuera-Ciapara I, Felix-Valenzuela L, Goycoolea FM (2006) Astaxanthin: a review of its chemistry and applications. Crit Rev Food Sci Nutr 46:185–196. https://doi.org/10.1080/10408690590957188

Hong M-E, Hwang SK, Chang WS, Kim BW, Lee J, Sim SJ (2015) Enhanced autotrophic astaxanthin production from *Haematococcus pluvialis* under high temperature via heat stress-driven Haber–Weiss reaction. Appl Microbiol Biotechnol 99(12):5203–5215. https://doi.org/10.1007/s00253-015-6440-5

Hu C, Cui D, Sun X, Shi J, Xu N (2020) Primary metabolism is associated with the astaxanthin biosynthesis in the green algae *Haematococcus pluvialis* under light stress. Algal Res 46:101768. https://doi.org/10.1016/j.algal.2019.101768

Huang L, Gao B, Wu M, Wang F, Zhang C (2019) Comparative transcriptome analysis of a long-time span two-step culture process reveals a potential mechanism for astaxanthin and biomass hyper-accumulation in *Haematococcus pluvialis* JNU35. Biotechnol Biofuels 12(1). https://doi.org/10.1186/s13068-019-1355-5

Huangfu J, Liu J, Sun Z, Wang M, Jiang Y, Chen ZY, Chen F (2013) Antiaging effects of astaxanthin-rich alga *Haematococcus pluvialis* on fruit flies under oxidative stress. J Agric Food Chem 61(32):7800–7804. https://doi.org/10.1021/jf402224w

Igielska-Kalwat J, Gościańska J, Nowak I (2015) Carotenoids as natural antioxidants. Postępy Hig Med Dośw 69:418–428. https://doi.org/10.5604/17322693.1148335

Iizuka M, Ayaori M, Uto-Kondo H, Yakushiji E, Takiguchi S, Nakaya K, Hisada T, Sasaki M, Komatsu T, Yogo M, Kishimoto Y, Kondo K, Ikewaki K (2012) Astaxanthin enhances ATP-binding cassette transporter A1/G1 expressions and cholesterol efflux from macrophages. J Nutr Sci Vitaminol 58(2):96–104. https://doi.org/10.3177/jnsv.58.96

Ishiki M, Nishida Y, Ishibashi H, Wada T, Fujisaka S, Takikawa A, Urakaze M, Sasaoka T, Usui I, Tobe K (2013) Impact of divergent effects of astaxanthin on insulin signaling in L6 cells. Endocrinology 154(8):2600–2612. https://doi.org/10.1210/en.2012-2198

Ji X, Cheng J, Gong D, Zhao X, Qi Y, Su Y, Ma W (2018) The effect of NaCl stress on photosynthetic efficiency and lipid production in freshwater microalga-*Scenedesmus obliquus* XJ002. Sci Total Environ 633:593–599. https://doi.org/10.1016/j.scitotenv.2018.03.240

Kamath BS, Srikanta BM, Dharmesh SM, Sarada R, Ravishankar GA (2008) Ulcer preventive and antioxidative properties of astaxanthin from *Haematococcus pluvialis*. Eur J Pharmacol 590:387–395. https://doi.org/10.1016/j.ejphar.2008.06.042

Kang CD, An JY, Park TH, Sim SJ (2006) Astaxanthin biosynthesis from simultaneous N and P uptake by the green alga *Haematococcus pluvialis* in primary-treated wastewater. Biochem Eng J 31(3):234–238. https://doi.org/10.1016/j.bej.2006.08.002

Karuppan R, Javee A, Gopidas SK, Pathmanapan A, Kattusamy K, Narayanan V, Subramanian SR, Subramani N (2022) Impact of cultivation parameters on astaxanthin accumulation in the green alga *Haematococcus lacustris* RRGK isolated from Himachal Pradesh, India. Energy Nexus 6:100083. https://doi.org/10.1016/j.nexus.2022.100083

Katsuda T, Shimahara K, Shiraishi H, Yamagami K, Ranjbar R, Katoh S (2006) Effect of flashing light from blue light emitting diodes on cell growth and astaxanthin production of *Haematococcus pluvialis*. J Biosci Bioeng 102(5):442–446. https://doi.org/10.1263/jbb.102.442

Khoo KS, Lee SY, Ooi CW, Fu X, Miao X, Ling TC, Show PL (2019) Recent advances in biorefinery of astaxanthin from *Haematococcus pluvialis*. Bioresour Technol 288:121606. https://doi.org/10.1016/j.biortech.2019.121606

Kidd P (2011) Astaxanthin, cell membrane nutrient with diverse clinical benefits and anti-aging potential. Altern Med Rev 16(4):355–364

Kim Z-H, Kim S-H, Lee H-S, Lee C-G (2006) Enhanced production of astaxanthin by flashing light using *Haematococcus pluvialis*. Enzyme Microb Technol 39(3):414–419. https://doi.org/10.1016/j.enzmictec.2005.11.041

Kim JH, Affan MA, Jang J, Kang MH, Ko AR, Jeon SM et al (2015) Morphological, molecular, and biochemical characterization of astaxanthin producing green microalga *Haematococcus* sp. KORDI03 (Haematococcaceae, Chlorophyta) isolated from Korea. J Microbiol Biotechnol 25:238–246. https://doi.org/10.4014/jmb.1410.10032

Kobayashi M, Kakizono T, Yamaguchi K, Nishio N, Nagai S (1992) Growth and astaxanthin formation of *Haematococcus pluvialis* in heterotrophic and mixotrophic conditions. J Ferment Bioeng 74:17–20. https://doi.org/10.1016/0922-338X(92)90261-R

Kobayashi M, Kurimura Y, Tsuji Y (1997) Light independent, astaxanthin production by the green microalga *Haematococcus pluvialis* under salt stress. Biotechnol Lett 19:507–509. https://doi.org/10.1023/A:1018372900649

Lababpour A, Hada K, Shimahara K, Katsuda T, Katoh S (2004) Effects of nutrient supply methods and illumination with blue light emitting diodes (LEDs) on astaxanthin production by *Haematococcus pluvialis*. J Biosci Bioeng 98(6):452–456. https://doi.org/10.1016/s1389-1723(05)00311-7

Lee C, Ahn J-W, Kim J-B, Kim JY, Choi Y-E (2018) Comparative transcriptome analysis of *Haematococcus pluvialis* on astaxanthin biosynthesis in response to irradiation with red or blue LED wavelength. World J Microbiol Biotechnol 34(7). https://doi.org/10.1007/s11274-018-2459-y

Lemoine Y, Schoefs B (2010) Secondary ketocarotenoid astaxanthin biosynthesis in algae: a multifunctional response to stress. Photosynth Res 106(1–2):155–177. https://doi.org/10.1007/s11120-010-9583-3

Li F, Cai M, Lin M, Huang X, Wang J, Zheng X, An Y (2019) Accumulation of astaxanthin was improved by the nonmotile cells of *Haematococcus pluvialis*. BioMed Res Int 2019:8101762. https://doi.org/10.1155/2019/8101762. PMID:30868075

Li L, Chen Z, Huang Q (2020) Exogenous γ-aminobutyric acid promotes biomass and astaxanthin production in *Haematococcus pluvialis*. Algal Res 52:102089. https://doi.org/10.1016/j.algal.2020.102089

Lin W-N, Kapupara K, Wen Y-T, Chen Y-H, Pan I-H, Tsai R-K (2020) *Haematococcus pluvialis*-derived astaxanthin is a potential neuroprotective agent against optic nerve ischemia. Mar Drugs 18(2):85. https://doi.org/10.3390/md18020085

López M, Arce L, Garrido J, Ríos A, Valcárcel M (2004) Selective extraction of astaxanthin from crustaceans by use of supercritical carbon dioxide. Talanta 64(3):726–731. https://doi.org/10.1016/j.talanta.2004.03.048

Lorenz RT, Cysewski GR (2000) Commercial potential for *Haematococcus* microalgae as a natural source of astaxanthin. Trends Biotechnol 18(4):160–167. https://doi.org/10.1016/s0167-7799(00)01433-5

Lu Y-P, Liu S-Y, Sun H, Wu X-M, Li J-J, Zhu L (2010) Neuroprotective effect of astaxanthin on H_2O_2-induced neurotoxicity in vitro and on focal cerebral ischemia in vivo. Brain Res 1360:40–48. https://doi.org/10.1016/j.brainres.2010.09.016

Lv H, Xia F, Liu M, Cui X, Wahid F, Jia S (2016) Metabolomic profiling of the astaxanthin accumulation process induced by high light in *Haematococcus pluvialis*. Algal Res 20:35–43. https://doi.org/10.1016/j.algal.2016.09.019

Ma R, Thomas-Hall SR, Chua ET, Eltanahy E, Netzel ME, Netzel G, Lu Y, Schenk PM (2018) Blue light enhances astaxanthin biosynthesis metabolism and extraction efficiency in *Haematococcus pluvialis* by inducing haematocyst germination. Algal Res 35:215–222. https://doi.org/10.1016/j.algal.2018.08.023

Miao F, Lu D, Zhang C, Zuo J, Geng Y, Hu H, Li Y (2008) The synthesis of astaxanthin esters, independent of the formation of cysts, highly correlates with the synthesis of fatty acids in

Haematococcus pluvialis. Sci China C Life Sci 51(12):1094–1100. https://doi.org/10.1007/s11427-008-0141-6

Nagaraj S, Arulmurugan P, Rajaram MG, Sundararaj R, Rengasamy R (2012) Enhanced production of astaxanthin at different physico-chemical parameters in the green alga Haematococcus pluvialis Flotow. Phykos 42(1):59–71

Niizawa I, Espinaco BY, Leonardi JR, Heinrich JM, Sihufe GA (2018) Enhancement of astaxanthin production from *Haematococcus pluvialis* under autotrophic growth conditions by a sequential stress strategy. Prep Biochem Biotechnol 48(6):528–534. https://doi.org/10.1080/10826068.2018.1466159

Orosa M, Franqueira D, Cid A, Abalde J (2001a) Carotenoid accumulation in *Haematococcus pluvialis* in mixotrophic growth. Biotechnol Lett 23(5):373–378. https://doi.org/10.1023/a:1005624005229

Orosa M, Valero JF, Herrero C, Abalde J (2001b) Comparison of the accumulation of astaxanthin in *Haematococcus pluvialis* and other green microalgae under N-starvation and high light conditions. Biotechnol Lett 23(13):1079–1085. https://doi.org/10.1023/a:1010510508384

Orosa M, Franqueira D, Cid A, Abalde J (2005) Analysis and enhancement of astaxanthin accumulation in *Haematococcus pluvialis*. Bioresour Technol 96(3):373–378. https://doi.org/10.1016/j.biortech.2004.04.006

Oslan SNH, Shoparwe NF, Yusoff AH, Rahim AA, Chang CS, Tan JS, Oslan SN, Arumugam K, Ariff AB, Sulaiman AZ, Mohamed MS (2021) A review on *Haematococcus pluvialis* bioprocess optimization of green and red stage culture conditions for the production of natural astaxanthin. Biomolecules 11(2):256. https://doi.org/10.3390/biom11020256

Østerlie M, Bjerkeng B, Liaaen-Jensen S (1999) Accumulation of astaxanthin all-E, 9Z and 13Z geometrical isomers and 3 and 3′ RS optical isomers in rainbow trout (Oncorhynchus mykiss) is selective. J Nutr 129(2):391–398. https://doi.org/10.1093/jn/129.2.391

Pang N, Gu X, Fu X, Chen S (2019) Effects of gluconate on biomass improvement and light stress tolerance of *Haematococcus pluvialis* in mixotrophic culture. Algal Res 43:101647. https://doi.org/10.1016/j.algal.2019.101647

Pan-utai W, Parakulsuksatid P, Phomkaivon N (2017) Effect of inducing agents on growth and astaxanthin production in *Haematococcus pluvialis*: organic and inorganic. Biocatal Agric Biotechnol 12:152–158. https://doi.org/10.1016/j.bcab.2017.10.004

Park JS, Mathison BD, Hayek MG, Massimino S, Reinhart GA, Chew BP (2011) Astaxanthin stimulates cell-mediated and humoral immune responses in cats. Vet Immunol Immunopathol 144(3–4):455–461. https://doi.org/10.1016/j.vetimm.2011.08.019

Pashkow FJ, Watumull DG, Campbell CL (2008) Astaxanthin: a novel potential treatment for oxidative stress and inflammation in cardiovascular disease. Am J Cardiol 101(10):S58–S68. https://doi.org/10.1016/j.amjcard.2008.02.010

Peng YJ, Hong-Ping LV, Wang SN et al (2017) Present research and prospect on natural astaxanthin. China Food Addit 158(4):193–197

Pereira S, Otero A (2020) *Haematococcus pluvialis* bioprocess optimization: effect of light quality, temperature and irradiance on growth, pigment content and photosynthetic response. Algal Res 51:102027. https://doi.org/10.1016/j.algal.2020.102027

Poonkum W, Powtongsook S, Pavasant P (2014) Astaxanthin induction in microalga *H. pluvialis* with flat panel airlift photobioreactors under indoor and outdoor conditions. Prep Biochem Biotechnol 45(1):1–17. https://doi.org/10.1080/10826068.2013.867871

Praveenkumar R, Lee K, Lee J, Oh Y-K (2015) Breaking dormancy: an energy- efficient means of recovering astaxanthin from microalgae. Green Chem 17:1226–1234. https://doi.org/10.1039/C4GC01413H

Ranga Rao A, Raghunath Reddy RL, Baskaran V, Sarada R, Ravishankar GA (2010) Characterization of microalgal carotenoids by mass spectrometry and their bioavailability and antioxidant properties elucidated in rat model. J Agric Food Chem 58(15):8553–8559. https://doi.org/10.1021/jf101187k

Ranga Rao A, Sindhuja HN, Dharmesh SM, Sankar KU, Sarada R, Ravishankar GA (2013a) Effective inhibition of skin cancer, tyrosinase, and antioxidative properties by astaxanthin and astaxanthin esters from the green alga *Haematococcus pluvialis*. J Agric Food Chem 61:3842–3851. https://doi.org/10.1021/jf304609j

Ranga Rao A, Baskaran V, Sarada R, Ravishankar GA (2013b) In vivo bioavailability and antioxidant activity of carotenoids from microalgal biomass-a repeated dose study. Food Res Int 54(1):711–717. https://doi.org/10.1016/j.foodres.2013.07.067

Ranga Rao A, Siew Moi P, Ravi S, Aswathanarayana RG (2014) Astaxanthin: sources, extraction, stability, biological activities and its commercial applications-a review. Mar Drugs 12:128–152. https://doi.org/10.3390/md12010128

Rao AR, Sindhuja HN, Dharmesh SM, Sankar KU, Sarada R, Ravishankar GA (2013) Effective inhibition of skin cancer, tyrosinase, and antioxidative properties by astaxanthin and astaxanthin esters from the green alga *Haematococcus pluvialis*. J Agric Food Chem 61(16):3842–3851. https://doi.org/10.1021/jf304609j

Recht L, Töpfer N, Batushansky A, Sikron N, Gibon Y, Fait A, Nikoloski Z, Boussiba S, Zarka A (2014) Metabolite profiling and integrative modeling reveal metabolic constraints for carbon partitioning under nitrogen starvation in the green algae *Haematococcus pluvialis*. J Biol Chem 289(44):30387–30403

Řezanka T, Nedbalová L, Sigler K, Cepák V (2008) Identification of astaxanthin diglucoside diesters from snow alga *Chlamydomonas nivalis* by liquid chromatography–atmospheric pressure chemical ionization mass spectrometry. Phytochemistry 69(2):479–490. https://doi.org/10.1016/j.phytochem.2007.06.025

Sadoud M, Mokhtar M, Bouamar S, Bouziane N, Belabbas M, Riazi A (2019) Nitrate starvation associated with strong light induces astaxanthin accumulation in a new green microalga *Haematococcus pluvialis* isolated from Algerian freshwater. South Asian J Exp Biol 9(2): 40–49. https://doi.org/10.38150/sajeb

Saha SK, McHugh E, Hayes J, Moane S, Walsh D, Murray P (2013) Effect of various stress regulatory factors on biomass and lipid production in microalga *Haematococcus pluvialis*. Bioresour Technol 128:118–124. https://doi.org/10.1016/j.biortech.2012.10.049

Sarada R, Tripathi U, Ravishankar GA (2002) Influence of stress on astaxanthin production in *Haematococcus pluvialis* grown under different culture conditions. Process Biochem 37:623–627. https://doi.org/10.1016/S0032-9592(01)00246-1

Sarada R, Vidhyavathi R, Usha D, Ravishankar GA (2006) An efficient method for extraction of astaxanthin from green alga *Haematococcus pluvialis*. J Agric Food Chem 54(20):7585–7588. https://doi.org/10.1021/jf060737t

Scibilia L, Girolomoni L, Berteotti S, Alboresi A, Ballottari M (2015) Photosynthetic response to nitrogen starvation and high light in *Haematococcus pluvialis*. Algal Res 12:170–181. https://doi.org/10.1016/j.algal.2015.08.024

Shah MMR, Liang Y, Cheng JJ, Daroch M (2016) Astaxanthin-producing green microalga *Haematococcus pluvialis*: from single cell to high value commercial products. Front Plant Sci 7. https://doi.org/10.3389/fpls.2016.00531

Shimidzu N, Goto M, Miki W (1996) Carotenoids as singlet oxygen quenchers in marine organisms. Fish Sci 62(1):134–137. https://doi.org/10.2331/fishsci.62.134

Solovchenko AE (2015) Recent breakthroughs in the biology of astaxanthin accumulation by microalgal cell. Photosynth Res 125:437–449. https://doi.org/10.1007/s11120-015-0156-3

Takaichi S (2011) Carotenoids in algae: distributions, biosyntheses and functions. Mar Drugs 9(6): 1101–1118. https://doi.org/10.3390/md9061101

Tam LT, Hoang DD, Ngoc Mai DT, Hoai Thu NT, Lan Anh HT, Hong DD (2012) Study on the effect of salt concentration on growth and Astaxanthin accumulation of microalgae *Haematococcus pluvialis* as the initial basis for two phase culture of astaxanthin production. Tap Chi Sinh Hoc 34(2). https://doi.org/10.15625/0866-7160/v34n2.964

Telli M, Şahin G (2020) Effects of gradual and sudden changes of salinity and light supply for astaxanthin production in *Haematococcus pluvialis* (Chlorophyceae). Fundam Appl Limnol 194(1):11–17. https://doi.org/10.1127/fal/2020/1300

Tjahjono AE, Hayama Y, Kakizono T, Terada Y, Nishio N, Nagai S (1994) Hyper-accumulation of astaxanthin in a green alga *Haematococcus pluvialis* at elevated temperatures. Biotechnol Lett 16(2):133–138. https://doi.org/10.1007/bf01021659

Turkez H, Geyikoglu F, Yousef MI (2012) Beneficial effect of astaxanthin on 2,3,7,8-tetrachlorodibenzo-p-dioxin-induced liver injury in rats. Toxicol Ind Health 29(7):591–599. https://doi.org/10.1177/0748233711434959

Urich K (1994) Comparative animal biochemistry. Springer, Germany

Vidhyavathi R, Venkatachalam L, Kamath BS, Sarada R, Ravishankar GA (2007) Differential expression of carotenogenic genes and associated changes in pigment profile during regeneration of *Haematococcus pluvialis* cysts. Appl Microbiol Biotechnol 75(4):879–887. https://doi.org/10.1007/s00253-007-0876-1

Wan M, Zhang J, Hou D, Fan J, Li Y, Huang J, Wang J (2014) The effect of temperature on cell growth and astaxanthin accumulation of *Haematococcus pluvialis* during a light–dark cyclic cultivation. Bioresour Technol 167:276–283. https://doi.org/10.1016/j.biortech.2014.06.030

Wan X, Zhou X-R, Moncalian G, Su L, Chen W-C, Zhu H-Z, Chen D, Gong Y-M, Huang F-H, Deng Q-C (2021) Reprogramming microorganisms for the biosynthesis of astaxanthin *via* metabolic engineering. Prog Lipid Res 81:101083. https://doi.org/10.1016/j.plipres.2020.101083

Wang F, Gao B, Wu M, Huang L, Zhang C (2019) A novel strategy for the hyper-production of astaxanthin from the newly isolated microalga *Haematococcus pluvialis* JNU35. Algal Res 39:101466

Wayama M, Ota S, Matsuura H, Nango N, Hirata A, Kawano S (2013) Three-dimensional ultrastructural study of oil and astaxanthin accumulation during encystment in the green alga *Haematococcus pluvialis*. PLoS One 8(1):e53618. https://doi.org/10.1371/journal.pone.0053618

Wei D, Peng J, Zhao Y, Zhao P, Xu J-W, Li T, Yu X (2018) A strategy for boosting astaxanthin accumulation in green microalga *Haematococcus pluvialis* by using combined diethyl aminoethyl hexanoate and high light. J Appl Phycol 31(1):171–181. https://doi.org/10.1007/s10811-018-1561-8

Wibrand K, Berge K, Messaoudi M, Duffaud A, Panja D, Bramham CR, Burri L (2013) Enhanced cognitive function and antidepressant-like effects after krill oil supplementation in rats. Lipids Health Dis 12(1):6. https://doi.org/10.1186/1476-511x-12-6

Yang Y, Kim B, Lee JY (2013) Astaxanthin structure, metabolism, and health benefits. J Hum Nutr Food Sci 1, 1003:1–1003:11

Yokoyama A, Adachi K, Shizuri Y (1995) New carotenoid glucosides, astaxanthin glucoside and adonixanthin glucoside, isolated from the astaxanthin-producing marine bacterium, *Agrobacterium aurantiacum*. J Nat Prod 58(12):1929–1933. https://doi.org/10.1021/np50126a022

Yoshida H, Yanai H, Ito K, Tomono Y, Koikeda T, Tsukahara H, Tada N (2010) Administration of natural astaxanthin increases serum HDL-cholesterol and adiponectin in subjects with mild hyperlipidemia. Atherosclerosis 209(2):520–523. https://doi.org/10.1016/j.atherosclerosis.2009.10.012

Zhang BY, Geng YH, Li ZK, Hu HJ, Li YG (2009) Production of astaxanthin from *Haematococcus* in open pond by two-stage growth one-step process. Aquaculture 295(3–4):275–281. https://doi.org/10.1016/j.aquaculture.2009.06.043

Zhang C, Liu J, Zhang L (2017) Cell cycles and proliferation patterns in *Haematococcus pluvialis*. Chinese J Oceanol Limnol 35:1205–1211. https://doi.org/10.1007/s00343-017-6103-8

Zhang W, Zhou X, Zhang Y, Cheng P, Ma R, Cheng W et al (2018) Enhancing astaxanthin accumulation in *Haematococcus pluvialis* by coupled light intensity and nitrogen starvation in column photobioreactors. J Microbiol Biotechnol 28(12):2019–2028. https://doi.org/10.4014/jmb.1807.07008

Zhu Y, Zhang Z, Xu X, Cheng J, Chen S, Tian J, Yang W, Crocker M (2021) Simultaneous promotion of photosynthesis and astaxanthin accumulation during two stages of *Haematococcus pluvialis* with ammonium ferric citrate. Sci Total Environ 750:141689. https://doi.org/10.1016/j.scitotenv.2020.141689

Chapter 4
Haematococcus Cultivation for Astaxanthin Production

Gamze Turan

Abbreviations

ACC	1-Aminocyclopropane-1-carboxylic acid
C/N	Carbon/nitrogen
CAGR	Compound annual growth rate
H. pluvialis	*Haematococcus pluvialis*
H/D	Height/diameter
LEDs	Light emitting diodes
USFDA	United States Food and Drug Administration
VVM	Volume of air per volume of liquid per minute

4.1 Introduction

Astaxanthin, a fat soluble pigment, is found in many red-colored micro- and macro-marine organisms, such as yeast, algae, salmon, shrimp, krill, trout, and crawfish species (Ambati et al. 2014). According to the 'Astaxanthin - Global Market Outlook (2023–2032)', the Global astaxanthin market was valued as US $758.2 Mn in 2022 and it is expected to grow to reach US $2001.5 Mn by the end of 2032 with a compound annual growth rate of 10.2% (https://www.transparencymarketresearch.com/astaxanthin-market.html). Due to its reduced production cost, synthetic astaxanthin accounts for 95% of market astaxanthin, but the naturally originated, algal *Haematococcus* astaxanthin has an increased preference in the market (Solymosi et al. 2015; Kumar et al. 2020).

G. Turan (✉)
Fisheries Faculty, Aquaculture Department, Ege University, Izmir, Turkey
e-mail: gamze.turan@ege.edu.tr

R. Raja et al. (eds.), *Haematococcus*, https://doi.org/10.1007/978-981-99-2901-6_4

On the dry weight basis astaxanthin accumulate up to 5% in the green microalga *Haematococcus* (Boussiba et al. 1999; Borowitzka 2019; Turan 2019, 2021). As a result, Haematococcus is a major source of astaxanthin for industries that produce astaxanthin (Orosa et al. 2001; Kumar et al. 2020). When the environment is unfavorable for normal algal cell growth, algal astaxanthin accumulates quickly in the resting algal cells (Boussiba et al. 1999). It is recognized that unfavorable growth circumstances, such as nitrate limitation, elevated temperature, light, and carbon dioxide concentration can cause increased algal astaxanthin synthesis and accumulation (Borowitzka 2019). In this chapter, the recent studies for commercial production of astaxanthin which is high valued bioactive compound from *Haematococcus* and the significant benefits of utilization of *Haematococcus* astaxanthin by various industries are summarized.

4.2 Astaxanthin

Astaxanthin (3, 3′-dihydroxy-β, β-carotene-4, 4′-dione), a very bright red color fat soluble pigment belonging to the family of xanthophylls (cartenoids) (Kumar et al. 2020). Astaxanthin is present in many red-colored freshwater and marine aquatic organisms. The aquatic natural resources of astaxanthin are yeast, algae, trout, salmon, shrimp, krill, and crawfish (Ambati et al. 2014; Kumar et al. 2020). Using astaxanthin as food and feed colorant has been approved by The United States Food and Drug Administration (USFDA) and European Commission (Ambati et al. 2014, Kumar et al. 2020). Astaxanthin has been also approved by USFDA as a food supplement in 1999 (Guerin et al. 2003; Kumar et al. 2020).

Because of its strong antioxidant activity, astaxanthin is also known as "super vitamin E" (Miki 1991; Kumar et al. 2020). Astaxanthin does not cause any hypervitaminosis A toxicity since astaxanthin cannot be transformed to vitamin A in humans (Jyonouchi et al. 1995; Fassett and Coombes 2011; Kumar et al. 2020). Addition to having antioxidant properties, astaxanthin also has anti-inflammatory properties and antidiabetic activity (Ranga Rao et al. 2013; Lim et al. 2018; Kumar et al. 2020). Because of these important properties, astaxanthin is used in the treatment of cancerous cells and cardiovascular diseases (CVD) (Ambati et al. 2014; Kumar et al. 2020).

4.3 *Haematococcus pluvialis*

The freshwater green alga *Haematococcus* is excellent natural source of astaxanthin which is produced and accumulated in the non-motile aplanospore cell stage of *Haematococcus* (Borowitzka 2010, 2013, 2019). Aplanospore cell formation and astaxanthin accumulation are induced by unfavorable growth conditions such as nitrate limitation, increased temperature, light, or carbon dioxide conditions

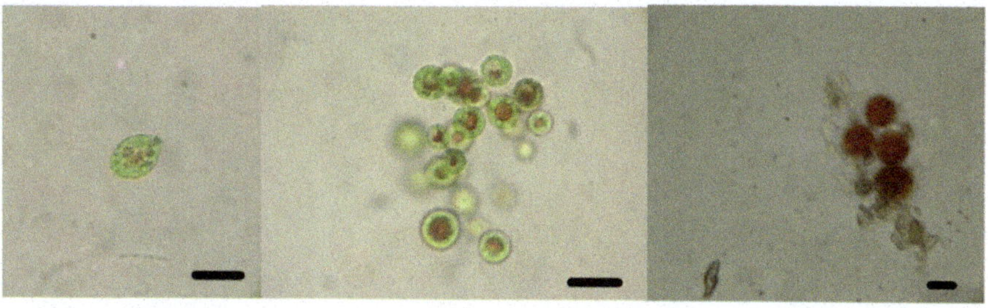

Fig. 4.1 Green stage cell (**a**), green to reddish color early stage palmelloid cells (**b**) and harvested red color aplanospores (Hematocysts) (**c**) of the Turkish isolate of *Haematococcus* sp. (Turan 2019). Scale bars represent: (**a**–**c**), 10 μm

(Borowitzka 1992, 1995; Boussiba and Vonshak 1991; Vidhyavathi et al. 2008). Astaxanthin and aplanospore cell formations generally occur together (Droop 1955); however, some astaxanthin formation may occur in the motile flagellate cell stage under specific nutrient limitation (Del Rio et al. 2005; Borowitzka 2019). *H. pluvialis* can also use organic carbon such as acetate either mixotrophically or heterotrophically (Kobayashi et al. 1992). However, astaxanthin formation was found to be very much higher in mixotrophic cultures in the light conditions (Orosa et al. 2001; Kang et al. 2005). There is a high degree of variation between *H. pluvialis* strains which were cultivated in different acetate concentrations and other organic carbon sources such as glucose, glycerol, and glycine (Borowitzka 1992, 1995, 2005, 2010, 2013, 2019).

The optimal growth temperature of *H. pluvialis* ranges between 10 and 20 °C. In *Haematococcus* batch cultivation systems, the motile flagellated green cell stage (Fig. 4.1) predominates during the exponential active phase of growth and the carotenogenic non-motile aplanospores (Fig. 4.1c) are formed during the stationary phase (Borowitzka 2005). This change in cell morphology—from the motile flagellated green cells to the extremely thick-walled, non-motile bright red color aplanospores—means that generally a two-stage production process is used (Borowitzka 2010; Turan 2019). Here, the first stage is optimized to produce growth on the flagellated "green stage cells" of *H. pluvialis* to maximize biomass and is usually carried out in closed photobioreactors to minimize contamination. This is then followed by a second stage, "red stage cells" of *H. pluvialis,* often in open ponds, or in ourdoor closed photobioreactors in which the cells are "stressed" (nitrate limitations, increased salt, temperature, and light conditions) to induce aplanospore and astaxanthin formation (Olaizola 2000a; Cysewski and Lorenz 2004; Borowitzka 2010, 2013, 2019).

Turan (2019) scaled up the Turkish isolate of *Haematococcus* sp. in greenhouse raceway ponds in batch mode for astaxanthin production. Cultivations of motile flagellated green stage cells (Fig. 4.1a), early stage palmelloid cells that are green to reddish color (Fig. 4.1b) and harvested red color non-motile aplanospores

Fig. 4.2 Cultivations of green stage cells (**a**), green to reddish color early stage palmelloid cells (**b**) and harvested red color aplanospores (Haematocysts) (**c**) of the Turkish isolate of *Haematococcus* sp. in raceway ponds (Turan 2019)

(Hematocysts) (Fig. 4.1c) were performed (Fig. 4.2) (Turan 2019) and the astaxanthin yield was found to be 3.6 ± 0.1% in dry weight and concluded that *Haematococcus* sp. has a high potential candidate for the commercial astaxanthin production among the four *Haematococcus* species—*H. pluvialis, H. Droebachiensis, H. capensis,* and *H. lacustris*. Laboratory and commercial cultures include 11 strains, such as the *H. pluvialis* cultivation in Turkey (Göksan and Gökpınar 2005; Göksan et al. 2010, 2011; Turan 2019).

4.4 *Haematococcus* Cultivation for Astaxanthin Production

Astaxanthin has been commercially available from *Haematococcus* species (Kumar et al. 2020) for over 35 years and production costs are up to US$100 *Haematococcus* per kg dry weight (Olaizola 2000a; Kumar et al. 2020). According to the research of Olaizola (2000b), *H. pluvialis* is an excellent candidate for commercial astaxanthin production in a closed circuit computer-controlled photobioreactor with a capacity of 25,000 L prepared for outdoor cultivation (Olaizola 2000b; Kumar et al. 2020). In another study conducted by Olaizola (2000b) for the feasibility of commercial astaxanthin production from *H. pluvialis*, the importance of various parameters affecting production during production was also determined (Olaizola 2000b; Kumar et al. 2020). The growth of H. pluvialis and production of astaxanthin in 30 (L) air-lift photobioreactors were also evaluated (Harker et al. 1996; Kumar et al. 2020). Under limited nitrogen conditions, the interaction between photochemical and physiological parameters with astaxanthin production was determined in

H. pluvialis cultures grown in open air conditions in photobioreactors (Torzillo et al. 2003, 2005; Kumar et al. 2020).

Different conditions such as inclusion of ten-fold concentrated culture medium as batch feed or illumination with blue light-emitting diodes (LEDs) in the cultivation of *Haematococcus* was also studied (Lababpour et al. 2005; Kumar et al. 2020). In a research study done by Katsuda et al. (2006), a significantly higher concentration of astaxanthin was produced as a result of illumination with flashing light compared to continuous light illumination (Katsuda et al. 2006; Kumar et al. 2020). Astaxanthin production and accumulation in *H. pluvialis* cells cultivated in a photochemostat with constant illumination and dilution ratio was also studied (Del Rio et al. 2005; Kumar et al. 2020). Del Rio et al. (2005) also determined the effect of nitrate concentrations on biomass, astaxanthin production and accumulation, and developed a mathematical model to relate growth rate to both light and specific nitrate availability (Del Rio et al. 2008; Kumar et al. 2020). This model was a very useful tool for efficient accumulation of *Haematococcus* astaxanthin in a single point system (Del Rio et al. 2008, Kumar et al. 2020).

It was found by Kang et al. (2005) that the photoautotrophic cultivation method was more effective compared to the heterotrophic cultivation technique for the production of astaxanthin from *H. pluvialis.* (Kang et al. 2005; Kumar et al. 2020). Another study showed that the effect of carbon to nitrogen ratio was highly pronounced on astaxanthin production and accumulation in *Haematococcus* compared to and light intensity (Kang et al. 2007; Kumar et al. 2020). The same authors also suggested that the fed-batch culture process following photoautotrophic cultivation has a high potential for astaxanthin production from *H. pluvialis* (Kang et al. 2010; Kumar et al. 2020).

Haematococcus was grown in photoautotrophic conditions in a bubble column by fed-batch culture treatment where limited nitrate applied increased the production and accumulation of astaxanthin by *H. pluvialis* (Ranjbar et al. 2008; Kumar et al. 2020). It was observed that tubular reactors are most suitable for the production of *H. pluvialis* biomass and astaxanthin due to higher light availability compared to bubble column (Lopez et al. 2006; Kumar et al. 2020). In the same study of Lopez et al. (2006), it was determined that the maximum astaxanthin production could be achieved under nitrate saturation conditions (Kumar et al. 2020).

Astaxanthin production in two-stage systems was found to be higher than in single-stage systems (Aflalo et al. 2007; Kumar et al. 2020). After determining the feasibility of a single-step system for the constant production of astaxanthin from *H. pluvialis,* it was demonstrated that astaxanthin could be produced efficiently in continuous outdoor culture systems with the appropriate amount of nitrate as well as the use of standard light within the culture system (Garcia-Malea et al. 2009; Kumar et al. 2020). A new method called "two-stage growth one-step process" designed by Zhang et al. (2009) cultivates four *H. pluvialis* strains for astaxanthin production (Zhang et al. 2009; Kumar et al. 2020). As the yield of astaxanthin correlates with the growth phases of *H. pluvialis,* the cellular phase was optimized during its transfer to the second production stage, yielding more than tenfold astaxanthin yields in the senescent green cell phase (Choi et al. 2011; Kumar et al. 2020).

Various parameters including the slope of the V-shaped base of a photobioreactor, the height/diameter (H/D) ratio, the volumetric flow rate of air (vvm), and the diameter of an air diffuser were optimized for long-term phototropic cultivation of *H. pluvialis* (Yoo et al. 2012; Kumar et al. 2020). In another study, besides the initial biomass density, the effect on cell death, morphology, growth, pigment content, and biomass productivity on the non-motile red cell stage growth of *H. pluvialis* were determined (Wang et al. 2013a; Kumar et al. 2020).). The effect of seasonal changes such as temperature and light intensity on the association between the initial biomass density, growth of *H. pluvialis,* and astaxanthin production was examined (Kumar et al. 2020). The effects of initial biomass density and nitrogen concentration ($\leq$4.4 mM) on *H. pluvialis* growth resulted in the highest astaxanthin productivity under a clear glass column photobioreactor (Wang et al. 2013b; Kumar et al. 2020).

An immobilized biofilm method, called the "Linked Cultivation" method was developed for the cultivation of *H. pluvialis* (Zhang et al. 2014; Kumar et al. 2020). It was observed the effect of inoculum density, medium volume, light intensity and nitrogen content on the astaxanthin accumulation in immobilized *H. pluvialis* (Zhang et al. 2014; Kumar et al. 2020). Cells grown in temperatures between 23 and 28 °C had better astaxanthin accumulation during light-dark cyclic cultivation of *H. pluvialis* (Wan et al. 2014; Kumar et al. 2020). In order to determine the effects of light on astaxanthin accumulation, fed-batch culture conditions of *H. pluvialis* (5 L photobioreactor) receiving different ratios of blue and white light were also examined, and the results showed that the mixture of blue and white light was better than a single light source(Sun et al. 2015; Kumar et al. 2020).

It was also found that high concentration of carbon dioxide stress and nuclear irradiation enhance the growth rate and astaxanthin accumulation of *H. pluvialis* with 5% increased growth rate and astaxanthin production in mutant *H. pluvialis* than the wild strain (Cheng et al. 2016; Kumar et al. 2020). Enhanced astaxanthin production was found whenever *H. pluvialis* cells were treated at the early red cell stage with 1-aminocyclopropane-1-carboxylic acid (ACC) where a combined potentiating effect of light intensity and ACC on better astaxanthin yield was observed (Vo et al. 2016; Kumar et al. 2020).

A biofilm cultivation method with high light intensities which makes a highly productive one-phase processing possible for astaxanthin production (Kiperstok et al. 2017; Kumar et al. 2020). In the study of Kobayashi et al. (1992), the astaxanthin of *H. pluvialis* produced in an Fe^{2+}-rich carbon (acetate) cultivation medium was applied in different photoperiod cycles and as a result, continuous illumination was the most favorable for astaxanthin formation. It was also observed that higher astaxanthin accumulation in *H. pluvialis* can be obtained with an input of carbon (acetate) plus Fe^{2+} compared to carbon (acetate) addition alone (Kobayashi et al. 1993; Kumar et al. 2020).

Effect of non-motile red cell or red cyst (aplanospore) cell inoculation and iron (II) supplementation on autotrophic astaxanthin production under outdoor summer conditions are also studied (Hong et al. 2016). The use of a conventional, reliable autotrophic culture system to minimize or eliminate potential predators and competitors by the addition of sodium chloride at the onset of the stationary phase of *H. pluvialis* growth was studied as well (Hong et al. 2016; Kumar et al. 2020). A higher yield of astaxanthin in *H. pluvialis* was obtained with increased light illumination, increased carbon dioxide concentration and devoid of a medium replacement step (Christian et al. 2018; Kumar et al. 2020). It was also suggested to add potassium acetate as an organic carbon source in mixotrophic cultivations for good astaxanthin production by *H. pluvialis* (Pan-utai et al. 2017; Kumar et al. 2020). A Type II model of light distribution for motile green cells of *H. pluvialis* cultivated under red LED lights was also developed (Gao et al. 2017; Kumar et al. 2020).The direct effect of electrical treatment on *H. pluvialis* growth was also studied and suggested to use 100 mA (voltage: 25 V) with optimum current (Kim et al. 2018; Kumar et al. 2020).

4.5 Conclusion

In this chapter, the effects of various variables, such as nitrate, nutritional supplements, light, carbon dioxide, temperature, etc. on the higher biomass yield, astaxanthin production and accumulation of *Haematococcus* under different growing conditions and cultivation systems are summarized. The green *Haematococcus* alga has many advantages since this strain is a renewable resource, produces high valuable compounds that are environmentally friendly, and it is also easy to scale up the synthesis processing simply by adjusting the growth conditions. However, the production cost of astaxanthin produced from *Haematococcus* is still very high. In order to reduce the production cost of *Haematococcus* astaxanthin, algal biomass production, astaxanthin production, and accumulation must be improved. Therefore, researchers should focus on discovering new *Haematococcus* species, including genetically engineered *Haematococcus* species, for better biomass yield and better ability to adapt to changing new environmental conditions. Advances in processing techniques are also absolutely necessary to reduce losses during processing.

Acknowlegments Dr. TURAN thankful to Ege University Fisheries Faculty Aquaculture Department for providing facility and financial support for this work.

Conflict of Interest The author has declared no conflict of interest.

References

Aflalo C, Meshulam Y, Zarka A, Boussiba S (2007) Determination of heat flux on dual bell nozzle by Monte Carlo method. Biotechnol Bioeng 98(1):300–305

Ambati RR, Moi PS, Ravi S, Aswathanarayana RG (2014) Astaxanthin: sources, extraction, stability, biological activities and its commercial applications–a review. Mar Drugs 12(1): 128–152

Borowitzka MA (1992) Comparing carotenogenesis in *Dunaliella* and *Haematococcus*: implications for commercial production strategies. In: Villa TG, Abalde J (eds) Profiles on biotechnology. Universidade de Santiago de Compostela, Santiago de Compostela, pp 301–310

Borowitzka MA (1995) The biotechnology of microalgal carotenoid production. In: Aubert JP, Martin PMV (eds) Microbes, environment, biotechnology. Abstracts, Colloques Internationaux de L'annee Pasteur. Institut Louis Malarde, Papeete, Tahiti, pp 149–151

Borowitzka MA (2005) Carotenoid production using microorganisms. In: Cohen Z, Ratledge C (eds) Single cell oils microbial and algal oils, 1st edn. AOCS Press, Urbana, pp 124–137

Borowitzka MA (2010) Carotenoid production using microorganisms, chapter 11. In: Cohen Z, Ratledge C (eds) Single cell oils, microbial and algal oils, 2nd edn. Elsevier, Amsterdam, pp 225–240

Borowitzka MA (2013) High-value products from microalgae-their development and commercialisation. J Appl Phycol 25(3):743–756

Borowitzka MA (2019) Microalgal metabolism and their utilisation. In: Malcata X, Pinto IS, Guedes AC (eds) Marine macro- and microalgae an overview. CNR Press, Boca Raton, pp 43–62

Boussiba S, Vonshak A (1991) Astaxanthin accumulation in the green alga *haematococcus* pluvialis. Plant Cell Physiol 32(7):1077–1082

Boussiba S, Bing W, Yuan JP, Zarka A, Chen F (1999) Changes in pigments profile in the green alga *Haematococcus pluvialis* exposed to environmental stresses. BiotechnolLett 21(7): 601–604

Cheng J, Zongbo Yang KL, Junhu Z, Kefa Cen S (2016) Enhancing the growth rate and astaxanthin yield of *Haematococcus pluvialis* by nuclear irradiation and high concentration of carbon dioxide stress. Bioresour Technol 204:49–54

Choi YE, Yun YS, Park JM, Yang JW (2011) Determination of the time transferring cells for astaxanthin production considering two-stage process of *Haematococcus pluvialis* cultivation. Bioresour Technol 102(24):11249–11253

Christian D, Zhang J, Sawdon AJ, Peng CA (2018) Enhanced astaxanthin accumulation in *Haematococcus pluvialis* using high carbon dioxide concentration and light illumination. Bioresour Technol 256:548–551

Cysewski GR, Lorenz RT (2004) Industrial production of microalgal cell-mass and secondary products—species of high potential: *Haematococcus*. In: Richmond A (ed) Microalgal culture: biotechnology and applied phycology. Blackwell Science, Oxford, pp 281–288

Del Rio E, Acien FG, Garcia-Malea MC, Rivas J, Molina-Grima E, Guerrero MG (2005) Efficient one-step production of astaxanthin by the microalga *Haematococcus pluvialis* in continuous culture. Biotechnol Bioeng 91(7):808–815

Del Rio E, Acien FG, Garcia-Malea MC, Rivas J, Molina-Grima E, Guerrero MG (2008) Efficiency assessment of the one-step production of astaxanthin by the microalga *Haematococcus pluvialis*. Biotechnol Bioeng 100(2):397–402

Droop MR (1955) Carotogenesis in *Haematococcus pluvialis*. Nature 175:42

Fassett RG, Coombes JS (2011) Astaxanthin: a potential therapeutic agent in cardiovascular disease. Mar Drugs 9(3):447–465

Gao X, Wang X, Li H, Roje S, Sablani SS, Chen S (2017) Parameterization of a light distribution model for green cell growth of microalgae: *Haematococcus pluvialis* cultured under red LED lights. Algal Res 23:20–27

Garcia-Malea MC, Gabriel Acien F, Del Rio E, Fernandez JM, Ceron MC (2009) Production of astaxanthin by *Haematococcus pluvialis*: taking the one-step system outdoors. Biotechnol Bioeng 102(2):651–657

Göksan T, Gökpınar Ş (2005) Vegetative growth characteristics of *Haematococcus pluvialis* Flotow (Chlorophyceae) at different light intensities, *E.U. J Fish Aquat Sci* 22:21–24. (In Turkish)

Göksan T, Ak İ, Gökpınar Ş (2010) An alternative approach to the traditional mixotrophic cultures of *Haematococcus pluvialis* flotow (Chlorophyceae). J Microbiol Biotechnol 20(9):1276–1282

Göksan T, Ak İ, Kılıç C (2011) Growth characteristics of the alga *Haematococcus pluvialis* Flotow as affected by nitrogen source, vitamin, light and aeration. Turk J Fish Aquat Sci 11:377–383. https://doi.org/10.4194/1303-2712-v11_3_06

Guerin M, Huntley ME, Olaizola M (2003) *Haematococcus* astaxanthin: applications for human health and nutrition. Trends Biotechnol 21(5):210–216

Harker M, Tsavalos AJ, Young AJ (1996) Autotrophic growth and carotenoid production of *Haematococcus pluvialis* in a 30 liter air-lift photobioreactor. J Ferment Bioeng 82(2):113–118

Hong ME, Choi YY, Sim SJ (2016) Effect of red cyst cell inoculation and iron(II) supplementation on autotrophic astaxanthin production by *Haematococcus pluvialis* underoutdoor summer conditions. J Biotechnol 218:25–33

Jyonouchi H, Sun S, Tomita Y, Gross MD (1995) Astaxanthin, a carotenoid without vitamin a activity, augments antibody responses in cultures including T-helper cell clones and suboptimal doses of antigen. J Nutr 125(10):2483–2492

Kang CD, Lee JS, Park TH, Sim SJ (2005) Comparison of heterotrophic and photo autotrophic induction on astaxanthin production by *Haematococcus pluvialis*. Appl MicrobiolBiotechnol 68(2):237–241

Kang CD, Lee JS, Park TH, Sim SJ (2007) Complementary limiting factors of astaxanthin synthesis during photoautotrophic induction of *Haematococcus pluvialis:* C/N ratio and light intensity. Appl Microbiol Biotechnol 74(5):987–994

Kang CD, Han SJ, Choi SP, Sim SJ (2010) Fed-batch culture of astaxanthin-rich *Haematococcus pluvialis* by exponential nutrient feeding and stepwise light supplementation. Bioprocess Biosyst Eng 33(1):133–139

Katsuda T, Shimahara K, Shiraishi H, Yamagami K, Ranjbar R, Katoh S (2006) Effect of flashing light from blue light emitting diodes on cell growth and astaxanthin production of *Haematococcus pluvialis*. J Biosci Bioeng 102(5):442–446

Kim JY, Lee C, Jeon MS, Park J, Choi YE (2018) Enhancement of microalga *Haematococcus pluvialis* growth and astaxanthin production by electrical treatment. Bioresour Technol 268: 815–819

Kiperstok AC, Sebestyén P, Podola B, Melkonian M (2017) Biofilm cultivation of *Haematococcus pluvialis* enables a highly productive one-phase process for astaxanthin production using high light intensities. Algal Res 21:213–222

Kobayashi M, Kakizono T, Nishio N, Nagai S (1992) Effects of light intensity, light quality, and illumination cycle on astaxanthin formation in a green alga, Haematococcus pluvialis. J Ferment Bioeng 74(1):61–63

Kobayashi M, Kakizono T, Nagai S (1993) Enhanced carotenoid biosynthesis by oxidative stress in acetate-induced cyst cells of a green unicellular alga, Haematococcus pluvialis. Appl Environ Microbiol 59(3):867–873

Kumar S, Gill BS, Verma A, Verma ML, Kushwaha R (2020) Chapter 7: Biotechnological production of high-valued algal astaxanthin and lutein under different growth conditions. In: Verma MD, Chandel AK (eds) Biotechnological production of bioactive compounds. Elsevier, Amsterdam, pp 191–220

Lababpour A, Shimahara K, Hada K, Kyoui Y, Katsuda T, Katoh S (2005) Fed-batch culture under illumination with blue light emitting diodes (LEDs) for astaxanthin production by *Haematococcus pluvialis*. J Biosci Bioeng 100(3):339–342

Lim KC, Yusoff FM, Shariff M, Kamarudin MS (2018) Astaxanthin as feed supplement in aquatic animals. Rev Aquacult 10:738–773. https://doi.org/10.1111/raq.12200

Lopez MCG-M, Sanchez EDR, Lopez JLC, Fernandez FGA, Sevilla JMF (2006) Comparative analysis of the outdoor culture of *Haematococcus pluvialis* in tubular and bubble column photobioreactors. J Biotechnol 123(3):329–342

Miki W (1991) Biological functions and activities of animal carotenoids. Pure Appl Chem 63(1): 141–146

Olaizola M (2000a) Commercial production of astaxanthin from *Haematococcus pluvialis* using 25,000-liter outdoor photobioreactors. J Appl Phycol 12(3–5):499–506

Olaizola M (2000b) Commercial production of astaxanthin from. Growth 12:499–506
Orosa M, Valero JF, Herrero C, Abalde J (2001) Comparison of the accumulation of astaxanthin in *Haematococcus pluvialis* and other green microalgae under N-starvation and high light conditions. Biotechnol Lett 23:1079–1085
Pan-utai W, Parakulsuksatid P, Pomkaivon N (2017) Effect of inducing agents on growth and astaxanthin production in Haematococcus pluvialis: organic and inorganic. Biocatalysis Agric Biotechnol 12:152–158. https://doi.org/10.1016/j.bcab.2017.10.004
Ranga Rao A, Sindhuja HN, Dharmesh SM, Sankar KU, Sarada R, Ravishankar GA (2013) Effective inhibition of skin cancer, tyrosinase, and antioxidative properties by astaxanthin and astaxanthin esters from the green alga *Haematococcus pluvialis*. J Agric Food Chem 61(16): 3842–3851
Ranjbar R, Inoue R, Shiraishi H, Katsuda T, Katoh S (2008) High efficiency production of astaxanthin by autotrophic cultivation of *Haematococcus pluvialis* in a bubble column photobioreactor. Biochem Eng J 39(3):575–580
Solymosi K, Latruffe N, Morant-Manceau A, Schoefs B (2015) Food colour additives of natural origin. In: Scotter MJ (ed) Colour additives for foods and beverages, Woodhead Publishing series in food science, technology and nutrition, pp 3–34. https://doi.org/10.1016/B978-1-78242-011-8.00001-5
Sun H, Kong Q, Geng Z, Duan L, Yang M, Guan B (2015) Enhancement of cell biomass and cell activity of astaxanthin-rich *Haematococcus pluvialis*. Bioresour Technol 186:67–73
Torzillo G, Göksan T, Faraloni C, Kopecky J, Masojıdek J (2003) Interplay between photochemical activities and pigment composition in an outdoor culture of *Haematococcus pluvialis* during the shift from the green to red stage. J Appl Phycol 15(2/3):127–136
Torzillo G, Göksan T, Işık O, Gökpinar S (2005) Photon irradiance required to support optimal growth and interrelations between irradiance and pigment composition in the green alga *Haematococcus pluvialis*. Eur J Phycol 40(2):233–240
Turan G (2019) Isolation and cultivation of a newly-discovered astaxanthin rich green microalga - *Haematococcus sp.* flotow strain from Homeros Valley (Bornova Creek, Izmir, Turkey). Turk J Fish&Aquat Sci 19(9):785–791. https://doi.org/10.4194/1303-2712-v19_9_07
Turan G (2021) Utilization of astaxanthin from Haematococcus for its use in aquaculture and poultry industries. In: Ravishankar GA, Ranga Rao A (eds) Global perspectives in astaxanthin: from industrial production to food, health, and pharmaceutical applications. Elsevier, Amsterdam. ISBN: 978-0-12-823304-7, pp 460–470
Vidhyavathi R, Venkatachalam L, Sarada R, Ravishankar GA (2008) Regulation of carotenoid biosynthetic genes expression and carotenoid accumulation in the green alga *Haematococcus pluvialis* under nutrient stress conditions. J Exp Bot 59:1409–1418
Vo TT, Lee C, Han SI, Kim JY, Kim S, Choi YE (2016) Effect of the ethylene precursor,1-aminocyclopropane-1-carboxylic acid on different growth stages of *Haematococcus pluvialis*. Bioresour Technol 220:85–93
Wan M, Zhang J, Hou D, Fan J, Li Y (2014) The effect of temperature on cell growth and astaxanthin accumulation of *Haematococcus pluvialis* during a light-dark cyclic cultivation. Bioresour Technol 167:276–283
Wang J, Han D, Sommerfeld MR, Lu C, Hu Q (2013a) Effect of initial biomass density on growth and astaxanthin production of *Haematococcus pluvialis* in an outdoor photobioreactor. J Appl Phycol 25(1):253–260
Wang J, Sommerfeld MR, Lu C, Hu Q (2013b) Combined effect of initial biomass density and nitrogen concentration on growth and astaxanthin production of *Haematococcus pluvialis* (Chlorophyta) in outdoor cultivation. Algae 28(2):193–202
Yoo JJ, Choi SP, Kim BW, Sim SJ (2012) Optimal design of scalable photo-bioreactor for phototropic culturing of *Haematococcus pluvialis*. Bioproc Biosyst Eng 35(1–2):309–315
Zhang BY, Geng YH, Li ZK, Hu HJ, Li YG (2009) Production of astaxanthin from *Haematococcus* in open pond by two-stage growth one-step process. *Aquaculture* 295(3–4):275–281
Zhang W, Wang JJ, Liu T (2014) Attached cultivation of *Haematococcus pluvialis* for astaxanthin production. *Bioresour Technol* 158C:329–335

Chapter 5
Bottlenecks in the Cultivation Processes of *Haematococcus pluvialis*

M. P. Sudhakar, D. Vidya, G. Dharani, and K. Arunkumar

5.1 Introduction

Haematococcus pluvialis, a green unicellular freshwater microalga grown in temperate regions becomes a valuable resource for bioactive molecules production (Oslan et al. 2021; Shah et al. 2016). *Haematococcus* algae are very rich in astaxanthin and lutein with high market value. The main product such as astaxanthin pigment production using *Haematococcus pluvialis* is gaining more importance in commercial aspects. Growing *Haematococcus pluvialis* in a lab scale level need precise conditions and environment (Yang et al. 2021). But growing outside the lab conditions is again a major challenge facing by the industries for the astaxanthin production (Jannel et al. 2020; Shah et al. 2016).

New formulations in the medium composition enhance the growth of *Haematococcus* sp. and astaxanthin production (Karuppan et al. 2022). However, the large freshwater footprint and the requirement of external nutrient supply are issues in its commercial scale cultivation (Nishshanka et al. 2021). Therefore, two-phase cultivation strategy is necessary for biomass accumulation (green stage) and secondary metabolite production such as astaxanthin (red stage) (Kiperstok et al. 2017). For commercial scale production one must see the external factors which are

M. P. Sudhakar (✉) · G. Dharani
Marine Biotechnology Division, National Institute of Ocean Technology, Chennai, Tamil Nadu, India
e-mail: sudhakar.niot@gov.in

D. Vidya · K. Arunkumar
Phycoscience Laboratory, Department of Plant Science, School of Biological Sciences, Central University of Kerala, Periye, Kerala, India

R. Raja et al. (eds.), *Haematococcus*, https://doi.org/10.1007/978-981-99-2901-6_5

affecting the growth condition of *Haematococcus* sp. Factors such as temperature, water quality, mites, insects, weather conditions, light intensity variation, bacterial contamination, salinity, pH, nutrients are the few important parameters has to be considered before startup of cultivation. In this chapter, *Haematococcus* sp. cultivation aspects and its difficulties in achieving the large scale for commercial aspects were emphasized and discussed.

5.2 Cultivation Process of *Haematococcus* Sp.

One of the most suggested techniques is the cultivation of microalgae in a multiple-stage process (the cultivation in two-stage systems), in which the first stage, optimum conditions are applied aiming at the maximization of biomass production, while in the second stage, stress conditions are applied aiming at the accumulation of the desired compounds (Markou and Nerantzis 2013). This type of research mainly focuses on the absolute effect of the specific stress conditions (nutrients, light, pH, and temperature) on the alteration of the biomass using photoautotrophic and heterotrophic cultivation (Sun et al. 2016). For commercial scale cultivation of *Haematococcus* sp. there are three types of cultivation usually adapted by industries, such as raceway ponds, circular ponds, and unstirred ponds. Open ponds are mostly used for the cultivation of microalgae (Yu et al. 2022) (Fig. 5.1).

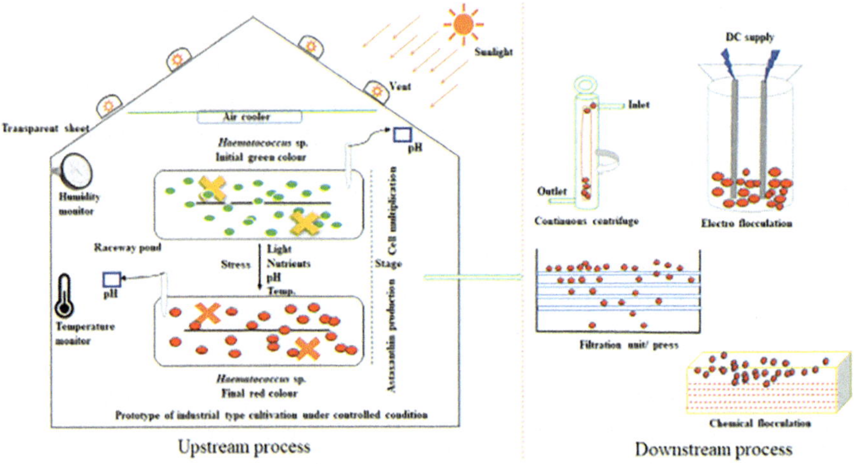

Fig. 5.1 Schematic representation of industrial scale *Haematococcus* sp. cultivation process

5.3 Contamination During Cultivation Process

5.3.1 Open Pond Farming

In open pond systems, microalgae cultivation seems to be simple and easy, with low-cost technology for mass cultivation. There are certain obstacles in the cultivation process and it has been monitored frequently inorder to get a higher yield of algal biomass. An open system favors natural sunlight and proper aeration for the microalgae. However, it is difficult to maintain the temperature and paddle wheel mixing. During the rainy season, the mixing of rainwater affects the pH of the culture system and increases the possibility of bacterial contamination. In an open pond system, there is a possibility of other microalgal blooms in the pond due to the open aeration system (Yu et al. 2022). In the case of *Haematococcus* sp. the cultivation process in an open system is very difficult to maintain the parameters such as temperature and sunlight. In the green stage, the microalgae are about to multiply rapidly with nutrient accumulation and growth. Axenic culture maintenance in an open pond system needs more precautions and routine clean-up activities. For industrial scale cultivation, only one culture can be maintained in an open pond system to avoid cross contamination of other microalgae.

5.3.2 Closed System Cultivation

Another type of microalgae cultivation system is the closed type. This type of cultivation is mostly preferred by industry to avoid external contamination and to get a high yield of biomass as well as products. The CO_2 mixing is properly maintained throughout the culture period with a controlling mechanism. In closed systems, different types of photobioreactors (PBRs) are used based on the type of microalgae and product development (Zhang et al. 2017). Photobioreactors help in the continuous supply of nutrients and other essential parameters such as light and CO_2 under controlled conditions. The biomass productivity and product formation will be higher than in open system cultivation. Semi-continuous production also favors microalgal growth and product harvest in-between the processes (Yu et al. 2022). The use of only one microalgae strain in the same PBRs, the formation of biofilms and fouling, which may inhibit the growth of microalgae, the possibility of rising production costs, and the requirement for a complex setup that is expensive are some of the disadvantages that come with industrial cultivation.

5.4 Bottlenecks in Cultivation Parameters

There are multiple factors affecting the cultivation process of *Haematococcus* sp. Factors such as nutrients, light, aeration, external contamination, harvesting, photobioreactors, etc. will be considered before the cultivation process. The life cycles of *Haematococcus* sp. in different stages are as follows: (1) Vegetative cell growth, (2) Encystment (vegetative to immature cyst cells), (3) Maturation (immature to mature cyst cells), and (4) Germination (mature cyst to vegetative cells). The above-mentioned stages are much important for industrial scale cultivation.

5.4.1 Nutrient Supplements and Its Effect on Growth

The biochemical composition of *H. pluvialis* is affected by the nutrients content and concentration of culture media used for stimulating the growth and biosynthetic pathway of the cells. Microalgae usually produce secondary metabolites under stress conditions. *Haematococcus* sp. under acetate and Fe^{2+} supplements accumulation produces astaxanthin (He et al. 2018). Oxidative stress (OH supplied by H_2O_2) is essential for promoting the formation of several secondary carotenoids and other carbon, nitrogen starvation also induces astaxanthin production (Perez-Garcia et al. 2011).

Studies revealed that acetate has a promising role in inducing the growth and astaxanthin accumulation in *H. pluvialis* (Su et al. 2014). But higher concentration of magnesium acetate resulted in the death of the cells due to the high stress condition. A 2.5 mM sodium orthovanadate concentration in the culture medium caused complete growth inhibition and induction of astaxanthin biosynthesis. Cells were exposed at high concentration of (100 mM) magnesium acetate which led to the dead phase in a short time. Due to high stress, the cells grew quickly and began to die in a relatively shorter period of time (Su et al. 2014). They also reported that treating with a high concentration of 2.5 mM sodium orthovanadate caused total growth inhibition, and the cells accumulated increased astaxanthin with total carotenoids 2.5-fold higher than the control. Moreover, during photoautotrophic cultivation of *H. pluvialis* NIES-144 low C/N ratios increased biomass but decreased astaxanthin production because of light limitation (Kang and Sim 2008).

5.4.2 Aeration and Light Properties

Aeration and light are the two important criteria for biomass accumulation and product synthesis. Under starvation condition *Haematococcus* sp. transforming its green vegetative stage to red aplanospores for astaxanthin production (Zhang et al. 2016). During cultivation, low light condition on early growth phase induces maximum cell growth to produce higher cell density. After attaining the maximum growth, higher light intensity induces the secondary carotenoid production. By adjusting the light cycle, the growth and astaxanthin accumulation in *Haematococcus* can be optimized. Kobayashi et al. (Kobayashi et al. 1993) proposed that, for carotenoid production 24:0 L/D cycles is more preferable than 12:12 L/D cycle. *H. pluvialis* maintained under continuous light and circadian L/D cycle is evaluated for its growth and carotenoid accumulation. Here, the algae cultured under circadian L/D cycle showed a significantly higher growth rate. But higher carotenoid production is by the algae cultured under continuous illumination of light regardless of the initial cell density.

Culturing of *H. pluvialis* in five different light intensities (40, 50, 60 μmol photons m^{-2} s^{-1}) by Imamoglu et al. revealed that low level light intensity up to 12th day of culturing attained maximum cell number of 9.50×10^5 cells mL^{-1}. A mild increase in light intensity affected the growth and 14.7% decrease in cell number is resulted. Further decrease in light intensity resulted in the reduction of cell number to 7×10^5 cells mL^{-1} (Imamoglu 2007). In nitrogen depleted conditions, increasing light intensity from 40 to 400 μmol photons m^{-2} s^{-1} supported the enhancement of dry cell weight of *H. pluvialis* to 2.19 ± 0.08 g L^{-1} (Zhang et al. 2018). While, light intensity from 40–50 μE m^{-2} s^{-1} is often used as optimal for the cultivation of H. pluvialis, reaching cell densities of 0.25×106 cells mL^{-1}. Aeration plays a major role in large scale cultivation of microalgae *H. pluvialis*. 1 to 5% of CO_2 supply is mandatory for the astaxanthin synthesis in phototrophic cultivation. Aeration with 15% CO_2 induced the astaxanthin synthesis to high level in shorter period of time (Christian et al. 2018). Studies reported that high level of aeration causes detrimental effect on growth of the algae by transforming the cells from motile stage to non-motile cyst stage in photo bioreactor. The level of aeration varies according to the type and size of photobioreactor used for cultivation (Vega-Estrada et al. 2005).

5.4.3 Cross Contamination and Its Effect on Process

H. pluvialis is very frequent to other microalgal contamination. Therefore, this microalga is usually cultivated in large scale exclusively in photobioreactors (PBRs) (Solovchenko and Chekanov 2014). The lack of a dependable mechanism for preventing and controlling several contaminations in culture will results decline in cell growth or product build up and have catastrophic implications. In order to boost the commercial viability of the production of *H. pluvialis*, eliminating contaminations should be the main focus. When *H. pluvialis* is mass-cultivated, biological contamination is a significant issue because the contaminants invariably enter the culture through water or air. Four different types of biological contaminants that affect algal cell growth during large-scale cultivation. They are zooplanktons, microorganisms such as prokaryote bacteria, eukaryotic fungi, and eukaryotic protozoa, other algae, and viruses (Zhu et al. 2020).

The cross contamination risk from other microbes or grazers may increase as a result of the mixotrophic approach in *H. pluvialis*. According to Wen et al., bacterial contamination still restricts the mixotrophic cultivation of *H. pluvialis* in an open raceway pond. According to Han et al., when the *H. pluvialis* green cells were injected and grew as biofilms, there were certain benefits over traditional cultivation techniques. One of the benefits was preventing protozoan contamination. This approach has only been proven indoors, thus it might be very difficult to use for outdoor production. Additionally, the expense of running a biofilm photobioreactor would be extremely expensive. Thus, choosing cultivation techniques that work with the facilities and production processes in use today is crucial to cutting operational costs and the risk of contamination (Oslan et al. 2021).

Contaminants not only fight for nutrients and light with cultivated microalgae, which hinder their growth, but also causes a rise in contaminants in algae powder. They compromise the quality of final goods and potentially compromise food safety. Contaminants in *H. pluvialis* astaxanthin biosynthesis, in particular, have an impact on the astaxanthin accumulation. Predators can eliminate 90% of the Haematococcus biomass in just 72 h. *H. pluvialis* cells readily stick to the wall of the closed horizontal tube photobioreactor when the culture system is polluted. Compared to the open culture system, the closed HTPBR had fewer biological pollutants (such as raceway pond). *P. sedebokerense* was managed by surfactants in the cultivation of *H. pluvialis* by Ding et al. In order to lessen the impact of contaminations on the *H. pluvialis* culture, Hwang et al. (Hong et al. 2015) suppressed the growth of *P. sedebokerense* by reducing pH. The *H. pluvialis* culture was successfully cleared of protozoa when treated with 3 g/L metronidazole. The effective concentration of sodium hypochlorite used in *H. pluvialis* for disinfection was 80–160 mg/L. Abscisic acid is a naturally occurring substance produced by algae, and copper sulfate and Triton-N have both been found to be effective against chytrids (Dawidziuk et al. 2017). The bioreactor and culture fluid were not shielded from contaminants by these methods, but they were employed to manage contaminants during the culture (Gutman et al. 2009).

5.4.4 Harvesting Difficulties and Solutions

Once the *H. pluvialis* reached the red stage, harvesting should be done for extracting astaxanthin. Generally, in large scale harvesting of microalgae the preferred method is sedimentation, gravitational settlement, centrifugation, and flocculation. But the cost of production increases while increasing the downstream processing technique with advanced automation (Solovchenko and Chekanov 2014). By using floatation and disk stack centrifugation methods, 95% recovery of cells can be achieved (Panis and Carreon 2016). Electrocoagulation floatation technique for harvesting yielded 90% recovery of cells from the culture media (Landels et al. 2019). Now economically viable and energy efficient methods for *H. pluvialis* is hardly available.

5.5 Conclusion

Haematococcus pluvialis, microscopic alga has its own growth cycle with two stages such as green and red stage. Cultivation at large scale or commercial level is quite difficult and the maintenance of mother culture for subsequent scale up process is really a tough challenge. Routine subculture and nutrient addition to keep the algae in a multiplication stage (log phase) is a tough challenge for industry. To compensate the expenses incurred during cultivation, an integrated product development using biorefinery concept in *Haematococcus* sp. cultivation is advised. One alga with multiple products for different applications is only a viable option for industrial cultivation. Any organic source as nutrients for cultivation is an added advantage for cost-effective production. However, lab scale is little easy to cultivate and optimize the parameters, but the industrial scale up of *H. pluvialis* under controlled condition is still not in a sustainable way of production. Continuous parameters optimization in a large scale cultivation under specified condition is necessary for growth of the microalgae and product accumulation. Through artificial intelligence (AI) system, optimizing the parameters at industry level *Haematococcus* sp. cultivation is in near future to meet the sustainable microalgae cultivation.

Acknowledgments The authors are thankful to National Institute of Ocean Technology and Central University of Kerala for their kind support and constant encouragement.

References

Christian D, Zhang J, Sawdon AJ, Peng CA (2018) Enhanced astaxanthin accumulation in *Haematococcus pluvialis* using high carbon dioxide concentration and light illumination. Bioresour Technol 256:548–551. https://doi.org/10.1016/j.biortech.2018.02.074

Dawidziuk A, Popiel D, Luboinska M, Grzebyk M, Wisniewski M, Koczyk G (2017) Assessing contamination of microalgal astaxanthin producer *Haematococcus* cultures with high-resolution melting curve analysis. J Appl Genet 58(2):277–285. https://doi.org/10.1007/s13353-016-0378-x

Gutman J, Zarka A, Boussiba S (2009) The host-range of Paraphysodermasedebokerensis, a chytrid that infects *Haematococcus pluvialis*. Eur J Phycol 44(4):509–514. https://doi.org/10.1080/09670260903161024

He B, Hou L, Dong M, Shi J, Huang X, Ding Y, Cong X, Zhang F, Zhang X, Zang X (2018) Transcriptome analysis in *Haematococcus pluvialis*: Astaxanthin induction by high light with acetate and Fe2+. Int J Mol Sci 19(1):175. https://doi.org/10.3390/ijms19010175

Hong ME, Hwang SK, Chang WS, Kim BW, Lee J, Sim SJ (2015) Enhanced autotrophic astaxanthin production from *Haematococcus pluvialis* under high temperature *via* heat stress-driven Haber–Weiss reaction. Appl Microbiol Biotechnol 99(12):5203–5215. https://doi.org/10.1007/s00253-015-6440-5

Imamoglu E (2007) Effect of different culture media and light intensities on growth of *Haematococcus pluvialis*. Int J Nat Eng Sci 1(3):5–9

Jannel S, Caro Y, Bermudes M, Petit T (2020) Novel insights into the biotechnological production of *Haematococcus pluvialis*-derived astaxanthin: advances and key challenges to allow its industrial use as novel food ingredient. J Mar Sci Eng 8(10):1–48. https://doi.org/10.3390/jmse8100789

Kang CD, Sim SJ (2008) Direct extraction of astaxanthin from *Haematococcus* culture using vegetable oils. Biotechnol Lett 30(3):441–444. https://doi.org/10.1007/s10529-007-9578-0

Karuppan R, Javee A, Gopidas SK, Pathmanapan A, Kattusamy K, Narayanan V, Subramanian SR, Subramani N (2022) Impact of cultivation parameters on astaxanthin accumulation in the green alga *Haematococcus lacustris* RRGK isolated from Himachal Pradesh, India. Energy Nexus 6: 100083. https://doi.org/10.1016/j.nexus.2022.100083

Kiperstok AC, Sebestyén P, Podola B, Melkonian M (2017) Biofilm cultivation of *Haematococcus pluvialis* enables a highly productive one-phase process for astaxanthin production using high light intensities. Algal Res 21:213–222. https://doi.org/10.1016/j.algal.2016.10.025

Kobayashi M, Kakizono T, Nagai S (1993) Enhanced carotenoid biosynthesis by oxidative stress in acetate-induced cyst cells of a green unicellular alga, Haematococcus pluvialis. Appl Environ Microbiol 59(3):867–873. https://doi.org/10.1128/aem.59.3.867-873.1993

Landels A, Beacham TA, Evans CT, Carnovale G, Raikova S, Cole IS, Goddard P, Chuck C, Allen MJ (2019) Improving electrocoagulation floatation for harvesting microalgae. Algal Res 39 (June):101446. https://doi.org/10.1016/j.algal.2019.101446

Markou G, Nerantzis E (2013) Microalgae for high-value compounds and biofuels production: A review with focus on cultivation under stress conditions. Biotechnol Adv 31(8):1532–1542. https://doi.org/10.1016/J.BIOTECHADV.2013.07.011

Nishshanka SH, Liyanaarachchi VC, Premaratne M, Ariyadasa TU, Nimarshana V (2021) Sustainable cultivation of *Haematococcus pluvialis* for the production of natural astaxanthin. MERCon 2021 - 7th international multidisciplinary Moratuwa engineering research conference, proceedings, pp 297–302. https://doi.org/10.1109/MERCON52712.2021.9525648

Oslan SNH, Shoparwe NF, Yusoff AH, Rahim AA, Chang CS, Tan JS, Oslan SN, Arumugam K, Bin Ariff A, Sulaiman AZ, Mohamed MS (2021) A review on Haematococcus pluvialis bioprocess optimization of green and red stage culture conditions for the production of natural astaxanthin. Biomolecules 11:256. https://doi.org/10.3390/biom11020256

Panis G, Carreon JR (2016) Commercial astaxanthin production derived by green alga *Haematococcus pluvialis*: A microalgae process model and a techno-economic assessment all through production line. Algal Res 18:175–190. https://doi.org/10.1016/j.algal.2016.06.007

Perez-Garcia O, Escalante FME, de-Bashan LE, Bashan Y (2011) Heterotrophic cultures of microalgae: metabolism and potential products. Water Res 45(1):11–36. https://doi.org/10.1016/j.watres.2010.08.037

Shah MMR, Liang Y, Cheng JJ, Daroch M (2016) Astaxanthin-producing green microalga *Haematococcus pluvialis*: from single cell to high value commercial products. Front Plant Sci 7:531. https://doi.org/10.3389/fpls.2016.00531

Solovchenko A, Chekanov K (2014) Production of carotenoids using microalgae cultivated in photobioreactors. In: Production of biomass and bioactive compounds using bioreactor technology, vol 9789401792233. Springer, Dordrecht, pp 63–91. https://doi.org/10.1007/978-94-017-9223-3_4

Su Y, Wang J, Shi M, Niu X, Yu X, Gao L, Zhang X, Chen L, Zhang W (2014) Metabolomic and network analysis of astaxanthin-producing *Haematococcus pluvialis* under various stress conditions. Bioresour Technol 170:522–529. https://doi.org/10.1016/j.biortech.2014.08.018

Sun Z, Li T, Zhou ZG, Jiang Y (2016) Microalgae as a source of lutein: chemistry, biosynthesis, and carotenogenesis. Adv Biochem Eng Biotechnol 153:37–58. https://doi.org/10.1007/10_2015_331

Vega-Estrada J, Montes-Horcasitas MC, Domínguez-Bocanegra AR, Cañizares-Villanueva RO (2005) Haematococcus pluvialis cultivation in split-cylinder internal-loop airlift photobioreactor under aeration conditions avoiding cell damage. Appl Microbiol Biotechnol 68(1):31–35. https://doi.org/10.1007/s00253-004-1863-4

Yang S, Zhao W, Mou H, Sun H (2021) Improving astaxanthin production of *Haematococcus pluvialis* by an efficient fed-batch strategy in a photobioreactor. Algal Res 60:102539. https://doi.org/10.1016/j.algal.2021.102539

Yu BS, Lee SY, Sim SJ (2022) Effective contamination control strategies facilitating axenic cultivation of *Haematococcus pluvialis*: risks and challenges. Bioresour Technol 344:126289. https://doi.org/10.1016/j.biortech.2021.126289

Zhang Z, Wang B, Hu Q, Sommerfeld M, Li Y, Han D (2016) A new paradigm for producing astaxanthin from the unicellular green alga *Haematococcus pluvialis*. Biotechnol Bioeng 113(10):2088–2099. https://doi.org/10.1002/bit.25976

Zhang L, Su F, Zhang C, Gong F, Liu J (2017) Changes of photosynthetic behaviors and photoprotection during cell transformation and astaxanthin accumulation in *Haematococcus pluvialis* grown outdoors in tubular photobioreactors. Int J Mol Sci 18(1):33. https://doi.org/10.3390/ijms18010033

Zhang WW, Zhou XF, Zhang YL, Cheng PF, Ma R, Cheng WL, Chu HQ (2018) Enhancing astaxanthin accumulation in *Haematococcus pluvialis* by coupled light intensity and nitrogen starvation in column photobioreactors. J Microbiol Biotechnol 28(12):2019–2028. https://doi.org/10.4014/jmb.1807.07008

Zhu Z, Jiang J, Fa Y (2020) Overcoming the biological contamination in microalgae and cyanobacteria mass cultivations for photosynthetic biofuel production. Molecules 25(22):5220. https://doi.org/10.3390/MOLECULES25225220

Part II
Biochemistry

Chapter 6
Biochemistry of *Haematococcus*

Thangavel Muthusamy and Sathyaramya Thangavel

6.1 Introduction

The freshwater microalgae with a high content of antioxidant astaxanthin compound are family of *Haematococcus pluvialis*. Mehariya et al. (2020) revealed the antimicrobial activity and antioxidant of *Haematococcus pluvialis* extractions and compounds using biochemical techniques (HPLC-MS), high-performance-mass spectrometry, liquid chromatography (HPLC-DAD), and GC-MS, as well as in vitro antimicrobial activity and antioxidant assays). The methylated fatty acid have no antimicrobial activity or reduced, the hydroxyl group (-OH) of the carboxyl group of fatty acid appears to be an important factor in their antimicrobial activity. According to studies, medium and long unsaturated fatty acid have number of double bonds with carbon chains is typically associated with fatty acid antibacterial activity. For the saturated fatty acids, however, fatty acids with 10 or 12 carbons in the chain were found to have the most antimicrobial activity. Total of 108 evidences suggest that inhibition acts on fatty acids as an antimicrobial through a variety of mechanisms, including uncoupling of oxidative phosphorylation, electron transport chain disruption, inhibition of enzyme activity, cell lysis, peroxidation, autoxidation, and impairment of nutrient uptake. In general more antimicrobial activity has with unsaturated fatty acids and they have superior activity of saturated fatty acid compared with the same carbon chain length.

T. Muthusamy (✉)
Research and Development Wing, Sree Balaji Medical College and Hospital (SBMCH), Bharath Institute of Higher Education and Research (BIHER), Chennai, Tamil Nadu, India

S. Thangavel
Department of Biological Chemistry, University of Louisville, Louisville, KY, USA

R. Raja et al. (eds.), *Haematococcus*, https://doi.org/10.1007/978-981-99-2901-6_6

6.2 Biochemical Composition of *Haematococcus pluvialis*

6.2.1 Carbohydrates and Proteins

The metabolic profile of the cell changes as it develops and progresses of its life cycle through several stages. The majority of the green stage cells of *H. pluvialis* contain larger protein content of H 30–46% by d.w (dry weight) (Grewe and Griehl 2012). Protein concentration in red stage cells ranges from 25% to 22% d.w., but it drops to 35% d.w. in palmella. Starch-type carbohydrates enable the cell to endure a protracted stress. The content rises in the green stage of d.w. from 16% to 18% and in the red cyst from 61% to 75% d.w. total carbs suddenly increase, by up to 64% d.w, the day of exposure on stress, as Recht et al. (2012) showed in their research, and then drop.

6.2.2 Lipids

Table 6.1 illustrates that the polyunsaturated fatty acids (PUFAs) deposited in chloroplasts with 11% of the primarily made up of lipid and overall lipid content in the green stage ranges from 20% to 25% (Shah et al. 2016). Unfavourable environmental circumstances lead to the build-up of lipids. These include low levels of nitrogen and phosphate, excessive salinity, intense light, and harsh temperatures. The main cause of the build-up of lipids in the cell is nitrogen constraint (Hagen et al. 2002). According to studies, nitrogen limitation of 0.16 g/L and strong light of 169 mol photons m2s both increased the quantity of astaxanthin esters and oleic acid (C18:1 monounsaturated fatty acid). When compared to the wild culture, the utilisation of adverse circumstances in this instance significantly increased the total lipid.

In the first stress conditions with continuous light intensity of 300 moL photons m3s1 and culture complete medium and no aeration were utilised. In the second variation, the same amount of light intensity as in the first case was used together with a medium devoid of nitrogen and aeration. Both 33.9% and 31.88% of dry weight were made up of total lipids. In the control culture the total content of lipids was 14.98% with a continuous aeration with a mixture of air, with a full medium (501–701 mL/min) and CO_2 (0.3 mL/min), and 12 h of illumination with a white lamp 89 moL photons m2s1. This study also revealed a notable rise in the amount of phospholipids in cultures kept under adverse circumstances. The first variety of stress conditions had the largest level of neutral lipids and glycolipids, at 19.79% and 7.79%, respectively, whereas the value of second variant was 17.59% and 7.59%. The content of phospholipid, which was equal to 9.79% in the second variety and 9.4% in the first, was higher. Glycolipids (3.80%), phospholipids (1.86%), and neutral lipids (9.20%) were found in the control culture. Three fatty acids, such as

Table 6.1 Comparison of fatty acid compositions (%) of different *H. pluvialis* strains

Fatty Acids	Kim et al. (2015)	Lorenz (1999)	Scodelaro Bilbao et al. (2016)	Lei et al. (2012)
C12:0 lauric	N/A	0.2	N/A	0.26
C14:0 myristic	0.2	0.4	1.99	0.64
C15:0 pentadecanoic acid	0.2	N/A	N/A	0.26
C16:0 palmitic	12.9	27	21.8	11.9
C16:1 palmitoleic	0.5	0.7	0.29	0.8
C16:2	0.3	N/A	N/A	N/A
C16:3	2.9	N/A	N/A	N/A
C16:4	2.9	N/A	N/A	N/A
C17:0 margaric	N/A	0.3	N/A	0.32
C17:1 margaroleic	N/A	1.4	N/A	0.01
C18:0 stearic	0.6	1.9	2.01	3.98
C18:1 oleic	5.01	24.8	15.9	10.9
C18:2 linoleic	25.8	19.9	24.8	12.99
C18:3 linolenic	28.9	11.9	11.9	1.99
C18:4 octadecatetraenoic	6.1	0.9	N/A	N/A
C20:0 arachidic	N/A	0.5	N/A	0.35
C20:1 gadoleic	0.4	0.4	N/A	0.9
C20:2 eicosadenoic	N/A	0.9	1.99	0.79
C20:3 eicosatrienoic gamma	N/A	0.4	N/A	0.20
C20:4 arachidonic	0.7	1.5	2.01	2.21
C20:5 eicosapentaenoic	0.5	0.7	0.77	0.88
C22:0 behenic	N/A	0.3	N/A	0.20
C24:0 lignoceric	0.4	0.3	0.44	0.3
C24:1 nervonic acid	0.2	0.2	N/A	0.20
Σ SFAs	16	40.1	22.9	20.1
Σ MUFAs	7	30.2	17.7	13.01
Σ PUFAs	80.2	39.6	40.3	20.11

N/A not available

linolenic acid, palmitic acid, and linoleic acid, are the most prevalent in *H. pluvialis* cells (Kim et al. 2015).

The strain affects the fatty acid profile. The monounsaturated fatty acid and saturated fatty acid contents of the *H. pluvialis* KORDI03 profile are both low (14 and 5.0%, respectively), while the polyunsaturated fatty acid concentration is 77% (Aflalo et al. 2007). SFA concentration ranged from 26.79% to 29.29% in the strain CCALA 1171, whereas MUFA content ranged from 17.99 to 21.10% and PUFA content ranged from 42.14% to 46.19% in Baha Blanca, Argentina (Butler et al. 2018). Cerón et al. (2007) found that the generation of fatty acids is significantly impacted by the nitrogen reduction. A culture maintained in a medium with a lower nitrogen content of 1.7 mM produced 7.60% of fatty acids in dry weight, whereas, higher nitrogen level of 4.6 mM exhibited a smaller amount of fatty acids. It was also shown in the aforementioned study that the nitrogen content affects the fatty acid composition. The total fatty acid content is 50% decreased by 2 mM when oleic acid levels rise when nitrogen content is decreased to 2 mM, accounting for 50% of the total fatty acid content (Cerón et al. 2007). Liang et al. (2015) revealed that biomass multiplication encourages only 12.8% when the total lipid content of microalga grown under control conditions. This study also demonstrated how stress variables affect lipid build-up in cells. The cultivation of biomass in a nitrogen-free medium led to a build-up of lipids at a rate of 46.71% dry weight. The content of lipid increased up to 45.79% dry weight in the second variant as a result of using a media no nitrogen and increasing the light intensity from 50 to 349 moL photons m2s1. The entire medium and a high light intensity of 349 moL photons m2 s1 produced the greatest results (55.69% d.w.) (Liang et al. 2015).

6.3 Biochemistry of Astaxanthin (AX)

Xanthophyll carotenoid like astaxanthin (Fig. 6.1), found in a variety of microalgae and yeast, United States and the European Union has been approved for use as a food colouring agent. It is utilised as a source of colour in the feed for salmon, trout, and shrimp. Ingesting seafood, extracting astaxanthin from the yeast species *Phaffia* microalga, or using *Haematococcus pluvialis* are the main sources of astaxanthin for humans. Similar to other carotenoids, astaxanthin has a high antioxidant effect by scavenging reducing lipid peroxidation, free radicals, boosting immune system performance, and controlling gene expression. Inhibiting transforming growth factor (TGF-1)-induced hepatic fibrosis and pro-fibrotic expression, astaxanthin suppresses activation of hepatic stellate cells. The high fat and high fructose diets given to mice, astaxanthin also reduces expression of hepatic endoplasmic reticulum stress proteins, enhances glucose metabolism, and reduces hepatocellular lipid build up (Bhuvaneswari and Anuradha 2012). However, it had no effect on the eight other CYP activities and is not thought to be a significant contributor to changed CYP activity. Astaxanthin has a mild inhibitory effect on human CYP2C19. ATX is part of the Xanthophil sub-group because of the additional keto- and hydroxyl-groups that are attached to the isoprene rings; this gives ATX the ability to orient into the optimal hydrophilic and hydrophobic position (as shown) and provides a lipid bilayer that provides superior protection against peroxidation. ATX belongs to the carotenoid group and shares a part of the same reduce expression of hepatic endoplasmic reticulum stress proteins biosynthetic pathway as carotene; it also possesses a polyene chain that quenches free radicals.

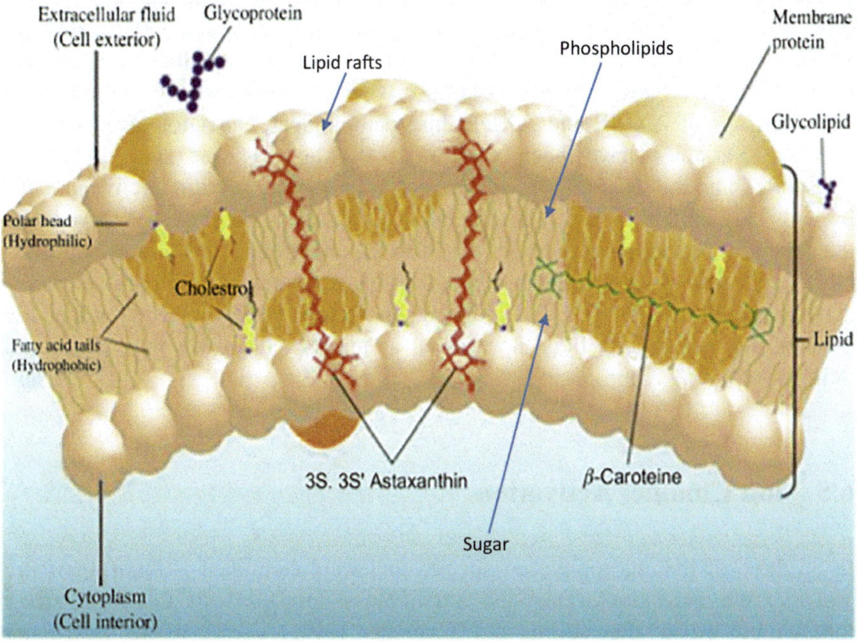

Fig. 6.1 Astaxanthin (ATX) chain (Adapted from J. Masojídek, G. Torzillo, 2014)

6.4 Biosynthesis of Astaxanthin

By adding hydroxyl and oxo moieties to the C3 and C4 positions of both ionone rings of carotene, respectively, AX is biosynthesised. Number of animals pigmentation is affected by the intense dark red xanthophyll, which has no provitamin A activity. Red crabs and salmon raised in pens and in the wild both require astaxanthin as part of their diet. Although new research suggests that astaxanthin has higher antioxidant qualities compared to carotene, there are still few astaxanthin useful in the nutraceuticals, food industries, and cosmetics (Eisenreich et al. 2001). In the trout aquaculture industries xanthophyll is mostly used as a colouring source, particularly in the salmon. Supplementing the fish diet with astaxanthin has also been linked to improved immunological function and fertility (Gwak et al. 2014). Haematococcus pluviais has been used for years to produce AX in a variety of environmental settings. With 1.4–2.9% of the biomass dry weight, these green microalgae represent AX's main natural resource (Varela et al. 2015). *H. pluvialis* was shown to be protected by the photosynthetic apparatus and the light filtering in AX. AX is a source of provitamin A, which boosts immunity, and it also has a strong antioxidant activity that provides protection against UV radiation and anti-inflammatory action.

In comparison to alpha-carotene, vitamin E, cantaxanthin, and even zeaxanthin, AX has stronger antioxidant properties (Gwak et al. 2014). Additionally, research studies have examined the effects of AX on human health (Gwak et al. 2014) and the potent suppression of skin cancer that AX mono (AXME) and diesters (AXDE) produced from biomass of the *H. pluvialis* provide (Nawrocki et al. 2015). Here, the effectiveness of AXME and AXDE against a UV 7,12-dimethylbenz(a)anthracene (DMBA)-induced skin cancer model in rats was investigated. To ascertain the characteristics of AXME and AXDE, histopathological, haematological, and biochemical variables have been examined. It was proposed that metabolites transported into the serum from AX derivatives may be more antioxidative in nature by tyrosinase inhibition, a non-antioxidative route, since prior research showed that the defensive action is not directly related to the antioxidant capabilities. Additionally, a multistep process that inhibits both ROS generation and tyrosinase enzyme activity may prevent UV-DMBA-induced skin cancer. Following these procedures, melanocyte and melanin build-up as well as unchecked growth may be avoided (Vidhyavathi et al. 2008).

6.5 Ion Channel Activation

In all three of the chlorophycean species that were examined—*Chlamydomonas*, Volvox, and *Haematococcus*—the amplitude is graded with the flash energy throughout a broad range (Braun and Hegemann 1999). A series of electrical events are triggered when the chlamydomonas eye is stimulated by light. Several distinct elements that have been connected to at least four conductances make up photocurrents. The first major component of the current is concentrated in the area around the eyespot in the plasma membrane which represents cation inflow over the area of beneath the eyespot. The ions that transport the photoreceptor current and initial discovery of photocurrents in *Haematococcus* have been a subject of discussion (Litvin et al. 1978). When the driving force is high enough, such as at low pH or high extracellular K+, H+, and K+ also contribute to the photoreceptor current, at least in *Chlamydomonas* (Holland et al. 1997). Although it is now obvious that H+ is the most easily transmitted ion, under practically all physiological circumstances, when proton concentrations are relatively low, Ca^{2+} contributes the most current.

SFA palmitic acid can be found in the FA profile of certain microalgal species. There is evidence that this SFA has antiviral properties; for instance, Jyonouchi et al. (2008) showed that evidence SFA has antiviral properties and *Dunaliella salina* and *Haematococcus pluvialis* produced palmitic acid which has anti-herpes simplex type-1 (HSV-1) activity. The palmitic acid has ability to bind to the CD4 receptor and prevent the entry of HIV-1 and infection. Palmitic acid was isolated from *Sargassum fusiforme* extract Lee et al. Mycobacterium TB and M. smegmatis were resistant to *Polysiphonia virgata* C. Agardh's (Rhodomelaceae) dichloromethane extraction. The findings showed that the main antimicrobial components of *Polysiphonia virgata* are FAs, such as myristic acid, lauric acid, linoleic acid, and oleic acid. Oleic acid shown considerable antibacterial activity against M. smegmatis with a MIC of 0.8 g, linoleic acid had a MIC of 1.56 g, and lauric acid had a MIC of 3.125 g. Lauric acid, myristic acid, and linoleic acid were also found to have 100% inhibition at MIC values of 25 g/mL, 90% and 76% inhibition, respectively, at 50 g/mL, and modest antibacterial activity at 50 g/mL, against multidrug-resistant (MDR) *M. tuberculosis* Jyonouchi et al. (2008).

An example of a successful and well-known biotechnological technique for the generation of bioactive compounds is the commercial cultivation of carotene from the halophile green microalga *Dunaliella salina*. This microalga's antibacterial effectiveness was tested against a variety of microbes (aureus, E. coli, S. aureus, A. niger, and *C. albicans*). GC-MS analysis of the *Dunaliella salina* extracts, including cyclocitral, 15 volatile chemicals and ionone, and phytol and FA neophytadiene, including linolenic, oleic acids and, palmitic, that had previously been reported as antimicrobials, were discovered (Hussein et al. 2006). The advantages of *Haematococcus pluvialis*, Xanthophyllomyces include easy cultivation and improved genetic techniques for strain improvement. Heterologous plasmids can be successfully integrated into the yeast's rDNA clusters several times. By complementing the relevant mutations in the Erwinia uredovora crt gene cluster, which was kept by recombinant *Escherichia coli* cells, cDNAs from *Xanthophyllomyces dendrorhous* were able to be produced. As in other fungi, the gene crtYB is responsible for both the conversion of geranylgeranyl pyrophosphate to phytoene and the synthesis of the -ionone rings from lycopene to -carotene. The gene's sequence diverges significantly from that of related genes from other fungi. With a 17% identity, the Neurospora crassa gene is the most comparable. The aminoterminal region of crtYB, which codes for the lycopene cyclase activity, and ORF7 in the carotene gene cluster of the myxobacterium *Myxococcus xanthus* share an almost identical degree of sequence similarity (18% identity).

By complementation, the phytoene dehydrogenase gene crtI was also cloned in *Escherichia coli*. A single gene product is sufficient for catalysing all four consecutive oxidation stages from phytoene to lycopene, according to the complementation study in the heterologous system. There are several enzymes that can convert -carotene into astaxanthin, and several of the genes that do this have been cloned from bacteria, algae, and plants. According to a metabolic route for *Xanthophyllomyces dendrorhous*, a keto group is first added to one of the -carotene rings, resulting in echinenone, which is subsequently changed into 3-hydroxyechinenone. The second ring is then used to carry out the same sequence of reactions. Although suitable *Escherichia coli* genotypes are available that should enable effective screening for astaxanthin producers after introducing the genes as cDNA copies, the concept is yet hypothetical for fungus. Neither the enzymes nor the related genes have been identified. However, the homologous genes crtW and crtZ from the marine *Agrobacterium aurantiacum* have been cloned, giving the model a respectable level of likelihood. Sequence-level analyses of unexpected cDNAs for crtI and crtYB point to alternative splicing of the mRNAs for both of those genes that would prevent the production of functional carotenogenic enzymes. Splicing to non-functional transcripts may constitute a regulatory mechanism permitting adaptability to environmental factors, albeit this hasn't been demonstrated at the biochemical or genetic level. In *Xanthophyllomyces dendrorhous*, astaxanthin appears to have the same roles that carotene does in mucoralean fungus. It guards against singlet oxygen's ability to destroy the yeast. This finding suggests that 1O2 and peroxyl radicals favourably control the synthesis of carotenoids. Additionally, there is early evidence that astaxanthin controls carotenoid production in a manner similar to how the end product inhibits carotene synthesis in *Phycomyces blakesleeanus* (Schroeder and Johnson 1995).

6.6 Anti-lipid Peroxidation Properties

Lipid peroxidation includes oxidising lipids and producing lipid peroxides as a result. Initiation, propagation, and termination are the three phases of the reaction. Aldehydes (such as malondialdehyde-MDA), hydroxyaldehydes (4-hydroxynonenal), and hydrocarbons (such as ethane) are produced through lipid peroxidation. These substances have the ability to change the hydrophobicity of cell membranes, including the physical characteristics of the asymmetry of the membranes being disturbed, lipid interior of the membranes being reduced,, the activity of membrane enzymes being inhibited, the activity of transporting proteins being inhibited being and the membranes becoming depolarised. The integrity of the intracellular membrane and plasma membrane may be lost as a result of these modifications. Astaxanthin's antioxidant capacity is pH-dependent. The phosphatidylcholine liposomes was zwitterionic at pH 7.5 (75%) and pH 7.9 (64%), Mano et al. (2018) demonstrated that the development of lipid peroxidation (reactive compounds of thiobarbituric acid) prevented by astaxanthin. Furthermore, at pH 6.2 (20%) and 6.8 (30%), it likewise only slightly suppresses the lipid peroxidation process. Mano and co. (2018). Supplementing with astaxanthin lessened the anti-lipid peroxidation impact in the case of rats with ulcers. The doses of astaxanthin 101, 251, and 501 g/kg b.w. per day. In blood the amount of reactive compounds thiobarbituric acid (measured as nmoL malondialdehyde per mg serum) reduced from 2.69 nmoL MDA/mg (control) to 2.03 nmoL MDA/mg, 1.94 nmoL MDA/mg, and 1.55 nm Kamath et al. (2008).

Interestingly, astaxanthin outperforms carotene in its ability to prevent lipid peroxidation. Carotene was twice as effective as astaxanthin at preventing the liposomal peroxidation caused by ADP and $Fe2^{+}$. According to Hussein, et al., the interaction of astaxanthin's terminating rings with hydrophilic polar sites of phospholipid membranes results in the formation of hydrophobicity of astaxanthin intramolecular which increases by hydrogen-bonded five-membered ring (2006). Olson (2004) investigated under ex vivo settings are evaluated by the impact of astaxanthin on lipid peroxidation in LDL. Volunteers received astaxanthin at various amounts (1.7, 3.5, 13.9, and 20.9 mg/day) for a period of 2 weeks. Normal group compared with (LDL from the group not drinking astaxanthin), experimental of LDL from the group that received astaxanthin were characterised by lower vulnerability to oxidation Iwamoto et al. (2000). Numerous illnesses, including diabetes and a number of neurodegenerative disorders, have chronic inflammation as their primary pathophysiological cause. Immune cells are particularly vulnerable to oxidative stress due to the high proportion of polyunsaturated fatty acids in their plasma membranes.

An excess of reactive oxygen species upsets the antioxidant equilibrium Chew et al. (2011). Inflammation is prevented by potent antioxidant, astaxanthin in biological systems and has the capacity to lower inflammation or control immune response brought on by peripheral illnesses (Chew and Park 2009). According to studies, the central nervous system's non-neuronal microglial cells can be controlled by astaxanthin. The tissue-specific resident macrophages known as microglial cells regulate homeostasis and participate in the immune response. Microglial cells emit IL-1, TNF-, IL-6, NO, and proinflammatory cytokines when they detect a threat, such as the presence of cell injury or pathogens. This response is excellent at neutralising threat in the initial stages, but because proinflammatory chemicals are toxic, persistent action of microglia can cause the CNS to be destroyed (Grimmig et al. 2017). Astaxanthin consumption reduced the release of Cox-2, IL-6, and nitric oxide/iNOS in microglia during the presence of bacteria, according to research by Kim et al. (2010). The frontal brain and hippocampus both experienced neurodegeneration and a decline in B, according to Park et al. (2009). This study examined how astaxanthin-treated mice performed in the Morris water maze (Park et al. 2009). Bowel diseases such as arteriosclerosis, inflammatory bowel disease, arteriosclerosis, rheumatoid arthritis, sepsis, brain inflammation, and gastric inflammation, can all be significantly improved by astaxanthin.

6.7 Cancer Prevention

Astaxanthin has the power to stop the development of cancer cells. Antioxidant substances prevent oxidative cell damage, which reduces mutagenesis and carcinogenesis. According to studies, astaxanthin not only prevents colon cancer cells from proliferating but also has the ability to trigger their apoptosis. Palloza et al. (2009) adapted an excerpt from *H. pluvialis*, which stopped human colon cancer cells from proliferating. The extract contained astaxanthin, which was in charge of inhibiting the cell cycle's progression and encouraging apoptosis (Palloza et al. 2009). Canthaxanthin and carotene were among the carotenoids with lower anticancer activity than astaxanthin (Chew and Park 2004). Mice treated with astaxanthin had higher levels of plasma interferon, natural killer cells, and immune cells (Nakano et al. 2008; Jyonouchi et al. 2008). By giving astaxanthin to rats and mice orally, it was possible to inhibit the development of tumour in their colons and mouths. Cell growth inhibition has been partially attributed to this action. According to research done by Song et al. (2011), astaxanthin has a high activity of anti-proliferative against tumour cells such as Lewis cells, CBRH-7919 hepatoma cells, and SHZ-88 breast cancer cells. It was found that astaxanthin concentrations and antiproliferative activities were significantly correlated. The CBRH-7919 cell line was the most susceptible to astaxanthin with a half maximum inhibitory dose of 39 M (Song et al. 2011).

6.8 Applications of Astaxanthin in Other Fields

Considering its high anti-inflammatory and antioxidant properties, this carotenoid has a positive effect. As a result, more and more studies are being done on how astaxanthin affects the neurological, cardiovascular, and ocular systems biologically. A significant risk cause for heart disease is hypertension. RO and nitrogen species that are produced in excess cause disorders such as arrhythmias, endothelial dysfunction, hypertension, and atherosclerosis. When astaxanthin was given orally to high blood pressure rats, the NO products were reduced, and the bp was dropped (Hussein et al. 2006). Astaxanthin intake reduced the levels of TG, TC, LDL-C, IL-6, CRP, and LPO, according to in vivo studies; this enhanced the antioxidant defence capacity and choroidal blood flow velocity. Also, astaxanthin boosted SOD activity and lowered the generation of IL-8, IFN-, LT-B4, PG-E2, and NO (Mosaad et al. 2016). Animal studies have shown that astaxanthin has a cardiovascular preventive effect, but human studies on the treatment of atherosclerotic cardiovascular disease have not been conducted.

In the treatment and prevention of eye conditions such as age-related cataracts, macular degeneration, keratopathy, and glaucoma, astaxanthin is useful. Every retina of the eye and systems of circulatory barriers are both crossed by antioxidant. As the only carotenoid, it integrates itself into eye cell membrane, shielding it from free radicals and harm. Since its carotenoid properties, astaxanthin prevented ischemia-induced retinal cell death (Otsuka et al. 2016). 8 weeks of astaxanthin supplementation in rats at a dose of 5 mg/kg/per day led to a reduction in retinal apoptosis, a reduction in the generation of NOS-2 and protein carbonyl, and an increase in retinoprotective characteristics (Curek et al. 2010). Because of its antioxidant and metabolism-regulating capabilities, anti-inflammatory, astaxanthin protects against a variety of eye disorders (Giannaccare et al. 2020). According to research by Fakhri et al. (2019), astaxanthin can cross the blood–brain barrier and prevent muscular degeneration pathways such as apoptosis, inflammation, and oxidative stress (2018). With supplementation, Bax and Cleaved-caspase-3 expression is decreased, which suppresses and lessens neuronal apoptosis and degenerative cell damage. Astaxanthin administration reduced NF-KB and TNF expression in postoperative rats, which reduced cerebral oedema and neurological impairment (Masoudi et al. 2017). Alzheimer's disease, Parkinson's disease, spinal cord and brain injuries, ageing, neuropathic pain, autism depression, and depression are among the neurological conditions that are thought to be treated by astaxanthin as a multi-target pharmaceutical agent (Fakhri et al. 2019).

6.9 Effect on DNA Repair

DNA damage results from skin exposure to UV light. Misconception in DNA repair, which can result in cancer mutations, are largely to blame for the medically damaging effects of exposure to UV radiation. Starting with the beginning of cell cycle arrest and the induced of DNA restore photoproducts, the DNA mechanisms created by UV-activated DNA changes are done by DNA damage (Timares et al. 2008). Mammalian cells use the nucleotide excision repair (NER) pathway as a crucial mechanism for the repair of damaged DNA (Cline and Hanawalt 2003).

Despite the lack of studies examining ASX's effects on the NER pathway, it has been claimed that ASX enhances cells' capacity to repair DNA. ASX was particularly effective at reducing DNA damage and affecting the speed of DNA repair (Santocono et al. 2006). Multiple defence mechanisms exist in human cells against UV-induced ROS either through damage repair or prevention. For instance, ASX promotes the expression of oxidative stress-responsive enzymes and prevents UV-induced DNA damage (Han et al. 2013). Additionally, it was shown that ASX protects against oxidative stress and DNA damage brought on by cyclophosphamide by activating Nrf2 and regulating the production of NQO1 and HO-1 (Tripathi and Jena 2010). Cyclophosphamide (CP), a cytotoxic alkylating agent, is widely employed in the highly effective treatment of numerous malignancies. However, it is extremely toxic to normal cells in both humans and lab animals, and it also induces genomic instability and DNA damage in addition to having toxic effects. Therefore, it's critical to guard against DNA damage caused by CP in clinical applications to normal cells. According to several studies, ASX reduced the oxidative stress and subsequent oxidative DNA damage caused by CP. Additionally, the AKT pathway is essential for controlling genome stability and DNA damage reactions. Double strand break (DSB) repair is hampered by AKT kinase activity suppression, according to studies (Xu et al. 2012). Recently, it was proposed that ASX's control of the AKT signal pathway could help maintain genomic integrity and prevent DNA damage (Table 6.2).

6.10 Conclusion

For its antioxidant and antibacterial properties, *Haematococcus pluvialis* is a frequently used chemotherapeutic agent for treating a variety of disorders. It has demonstrated the effectiveness against acute leukaemia, lymphoma, and several other solid human malignancies. However, evidence has shown that astaxanthin chemotherapy works as an antimicrobial through a variety of mechanisms, including disruption of the impairment of nutrient uptake, electron transport chain, uncoupling of oxidative phosphorylation, cell lysis, inhibition of enzyme activity, impairment of nutrient uptake, and autoxidation and peroxidation. Here, it was attempted to demonstrate how astaxanthin, when given to mice on a high fat, high-fructose diet,

Table 6.2 Study on skin and astaxanthin human intervention summary

Intervention	Design method	Control	Population (*n*)	Time	Output	Dosage
ASX capsules administration	Randomized double-blind, controlled study	Placebo	Healthy female subjects (14/diet group)	7 weeks	↓ biomarker DNA damage; ↑ of T cells, NK cells, IL-6 and B cells	3 or 9 mg
ASX capsules administration	Monitoring of oxidative stress and skin aging	Nil	31 middle-aged volunteers	5 weeks	↓ MDA; ↓ RSSC	5 mg
ASX capsules administration	Randomized, double-blind, parallel-group, placebo-controlled	Placebo	65 healthy female subjects	17 weeks	↓ wrinkle parameters; ↓ IL-α	9 or 13 mg
ASX capsules administration	Pilot study	Nil	3 healthy female subjects	2 weeks	↓ wrinkle parameters	0.7 mg/g of ASX cream
ASX on topical application	Pilot study	Nil	3 healthy male subjects	N/S	↓ erythema	N/S
ASX capsules administration	Randomized, single-blind, placebo-controlled	Placebo	50 female healthy subjects	5 weeks	↓ wrinkle parameters	1.9 mg
ASX with Oral and topical treatment	N/A	N/A	30 female healthy subjects	7 weeks	↓ wrinkle parameters	5.9 mg
Oral forms of two (tablets collagen; ASX capsules)	Randomized, double-blind placebo-controlled	Placebo	44 healthy female volunteers	13 weeks	↑ viscoelastic parameters; ↓ TEWL; ↑ procollagen type I; ↓ MMP-1 and MMP-12	1.9 mg
Capsules of ASX combined with topical application	Noncontrolled open-label	None	30 healthy female subjects	9 weeks	↓ age spot ↓ wrinkles; size; ↑ elasticity; ↑ skin texture	5 mg and 1.5 mL (79.9 μM solution)
Administration of ASX capsules	Double-blind controlled (randomized)	Placebo	36 healthy male subjects	7 weeks	↓ wrinkles; ↑ elasticity; TEWL; ↑ moisture content; ↓ sebum oil	7 mg

Abbreviations: ↑ increase, ↓ decrease, *NK* natural killer, *ASX* astaxanthin, *MDA* malondialdehyde, *IL-6* interleukin-6, *RSSC* residual skin surface components, *TEWL* transepidermal water loss, *MMP* matrix metalloproteinase, *N/S* not specified also enhances glucose metabolism, reduces the expression of hepatic endoplasmic reticulum stress proteins, and reduces hepatocellular lipid build up. Various illnesses, including a number of neurodegenerative and diabetes disorders, have chronic inflammation as their primary pathophysiological cause. Immune cells are especially vulnerable to oxidative stress due to the high proportion of polyunsaturated fatty acids in their plasma membranes. An excess of reactive oxygen species upsets the antioxidant equilibrium. A potent antioxidant, astaxanthin prevents inflammation in biological systems. In peripheral illnesses carotenoid has the ability to control immune response and lessen inflammation.

References

Aflalo C, Meshulam Y, Zarka A, Boussiba S (2007) On the relative efficiency of two- vs. one-stage production of astaxanthin by the green alga *Haematococcus pluvialis*. Biotechnol Bioeng 98: 300–305

Bhuvaneswari S, Anuradha CV (2012) Astaxanthin prevents loss of insulin signaling and improves glucose metabolism in liver of insulin resistant mice. Can J Physiol Pharmacol 90:1544–1552

Braun FJ, Hegemann P (1999) Two light-activated conductances in the eye of the green alga *Volvox carteri*. Biophys J 76:1668–1678

Butler T, McDougall G, Campbell R, Stanley M, Day J (2018) Media screening for obtaining *Haematococcus pluvialis* red motile macrozooids rich in astaxanthin and fatty acids. Biology 7: 2

Cerón MC, García-Malea MC, Rivas J, Acien FG, Fernandez JM, Del Río E, Guerrero MG, Molina E (2007) Antioxidant activity of *Haematococcus pluvialis* cells grown in continuous culture as a function of their carotenoid and fatty acid content. Appl Microbiol Biotechnol 74:1112–1119

Chew BP, Park JS (2004) Carotenoid action on the immune response. J Nutr 134:257–261

Chew BP, Park JS (2009) Carotenoids: nutrition and health. In: Britton G, Liaanen-Jensen S, Pfander H (eds) Carotenoids against disease: part C: the immune system and disease, vol 5. Birkhauser Press, Basel, pp 363–382

Chew BP, Mathison BD, Hayek MG, Massimino S, Reinhart GA, Park JS (2011) Dietary astaxanthin enhances immune response in dogs. Vet Immunol Immunopathol 140:199–206

Cline SD, Hanawalt PC (2003) Who's on first in the cellular response to DNA damage? Nat Rev Mol Cell Biol 4:361–372

Curek GD, Cort A, Yucel G, Demir N, Ozturk S, Elpek GO, Savas B, Aslan M (2010) Effect of astaxanthin on hepatocellular injury following ischemia/reperfusion. Toxicology 267:147–153

Eisenreich W, Rohdich F, Bacher A (2001) Deoxyxylulose phosphate pathway to terpenoids. Trends Plant Sci 6:78–84

Fakhri S, Aneva IY, Farzaei MH, Sobarzo-Sánchez E (2019) The neuroprotective effects of astaxanthin: therapeutic targets and clinical perspective. Molecules 24:2640

Giannaccare G, Pellegrini M, Senni C, Bernabei F, Scorcia V, Cicero AFG (2020) Clinical applications of astaxanthin in the treatment of ocular diseases: emerging insights. Mar Drugs 18:239

Grewe CB, Griehl C (2012) The carotenoid Astaxanthin from *Haematococcus pluvialis*. In: Walter C (ed) Microalgal biotechnology: integration and economy, vol 1. De Gruyter, Berlin, pp 129–144

Grimmig B, Kim SH, Nash K, Bickford PC, Douglas Shytle DR (2017) Neuroprotective mechanisms of astaxanthin: a potential therapeutic role in preserving cognitive function in age and neurodegeneration. Geroscience 39:19–32

Gwak Y, Hwang YS, Wang B, Kim M, Jeong J, Lee CG et al (2014) Comparative analyses of lipidomes and transcriptomes reveal a concerted action of multiple defensive systems against photooxidative stress in *Haematococcus pluvialis*. J Exp Bot 65:4317–4334

Hagen C, Siegmund S, Braune W (2002) Ultrastructural and chemical changes in the cell wall of *Haematococcus pluvialis* (Volvocales, Chlorophyta) during aplanospore formation. Eur J Phycol 37:217–226

Han D, Li Y, Hu O (2013) Astaxanthin in microalgae: pathways, functions and biotechnological implications. Algae 28:131–147

Holland EM, Harz H, Uhl R, Hegemann P (1997) Control of phobic behavioral responses by rhodopsin-induced photocurrents in *Chlamydomonas*. Biophys J 73:1395–1401

Hussein G, Goto H, Oda S, Sankawa U, Matsumoto K, Watanabe H (2006) Antihypertensive potential and mechanism of action of astaxanthin: III. Antioxidant and histopathological effects in spontaneously hypertensive rats. Biol Pharm Bull 29:684–688

Iwamoto T, Hosoda K, Hirano R (2000) Inhibition of low density lipoprotein oxidation by astaxanthin. J Atheroscler Thromb 7:216–222

Jyonouchi H, Sun S, Iijima K, Gross MD (2008) Antitumor activity of astaxanthin and its mode of action. Nutr Cancer 36:59–65

Kamath BS, Srikanta BM, Dharmesh SM, Sarada R, Ravishankar GA (2008) Ulcer preventive and antioxidative properties of astaxanthin from *Haematococcus pluvialis*. Eur J Pharmacol 590: 387–395

Kim YH, Koh HK, Kim DS (2010) Down-regulation of IL-6 production by astaxanthin *via* ERK-, MSK-, and NFkappaB- mediated signals in activated microglia. Int Immunopharmacol 10: 1560–1572

Kim JH, Affan MA, Jang J, Kang MH, Ko AR, Jeon SM (2015) Morphological, molecular, and biochemical characterization of astaxanthin-producing green microalga *Haematococcus* sp. KORDI03 (Haematococcaceae, Chlorophyta) isolated from Korea. J Microbiol Biotechnol 25:238–246

Liang C, Zhai Y, Xu D, Ye N, Zhang X, Wang Y, Zhang W, Yu J (2015) Correlation between lipid and carotenoid synthesis and photosynthetic capacity in *Haematococcus pluvialis* grown under high light and nitrogen deprivation stress. Grasas Aceites 66:77

Litvin FF, Sineshchekov OA, Sineshchekov VA (1978) Photoreceptor electric potential in the phototaxis of the alga *Haematococcus pluvialis*. Nature 271:476–478

Mano CM, Guaratin T, Cardozo K, Colepicolo P, Bechara E, Barros M (2018) Astaxanthin restrains nitrative-oxidative peroxidation in mitochondrial-mimetic liposomes: a pre-apoptosis model. Mar Drugs 16:126

Masoudi A, Dargahi L, Abbaszadeh F, Pourgholami MH, Asgari A, Manoochehri M, Jorjani M (2017) Neuroprotective effects of astaxanthin in a rat model of spinal cord injury. Behav Brain Res 329:104–110

Mehariya S, Sharma N, Iovine A, Casella P, Marino T, Larocca V, Molino A, Musmarra D (2020) An integrated strategy for nutraceuticals from *Haematococcus pluvialis*: from cultivation to extraction. Antioxidants 9:825

Mosaad YO, El Khalik A, Gobba N, Hussein MA (2016) Astaxanthin; a promising protector against gentamicin-induced nephrotoxicity in rats. Curr Pharm Biotechnol 17:1189–1197

Nakano M, Onodera A, Saito E, Tanabe M, Yajima K, Takahashi J, Ngyuen VC (2008) Effect of astaxanthin in combination with alpha-tocopherol or ascorbic acid against oxidative damage in diabetic ODS rats. J Nutr Sci Vitaminol 54:329–334

Nawrocki WJ, Tourasse NJ, Taly A, Rappaport F, Wollman FA (2015) The plastid terminal oxidase: its elusive function points to multiple contributions to plastid physiology. Annu Rev Plant Biol 66:49–74

Olson JA (2004) Carotenoids: absorption, transport, and metabolism of carotenoids in humans. Pure App Chem 66:1011–1016

Otsuka T, Shimazawa M, Inoue Y, Nakano Y, Ojino K, Izawa H, Tsuruma K, Ishibashi T, Hara H (2016) Astaxanthin protects against retinal damage: evidence from *in vivo* and *in vitro* retinal ischemia and reperfusion models. Curr Eye Res 41:1465–1472

Palloza P, Torelli C, Boninsegna A, Simone R, Catalano A, Mele MC, Picci N (2009) Growth inhibitory effects of the astaxanthin-rich alga *Haematococcus pluvialis* in human colon cancer cells. Cancer Lett 283:108–117

Park SK, Kim K, Page GP, Allison DB, Weindruch R, Prolla TA (2009) Gene expression profiling of aging in multiple mouse strains: identification of aging biomarkers and impact of dietary antioxidants. Aging Cell 8:484–495

Recht L, Zarka A, Boussiba S (2012) Patterns of carbohydrate and fatty acid changes under nitrogen starvation in the microalgae *Haematococcus pluvialis* and *Nannochloropsis* sp. Appl Microbiol Biotechnol 94:1495–1503

Santocono M, Zurria M, Berrettini M, Fedeli D, Falcioni G (2006) Influence of astaxanthin, zeaxanthin and lutein on DNA damage and repair in UVA-irradiated cells. J Photochem Photobiol B 85:205–215

Shah MMR, Liang Y, Cheng JJ, Daroch M (2016) Astaxanthin-producing green microalga *Haematococcus pluvialis*. Single cell to high value commercial products. Front Plant Sci 7:531

Song XD, Zhang JJ, Wang MR, Liu WB, Gu XB, Lv CJ (2011) Astaxanthin induces mitochondria-mediated apoptosis in rat hepatocellular carcinoma CBRH-7919 cells. Biol Pharm Bull 34:839–844

Timares L, Katiyar SK, Elmets CA (2008) DNA damage, apoptosis and langerhans cells-activators of UV-induced immune tolerance. Photochem Photobiol 84:422–436

Tripathi DN, Jena GB (2010) Astaxanthin intervention ameliorates cyclophosphamide-induced oxidative stress, DNA damage and early hepatocarcinogenesis in rat: role of Nrf2, p53, p38 and phase-II enzymes. Mutat Res 696:69–80

Varela JC, Pereira H, Vila M, León R (2015) Production of carotenoids by microalgae: achievements and challenges. Photosynth Res 125:423–436

Vidhyavathi R, Venkatachalam L, Sarada R, Ravishankar GA (2008) Regulation of carotenoid biosynthetic genes expression and carotenoid accumulation in the green alga *Haematococcus pluvialis* under nutrient stress conditions. J Exp Bot 59:1409–1418

Xu N, Lao Y, Zhang Y, Gillespie DA (2012) Akt: a double-edged sword in cell proliferation and genome stability. J Oncol 2012:951724

Chapter 7
Lipids, Lipidomics, and Biosurfactants of *Haematococcus*

S. Pujithaa, M. Rajamiriyam, and M. K. Kalaivani

Lipids are fatty oil compounds that are soluble in organic solvents and insoluble in polar solvents. Lipids are natural compounds that include sterols, fat soluble vitamins, monoglycerides, diglycerides, and phospholipids. The major function of lipids includes cell signaling, energy storing and acts as a structural component of cell membrane. Lipids also contribute to various diseases such as diabetes, obesity, cancer, liver degeneration, and neurodegenerative disorders. Because of their diverse nature, lipids are considered to be difficult molecules to study. Cellular lipids are highly complex and dynamic molecules that constantly change according to physiological, pathological, and environmental conditions. As lipids play a major physiological role and affect a wide range of pathologies, studying the lipids and its metabolic pathways provides great potential in inventing new biomarkers and can also help in creating novel therapeutic interventions (Sargent et al. 2003). Therefore, exploration of lipids will be a key component of future research fields. In the year 2003, the emergence of lipidomics saw the plausibility of studying the lipids on a larger scale.

7.1 Lipidomics

Lipidomics refers to analyzing lipids in bulk from biological samples which, can include tissues, body fluid, plants, and processed food. Historically, lipids were studied using chromatographic techniques such as TLC and GC. However, these techniques merely separate the various lipid classes. But in recent years, lipidomics has undergone rapid transformation as a result of technical improvement. Advanced

S. Pujithaa · M. Rajamiriyam · M. K. Kalaivani (✉)
Research and Development Wing, Sree Balaji Medical College and Hospital (SBMCH), BIHER, Chennai, India
e-mail: kalaivani.mb@bharathuniv.ac.in

R. Raja et al. (eds.), *Haematococcus*, https://doi.org/10.1007/978-981-99-2901-6_7

technologies for examining lipids more effectively include mass spectroscopy, NMR spectroscopy, fluorescence spectroscopy, and computational techniques. Lipidomics involves identifying and quantifying the various species of cellular lipids and their interaction with other biomolecules and metabolites. Lipidomics studies the structure, functional specificity, interaction, and dynamics of cellular lipids and also studies the changes that occur in those cellular lipids due to the perturbation in the system in which the lipid was present (Wenk 2010).

7.2 Lipidomics Workflow and Techniques

7.2.1 Sample Preparation

Before any lipidomics analysis, the samples must be carefully collected and preserved. Any discrepancy in sample condition, preprocessing, and storage might affect the quality of lipids being extracted (Fig. 7.1). Most lipidomic techniques utilize biological samples and the extraction techniques exploit the solubility of lipid

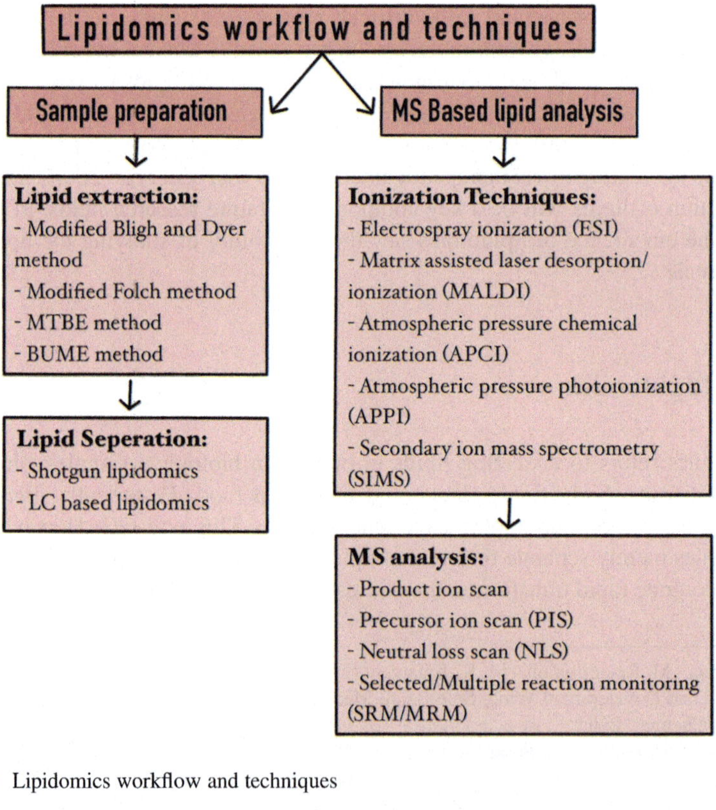

Fig. 7.1 Lipidomics workflow and techniques

hydrocarbon chains in organic solvents. Due to the complexity of the lipid, single universal extraction method cannot be used for all classes of lipid.

7.2.2 *Lipid Extraction*

Thus various different lipid extraction techniques are used (Honeycutt et al. 1995), which include the following.

7.2.2.1 Modified Bligh and Dyer Method

The modified Bligh and Dyer method, which employs methanol and chloroform, is regarded as the gold standard method for lipid extraction. This method begins by combining the sample with water, methanol, and chloroform in a specific ratio of 0.9:1:1. Following which, a biphasic system is formed spontaneously which separates proteins, carbohydrates, and phospholipids into water rich upper phase and most lipids into chloroform rich lower phase. This method is still a widely used method for lipid estimation but it has a serious disadvantage due to the usage of chloroform, which is considered to be highly toxic and carcinogenic in nature. Thus using chloroform on a larger scale is dangerous in terms of safety (Table 7.1).

7.2.2.2 Modified Folch Method

The Folch method is quite similar to that of the Bligh and Dyer method. The sole distinction is the 2:1 ratio of chloroform to methanol, which is followed by the addition of water or 0.9% NaCl in the subsequent step. It has similar merits and demerits to that of Bligh and Dyer method.

7.2.2.3 MTBE Method

The Methyl-Tert-Butyl-Ether, abbreviated as MTBE, is a method of lipid extraction that produces faster and automated generation of total lipids from the biological

Table 7.1 Lipid extraction methods

Sl. no	Extraction method	Chemicals used	Ratio
1	Modified Bligh and Dyer	Water, methanol, chloroform	0.9:1:1
2	Modified Folch method	Chloroform, methanol	2:1
3	MTBE method	Methly-Tert-butyl-ether	–
4	BUME method	Butanol, methanol	3:1
		Hepatane, ethyl acetate	3:1

samples. Due to the low density of MTBE, this approach is more beneficial than the two previously mentioned methods. Thus, during phase separation, the upper organic phase contains the lipids, this simplifies the collection process and lipid loss due to dripping. But the drawback of this method is that a significant amount of aqueous component can carry some water-soluble contaminants.

7.2.2.4 BUME Method

The Butanol-Methanol (BUME) method delivers high lipid recovery from the extraction process. A 96-well plate can be used to carry out this procedure in 60 min utilizing a no-chloroform-free total lipid extraction from blood plasma. BUME method uses butanol and methanol in 3:1 ratio followed by a two-phase separation using heptane: ethyl acetate in 3:1 along with 1% acetic acid as buffer. This method relatively eliminates the water soluble contaminants to a reasonable extent. However, it has the disadvantage of evaporating butanol from the organic phase containing lipid.

7.2.3 Lipid Separation

Some additional optional steps can be applied to the lipid extracted before the mass spectrometry analysis, which simplifies the complexity of the lipid extracted. This can be done using a physical approach which separates polar from non-polar lipids by using the liquid–liquid partitioning or solid phase partitioning or through a chemical approach which uses base hydrolysis method to separate sphingolipids from phospholipids/glycerolipids. There is also the alternative of derivatizing the extract by tagging the specific functional group of the lipids. The recovered lipid solutions are next examined using one of two different techniques as follows.

7.2.3.1 Shotgun Lipidomics

This method utilizes the physical and chemical properties of the lipid, which facilitates high throughput analysis of the cellular lipids directly from the organic extract of the sample.

7.2.3.2 LC Based Lipidomics

The Liquid Chromatography (LC) based analysis separates the lipids extracted and reduces the complexity. This method is more sensitive than the shotgun method.

7.2.4 MS-Based Lipid Analysis

Mass spectrometry (MS) is an analytical technique that measures the mass-to-charge ratio of the ions present in the given sample. This technique measures the exact molecular weight of the sample and also helps in elucidating the structure and chemical properties of different molecules present in the given sample. Recent advances in MS have allowed for extensive lipid profiling of the biological samples (Yang and Han 2016). The MS technique involves three important steps:

- Ionization techniques.
- MS analysis.
- Data detection.

7.2.4.1 Ionization Techniques

Some of the frequently used ionization techniques in MS for lipidomics are as follows.

7.2.4.2 Electrospray Ionization (ESI)

ESI is a soft ionization technique that transfers ions from liquid to gaseous phase with the assistance of electrical energy. When a strong electric field is applied to the lipid in the sample, an electrospray of lipids is produced, which then passes through a capillary tube and creates fine aerosol from which ions are formed by desolvation.

7.2.4.3 Matrix-Assisted Laser Desorption/Ionization (MALDI)

MALDI uses laser energy absorbing matrix to create ions. This is also a soft ionization technique. The lipid molecules are embedded into a matrix that absorbs the laser at a wavelength similar to that of the laser. The matrix along with the lipid molecule is irradiated by the pulsed laser which triggers ablation and desorption of the lipid and matrix. This facilitates the ionization of the lipid molecule in the hot plume of ablated gas.

7.2.4.4 Atmospheric Pressure Chemical Ionization (APCI)

This technique utilizes the gas-phase ion molecule reaction that takes place at atmospheric pressure. The solvent and the sample flows through the injector along with gas assistance. The heater in the injector vaporizes the solvent and the sample and ejects it through the outlet. After which an electro-acupuncture corona discharge is used to ionize the solvent that comes through the outlet.

7.2.4.5 Atmospheric Pressure Photoionization (APPI)

This method utilizes photochemical action to ionize samples in the gas phase. APPI even detects weak polar and non-polar compounds. In this technique the molecules are ionized using a vacuum ultraviolet light at atmospheric pressure. This is done by either direct absorption or by ionizing the dopant molecule which leads to chemical ionization of the target molecule.

7.2.4.6 Secondary Ion Mass Spectrometry (SIMS)

SIMS is one of the most sensitive surface analytical techniques, where the solid surface of the molecule is bombarded with a primary ion beam like silver and gold ions. The ejected secondary ions are then collected and sent for mass spectrometric analysis.

7.2.4.7 Desorption Electrospray Ionization (DESI)

This technique combines electrospray and desorption ionization methods. In this method an electrically charged mist is directed into the sample with the help of air pressure. The subsequent splashed droplets carry the desorbed and ionized analytes. The ions then proceed through the interface of air pressure into the mass spectrometer.

7.2.4.8 MS Analysis

Depending on the type of analysis required, the ionized molecules are then sent for either full mass spectrometric analysis (MS) or tandem mass spectrometric analysis (MS/MS). The following are some of the typical tandem mass spectrometric methods used in lipidomics

7.2.4.9 Product Ion Scan

In product ion scan the first analyzer mass selects the precursor ions, which are later induced to dissociate into product ions. After which a second analyzer analyzes the mass of those product ions.

7.2.4.10 Precursor Ion Scan (PIS)

In this technique, the precursor ions are selected on the basis of mass by the first mass analyzer. Some of the fragment ions from the precursor ions are selected and the second analyzer examines only the selected fragment ions. This selected fragment represents the common precursor ions. Thus, all the precursor ions are monitored through the fragment ions, respectively.

7.2.4.11 Neutral Loss Scan (NLS)

In NLS, all the precursor ions are scanned by the first analyzer like all other ion scan types. The second analyzer focuses on the precursor ions in which there is common neutral loss that happens during the ion dissociation. Thus all the precursor fragments that undergo specific neutral loss are monitored.

7.2.4.12 Selected/Multiple Reaction Monitoring (SRM/MRM)

This approach for quantifying specific samples from complex mixtures is extremely sensitive and precise. This is non scanning mass spectrometric analysis that is used only in targeted analysis. This technique utilizes two mass analyzers, one acts as mass filters and another monitors that specific fragment of the specified precursor ion. This specific fragment and precursor ion pair are called transitions. The technique in which parallel data acquisition from multiple selected transitions is called multiple reaction monitoring (MRM).

7.2.4.13 Data Detection

If the internal standard and calibration curve for each distinct species are missing, a de-isotoping operation is required after the entire data collecting process through several MS data analysis approaches. Following which identification and quantification of the individual lipid species is carried out by means of analytical approaches (Wu et al. 2014; Ekroos 2012). Further these quantitative and qualitative data are explored through bioinformatics data mining to understand the underlying lipid metabolism and its dysregulation.

7.2.5 Applications of Lipidomics

1. Understanding the changes in cellular lipids caused by diet can help lipidomics be a useful tool in nutritional lipidomics. Nutritional lipidomics also assign

function to lipids like signaling compound, nutrient sensor to find the interaction between the nutrient and its metabolism. Also lipidomics can be used to analyze and monitor the dietary intake and can provide nutritional advice to improve the quality of health.

2. Pharmaceutical industries and pharmaceutical research utilizes lipidomics for drug discovery, drug target identification, screening, toxicity testing, pre-clinical drug testing, predicting, and monitoring drug responses.
3. Lipidomics can help in metabolic disorders such as CVD, diabetes, non-alcoholic fatty liver disease, etc. Most metabolic disorders are highly related to lipids thus lipidomics plays a significant role in predicting the associated risk factors and also helps in therapeutic monitoring. Studying the high-density lipid (HDL) and identifying the phospholipids as a major bioactive compound of HDL through lipidomics has unraveled its connection to vascular diseases.
4. As our brain contains large amounts of lipids, thus normally lipid signaling, metabolism and homeostasis are highly connected to neurological disorders. Hence discovering and developing biomarkers for prognosis and diagnosing the disorders through lipidomics can help in preventing neurological disorders.
5. Lipid plays a major role in cell growth and cellular metabolism like lipids form the major structural component of cell membrane, some triglycerides are involved in energy ATP production, energy storage, and metabolism and some of the bioactive lipids act as secondary messengers and hormones. These functional activities are used by cancer cells for tumor development, cell proliferation, survival, and cell signaling. Thus using lipidomics we can identify cancer biomarkers and by monitoring the lipid levels in blood and other body fluids can aid in monitoring the efficacy and efficiency of the anticancer treatments.
6. Lipidomics analysis in ophthalmology can aid in the diagnosis and understanding eye diseases. Lipidomics has helped in identifying new biomarkers for eye diseases and helped in studying the tear film stability which helps in prognosis, diagnosis, and treatment of some ocular surface disease. Additionally, lipidomics has facilitated the development of various lipid-based treatments for ocular surface inflammatory diseases.
7. Lipids are considered to be a major factor in developing and processing of foods, which have a heavy toll on human health. Thus lipidomics can be applied to assess and analyze the food products and also in quality judgment of the food.
8. Lipidomics are also used as a tool to evaluate the effects of environmental stress through in vitro and in vivo models. This can be specifically used against organic pollutants and heavy metals in animals and humans.
9. Plant lipidomics research which involves identifying and quantifying lipids from plant tissues can unveil the plant's response to environmental stresses and can help in identifying the genes involved in the lipid metabolisms of plants.
10. Microbial lipidomics can help in studying the microbial lipids, their metabolism and their functions for drug discovery. Also lipidomics can help in optimizing the microbial fermentation conditions.

11. MS-based lipidomics approaches have helped in exploring the bioactive properties of the glycolipids present in microalgae. Understanding the polar lipid molecules found in the macroalgae has been made easier by lipidomics. These polar lipids have important nutraceuticals and bioactive properties.

7.2.5.1 Biosurfactant

The word "biosurfactant" was used to describe surface-acting substances with a natural origin, such as bacteria, plants, and animals, which improve surface-surface interactions by forming micelles (Rahman and Gakpe 2008). Additionally, the application of the biosurfactants reduces the tension of the interfacial surface between the surface and solution, water/oil, or water/air interfaces. In other words, the addition of surfactants lowers surface tension till the formation of structures such as micelles, vesicles, and bilayers; this critical point is known as the critical micelle concentration (Kumar et al. 2021). The biosurfactants produced by microbes have been used frequently as a raw material in a variety of industries, including waste management, agriculture, pharmaceutical sector, for lubrication, foaming, wetting, stabilizing dispersions, emulsions, and formulation. Biosurfactants, a surface-acting agent used to increase the production of micelles from natural sources such as plants, microorganisms, and animals through surface–surface interaction. The surfactants which are most commonly used known as chemical origin, but their high persistence power, toxic nature, and low degradation rate control their routine use in the cosmetics, food, and pharmaceutical industries (Kumar et al. 2021).

Due to their unique properties, including low toxicity, selectivity, and relative ease of manufacture, surface-active biomolecules have recently piqued widespread attention. The biosurfactant which contain special properties will be employed in a wide range of sectors, including organic chemicals, petrochemicals, petroleum, agrochemicals, miming, metallurgy fertilizers, foods, beverages, cosmetics, pharmaceuticals, and many others. Biosurfactant can be used as emulsifiers as well as de-emulsifiers, forming agents, functional food ingredients, wetting agents, spreading agents, and detergents. The biosurfactants play major role in oil industry and heavy crude oil bioremediation by interfacial surface tention reducing ability (Vijayakumar and Saravanan 2015).

Some microorganisms (mostly bacteria, yeasts, and fungus) may synthesize the biological surfactants extracellularly and as components of the cell membrane. The same functionality of synthetic surfactants was compared and the biosurfactants which have ability of being to work in extreme salinity conditions, pH and temperature, and also being nontoxic or very low level of toxicity, and biodegradable. Furthermore, despite being regarded as having a biological origin, biosurfactants can be produced using renewable substrates and structurally modified using genetic engineering and biochemical methods. With these innovative formulations, biosurfactant can reach different markets.

The advantages of the biosurfactants in several facets of life have been investigated and documented. Properties of biosurfactants such as emulsifying (liquid-liquid mixture), wetting, stabilizing, and spreading, cleaning, and foaming can accumulate at different interfaces such as liquid/gas, liquid/liquid, and liquid/solid thereby reducing internal surface tension. They are frequently utilized in detergent, petroleum, cosmetics, agriculture, food, and pharmaceutical industries with these properties (Gürkök and Özdal 2021).

The interaction between metals and biosurfactants results in a range of complexes and the removal of heavy metals off the surface of the soil, which may increase the metal ion concentration and bioavailability. In the last several years, there has been increased interest in the microbial-based synthesis of surfactants and their widespread use in the breakdown of herbicides and insecticides have garnered more attention.

A wide variety of chemicals that contain bacteria can be used as a source of carbon and energy for development. Microorganisms, however, diffuse a variety of compounds as biosurfactants when carbon is an insoluble hydrocarbon, while some yeast and bacteria diffuse biosurfactants that could also dissolve hydrocarbons in the medium.

7.2.5.2 Structure of Biosurfactant

Biosurfactant are amphiphilic molecules comprising of hydrophilic head and hydrophobic tail. The presence of hydrophilic head and hydrophobic tail helps to form the water-oil interface which helps to reduce the surface tension (Fig. 7.2).

7.3 Biosurfactant Types

According to their molecular weight, chemical makeup, and microbial origin, biosurfactants can be categorized as follows (Fig. 7.3):

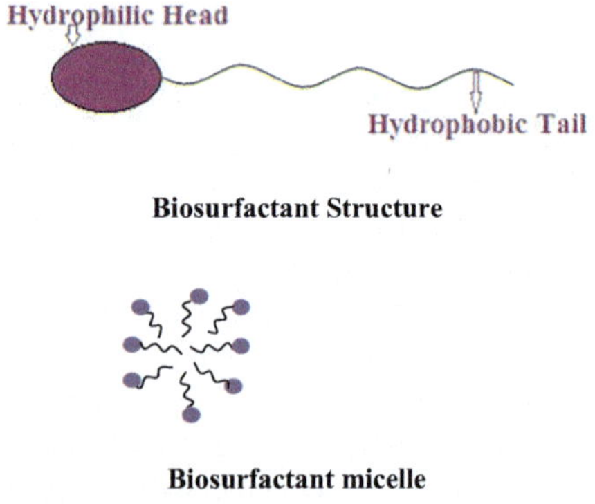

Fig. 7.2 Structure of biosurfactant and micelle formation

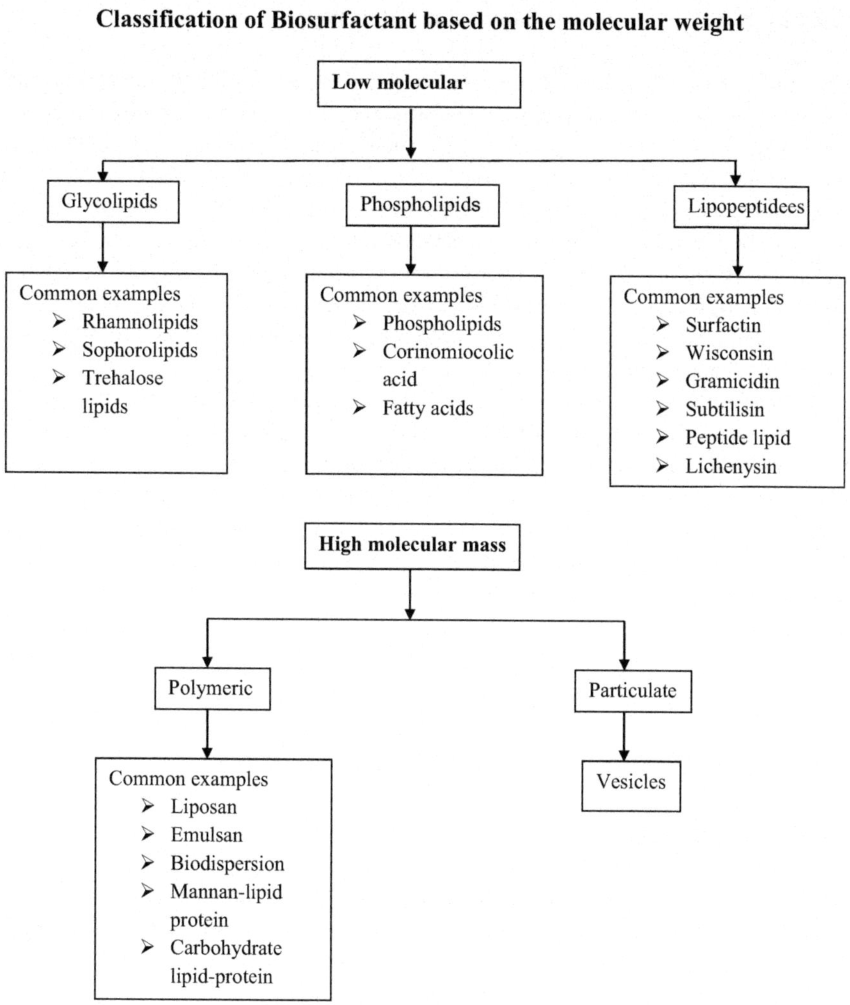

Fig. 7.3 Biosurfactant classification based on the molecular weight and some of their examples

7.3.1 On the Basis of Molecular Weight

7.3.1.1 Biosurfactants with Low Molecular Weight

This type of biosurfactant will reduce the surface and interfacial tension at the water/air interfaces. The majority of low molecular weight biosurfactants are glycolipids or lipopeptides. The glycolipids that have been studied the most are rhamnolipids, trehalolipids, and sophorolipids, which are disaccharides that are acetylated with hydroxyl fatty acids or long-chain fatty acids (Shah et al. 2016a, b).

7.3.1.2 Biosurfactants with High Molecular Weight

The ability to create a stable bio-emulsion is correlated with the high molecular weight of the biosurfactant. They act at low concentrations and are highly effective at stabilizing oil in water emulsions. When categorizing biosurfactants based on their polar groupings and the hydrophobic moiety long-chain fatty acids or fatty acids derivatives, the majority are either neutral or anionic, while the hydrophilic component can be phosphate, amino acid, carbohydrates, or cyclic peptide. According to Desai and Banat (1997), the structure of biosurfactant typically consists of a hydrophilic moiety made up of amino acids or peptides, anions or cations, mono-, di, or polysaccharides, and a hydrophobic moiety made up of derivatives of unsaturated, saturated, or fatty acids.

Similar to their chemical equivalents, biosurfactants can be categorized according to their molecular weight, critical micelle concentration (C.M.C.), the type of microorganisms they produce, and how they work. The most frequently reported low molecular weight compounds are glycolipids, phospholipids, and lipopeptides. However, high molecular weight is a characteristic of polysaccharides, lipopolysaccharides, and a complex mixture of biopolymers. Figure 7.1 provides a full classification and a few typical cases.

7.4 Categorized According to Molecular Weight

7.4.1 Based on the Chemical Composition

7.4.1.1 Glycolipids

The most well-known biosurfactants are glycolipids, which typically are carbohydrates that have an ester group attached to them along with long-chain aliphatic or hydroxy aliphatic acids (Vijayakumar and Saravanan 2015).

The most popular glycolipids are as follows.

1. Rhamnolipids.
2. Trehalolipids.
3. Sophorolipids.

7.4.1.2 Rhamnolipids

Rhamnolipids are glycolipids that contain one or two rhamnose molecules linked to one or two b-hydroxydecanoic acid molecules. The first rhamnose-containing glycolipids were discovered in *Pseudomonas aeruginosa* (Desai and Banat 1997).

7.4.1.3 Trehalolipids

The presence of trehalose ester allowed researchers to observe the serpentine group, one of the glycolipids, in large numbers on the cell surface of the *Mycobacterium* genus. Species of *Mycobacterium*, *Corynebacterium*, and *Nocardia* linked to mycolic acids at C-6 and C-69 by the disaccharide trehalose. Mycolic acids are long-chain, a-branched, b-hydroxy fatty acids. Trehalolipids from various organisms differ in the number of carbon atoms, degree of unsaturation, and size and structure of mycolic acid (Rahman and Gakpe 2008).

7.4.1.4 Sophorolipids

The sophorolipid class of glycolipids, which *Torulopsisbombicola*, *T. petrophilum*, and *T. apicola* primarily produce, consists of a long-chain hydroxy fatty acid connected to a dimeric carbohydrate called sophorose. These biosurfactants contain at least six to nine different hydrophobic sophorosides. Additionally, it was stated that water soluble sophrolipids were a mixture of several yeasts. As demonstrated by Cutler and Light, *Candida bogoriensis* generates glycolipids in which sophorose is connected to docosanoic acid diacetate (Gayathiri et al. 2022).

7.4.1.5 Lipopeptides and Lipoprotein

With cyclic lipopeptides, lipopeptides have a lipid component and 8–17 amino acids. They also exhibit significant diversity in both their amino acid and lipid makeup. *Bacillus* sp. produces lipopeptides known as surfactines, which have seven amino acids bound to carboxyl and hydroxyl groups in a 14-carbon acid (Rahman and Gakpe 2008).

7.4.1.6 Fatty Acid and Phospholipids

Bacteria, yeast, and other microorganisms produce biosurfactants fatty acids and phospholipids, which are in high demand due to their effectiveness and diversity. These biosurfactants will create alkane microemulsions, and proteinogenic or non-proteinogenic amino acids will create lipoamino acids (Gayathiri et al. 2022). Fatty acids produced by alkanes are thought of as a surfactant because they cause microbial oxidation. The proportion of hydrophilic and lipophilic fatty acids is definitely correlated with the length of the hydrocarbon chain. The most useful saturated fatty acid for lowering surface and interfacial tensions was discovered to be in the C12–C14 range (Rahman and Gakpe 2008).

7.4.1.7 Polymeric Biosurfactant

The majority of these biosurfactants are proteins that incorporate polymeric heterosaccharides. The best researched polymeric biosurfactants are polysaccharide protein complexes, mannoprotein, liposan, and emulsan (Desai and Banat 1997). Many bacterial species from various genera have been isolated using extracellular polymeric surfactants, which contain a complex combination of biopolymers as basic components, including proteins, polysaccharides, and lipopolysaccharides (Gayathiri et al. 2022).

7.4.1.8 Particulate Biosurfactant

Some of the bacteria produced microemlusion, which forms extracellular membrane vesicles that partition hydrocarbons. The microemlusions were created by microbial cells absorbing the alkane and are crucial in many processes (Desai and Banat 1997). In the marine environment, the glycolipids as a group have been extensively studied because the glycolipids generated by a wide range of bacteria and these bacteria were isolated from various marine matrices. They have been isolated from both animal and contaminated soil, including seapenpteroeides, annelids, and fish guts, as well as Antarctic and Arctic sediments. The glycolipopeptide biosurfactants were generated by the coral-associated *Bacillus* sp. (Floris et al. 2018).

Tables 7.2 and 7.3 illustrate the numerous researched types of biosurfactant together with the source of separation of their corresponding microbial producer (Floris et al. 2018) (Table 7.4).

Table 7.2 Biosurfactant from natural sources

Source of isolation	Biosurfactant	Bacterial strains
Sea water	Glycoprotein	*Pseudomonas* sp.
Sand, soil, samples and sea water	Not determined	*Catenovulum* sp., *Pseudomonas* sp., *Serratia* sp., *Rhodocccus* sp., *Glaciecola* sp.
Oil spilled, sample, sea water	Not determined	*Accenitobacter* sp., *Bacillus* sp., *Pseudomonas* sp.
Shoreline marine sediment enrichments	Glycolipid	*Pseudomonas* sp., *Idiomarina* sp., *Rhodocccus* sp.
Black Sea, hot water springs	Not determined	*Bacillus* sp.
Sea water samples	Not determined	*Aeromonas hydrophila, Pseudomonas aeruginosa*
Sea water samples	Lipopeptide	*Hallobacterium salinarum*
Sea water sediment, shells samples	Not determined	*Bacillus* sp., *Accenitobacter* sp., *Pseudomonas* sp.
Sea water enrichments	Not determined	*Lactobacillus casei, C. xervsis, Escherichia coli, Klebsillaozaenae, Citrobacter intermedius, B. megaterium*

Table 7.3 Biosurfactant from biological matrices

Source of isolation	Biosurfactant	Bacterial strains
Callyspongia diffusa	Lipopeptide	*Ncardiopsis alba*
Dendrilla nigra	Glycolipid	*Brachybacterium paraconglomeratum*
Sparus aurata	Glycolipid/ Rhamnolipid	*Acinetobacter* sp., *Aeromonas* sp., *Sphingomonas* sp., *Pseudomonas* sp.
Dendrilla nigra	New lipopeptide	*Brevibacterium aureunt*
Acropora digitifera	Lipopeptide/new biosurfactants	*Bacillus flexus, Phychrobacter* sp., *Bacillus pumilus, Psychrobacter* sp., *Bacillus anthracis*
Pterocides spinosum	Glycolipid	*Vibrio* sp., *Brevibacterium* sp.
Callyspongia diffusa	Lipopeptide	*Bacillus* sp., *Halomonas* sp., *Alcaligenes* sp.
Sarchophyton glaucum	Glycolipopeptide	*Bacillus* sp. E34

7.4.2 *Properties of Microbial Biosurfactants*

7.4.2.1 Surface and Interfacial Activity

At very low concentration, biosurfactants produce lower surface tension and it gives high effectiveness. The efficiency was compared to conventional surfactants.

7.4.2.2 Temperature, pH, and Ionic Strength

Many biosurfactants were used under severe condition. For example, the lipopeptide from *Bacillus licheniformis* temperature will be stable at 5 °C for up to 140 h, with a pH range of 5–12. It also can withstand salt concentration up to 10%; and 2% NaCl concentration for is sufficient inactive conventional surfactants.

7.4.2.3 Biodegradability

Bacteria and other microorganisms in soil or water will easily degrade the biosurfactant. As a result, they are suitable for waste treatment and bioremediation applications.

Table 7.4 Classification based on the microbial origin: major categories of microbial surfactants with their microbial source biosurfactants

Biosurfactant		Organism
Glycolipids	Rhamnolipids	*P. aeruginosa* *Pseudomonas* sp.
	Trehalolipids	*R. erythropolis* *N. erythropolis*
	Sophorolipids	*Mycobacterium* sp. *T. bombicola* *T. apicola*
	Cellobiolipids	*T. petrophilum* *U. zeae* *U. maydis*
Lipopeptides and lipoproteins	Peptide-lipid	*B. licheniformis*
	Serrawettin	*S. marcescens*
	Viscosin	*P. fluorescens*
	Surfactin	*B. subtilis*
	Subtilisin	*B. subtilis*
	Gramicidins	*B. brevis*
	Polymyxins	*B. polymyxa*
Fatty acids, neutral lipids, and phospholipids	Fatty acids	*C. lepus*
	Neutral lipids	*N. erythropolis*
	Phospholipids	*T. thiooxidans*
Polymeric surfactants	Emulsan	*A. calcoaceticus*
	Biodispersan	*A. calcoaceticus*
	Mannan-lipid-protein	*C. tropicalis*
	Liposan	*C. lipolytica*
	Carbohydrate-protein-lipid	*P. fluorescens* *D. polymorphis*
	Protein PA	*P. aeruginosa*
Particulate biosurfactants	Vesicles and fimbriae	*A. calcoaceticus*
	Whole cells	Variety of bacteria

7.4.2.4 Low Toxicity

Due to the increasing concern biosurfactant have received greater attention regarding the allergic effect of artificial product on the part of the population. Furthermore, due to their low toxicity, they are allowed for use in cosmetic, food, and pharmaceutical products.

7.4.2.5 Availability

Because biosurfactants can be extensively produced, even industrial waste are good resources to increase its availability.

7.4.2.6 Specificity

Due to their complex organic molecular structure and unique functional group, biosurfactants frequently have very distinct properties. It has considerable specificity to pollutant of detoxification as well as applications in the cosmetic, food, and pharmaceutical industries.

7.4.2.7 Biocompatibility and Digestibility

Because of their biocompatibility and digestibility, they can be employed in the cosmetic, food, and pharmaceutical industries (Table 7.5).

7.4.2.8 The Use of Biosurfactants and Their Significance

Application of Soil Quality Improvement

The presence and availability of appropriate concentrations of micronutrients and macronutrients in the soil support crop ecology, growth, and productivity. Physiological processes in plants are directly influenced by trace elements present in the soil. In fact, excess or deficiency of any of these elements has resulted in poor quality as well as various plant growth retardation. The high level of global temperature, increase in salinity, imbalance of soil pH, continuous changes in climatic condition, or efficacy of soil, which has adversely affected by deposition of environmental contaminants, resulting in poor availability to the plants, and ultimately, low crop production and poor food quality. Furthermore, the microbial biosurfactant influence plant growth by synthesizing phytohormones, which significantly modulates and activates resistant mechanisms. Hence, the soil which contains micronutrient has high efficiency and is extremely suitable to enhance plant output due bioaugmentation-producing bacteria. The biosurfactants have high impact on the native microflora, which is indirectly and directly responsible for promotion in plant growth or soil helps to remove contaminants from plant roots or soil, and also helps to mitigating biotic and abiotic stresses. Liao et al. 2016 examined the biosurfactant (lecithin and rhamnolipid) and, as well as chemical surfactant (tween 80) in pot experiment with maize. The biomass of maize was not significantly impacted after application in crude oil-contaminated soil, however, biosurfactants such as rhamnolipid and lecithin improved the microbial population, resulting in higher petroleum hydrocarbons from the contaminated soil.

Table 7.5 The bioremediation of oil polluted environments which contain potential application for various types of biosurfactant with their producing microorganisms (de Cássia F S Silva et al. 2014)

Biosurfactant types	Microorganisms	Applications
Rhamnolipid	*Pseudomonas aeruginosa BS20*	Bioremediation of hydrocarbon-contaminated sites
	Pseudomonas alcaligenes	Environmental applications
	Pseudoxanthomonas sp. PNK-04	Environmental applications
	*Pseudomonas cepacia*CCT6659	Bioremediation of marine and soil environments
	Pseudomonas aeruginosa S2	Bioremediation of oil-contaminated sites
Glycolipid	*Nocardiopsislucentensis*MSA04	Bioremediation of marine environment
	Pseudozyma hubeiensis	Bioremediation of marine oil pollution
	*R. wratislaviensis*BN38	Bioremediation applications
	*C. guilliermondii*UCP0992	Removal of petroleum derivate motor oil from sand
Trehalose tetra ester	*Micrococcus luteus* BN56	Bioremediation of oil-contaminated environments
Mannosylerythritol lipid	*C. sphaerica*UCP0995	Oil removal
Sophorolipids	*C. lipolytica*UCP0988	Oil recovery
	Candida bombicola	Environmental applications
Protein-carboydrate-lipid complex	*C. glabrata* UCP1002	Oil removal
	C. tropicalis UCP0996	Removal of petroleum and motor oil adsorbed to sand
	C. sphaerica UCP0995	Bioremediation processes
	C. glabrata UCP1002	Oil recovery from sand
Glucolipid and trehalose lipid	*Rhodococcus erythropolis*3C-9	Oil spill cleanup operations
	Calyptogena soyoae	Bioremediation of marine environment

Application of Biosurfactant in Pharmaceutical Industry

7.4.2.9 Antioxidant Properties

The use of microbial biosurfactants as antioxidant agents has a significant impact on the pharmaceutical and food industries these days. Free radicals are generated in our body during several physiological process and are neutralized by antioxidant molecules. Extremely active free radicals lead to severe oxidative damage or oxidative stress (Kumar et al. 2021). The biosurfactant of microbiological origin can act on

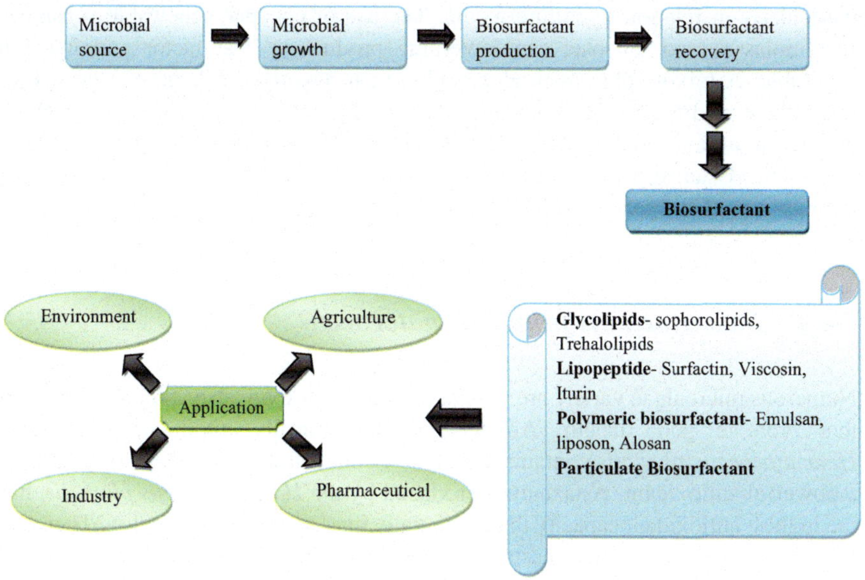

Fig. 7.4 Schematic representation of microbial resource utilized for the production of biosurfactant and their potential application (Kumar et al. 2021)

surface physiological features, and, as a result can prevent bacterial adhesions on the surface. Similarly, the flow of oxidative chain reactions was also arrested by providing antioxidant activities. Due to its advantages over the synthetic antioxidants, such as their antioxidant characteristics, low toxicity, antibacterial activities, and biodegradability, biosurfactants have gained significant industrial attention (Abdollahi et al. 2020) (Fig. 7.4).

7.4.3 Biosurfactants from Algae

Microalgae are employed to make biosurfactants since they are significant producers of neutral lipids, phospholipids, and glycolipids. Additionally, as a source of biocompounds, microorganisms have numerous applications in the food and pharmaceutical industries, including the production of bio-pigments and essential fatty acids, the direct incorporation of biomass into animal and human feed, the production of biofuels like biodiesel, H_2, biogas, and bioethanol, and the biofixation of carbon dioxide (Bellou et al. 2014).

The fact that many of the microorganisms used in biosurfactant production fall under the generally recognized as safe (GRAS) classification is a major benefit for growing microalgae. Such verified organisms can be applied in the pharmaceutical and food industries and pose no risk of pathogenicity or toxicity (Soccol et al. 2013). The investment costs were rising due to the low output of commercial biosurfactants, particularly in relation to culture media. The use of cheaper, renewable substrates like molasses and glucose, can lower the production cost factor and make it profitable. A mixotrophic microalgae culture can greatly accelerate the growth of microalgae, and produce cell densities that are three to ten times higher than those attained in an autotrophic culture. Moreover, while respiration can supply cellular energy needs, an organic substrate in the culture medium can prevent nighttime biomass loss (Elisngela et al. 2015).

7.4.4 Biosurfactants from Haematococcus

Numerous microalgae variety are used as nutrient-rich food, also contain substances that promote good health. Among the economically important microalgae, *Haematococcus pluvialis* contains the highest concentration of natural astaxanthin, a powerful antioxidant. Astaxanthin produced by the *Haematococcus pluvialis*, has the highest antioxidant capacity than artificially manufactured astaxanthin. There are significant uses for the natural astaxanthin in the cosmetics, nutraceuticals, aquaculture, and food industries (Shah et al. 2016a, b). Previous research by Raman et al. (2008) showed that under high light and sodium acetate stresses, *Haematococcus pluvialis* accumulated enormous amounts of astaxanthin, due to increased activity of the transcription genes in astaxanthin biosynthetic pathway. Among high-value microalgal compounds, the red pigment astaxanthin has the highest demand in the market due to its antioxidant properties. It belongs to the xanthophyll family and can be found in some microalgae, bacteria, yeasts, and marine animals. Among high-value microalgal compounds, the red pigment astaxanthin has the highest demand in the market due to its antioxidant properties. It belongs to the xanthophyll family and can be found in some microalgae, bacteria, yeasts, and marine animals.

Astaxanthin has the highest market demand among high-value microalgal compounds due to its antioxidant qualities. This pigment belongs to the xanthophyll family and can be found in some microalgae, marine animals, yeasts, and bacteria. Also farmed salmon, poultry, and trout are key sources for astaxanthin. In humans, the decrease in occurrence of neurodegenerative diseases, ocular diseases, cancer, and skin diseases and improvement of immunological response has been demonstrated by diet supplemented with astaxanthin. It has reportedly been shown to have anti-inflammatory and LDL cholesterol-inhibiting properties. In comparison to other species that produce astaxanthin naturally, such as the microalga *Chromochloris zofingiensis* and the red yeast *Xanthophyllomyces dendrorhous*, *H. pluvialis* may synthesize and deposit the pigment in significantly greater quantities (Le-Feuvre et al. 2020).

7.4.4.1 Astaxanthin Production by Green Microalgae, *Haematococcus pluvialis*

There are more than 7000 different kinds of "green microalgae" that thrive in various environments. Unicellular freshwater microalga, *Haematococcus pluvialis* (class Chlorophyceae, order Volvocales) is ubiquitous distributed all over the world. It is regarded as the primary producer of the best commercial natural astaxanthin. It belongs to the same family of carotenes as lycopene, lutein, and beta-carotene (Rao et al. 2010). According to Pérez-López et al. (2014) and Cyanotech (2015), astaxanthin is produced through the carotenoid pathway from glyceraldehyde-3-phosphate and pyruvate. Depending on the cultivation circumstances and/or the end products of photosynthesis, both of these substances are glycolytic. These two crucial metabolic intermediates then enter the non-mevalonate pathway (MEP). Following the integration, IPP, a crucial intermediary in the synthesis of all carotenoid pigments, including astaxanthin begins to be produced. Astaxanthin has a wide range of uses in the cosmetic, food, aquaculture, feed, nutraceutical, and pharmaceutical industries because of its ability to scavenge free radicals. As an antioxidant, astaxanthin is 100 times more effective than α-tocopherol and 65 times more potent than vitamin C. Over 95% of the astaxanthin available in the market is synthetically manufactured. Most notably just 1% of the astaxanthin produced by *Haematococcus pluvialis* is used for commercial purposes.

One of the most recent improvements in the cultivation of *Haematococcus pluvialis* for the synthesis of astaxanthin is the use of a two-stage mixotrophic culture system and an attached cultivation system that uses an immobilized biofilm (Zhang et al. 2014). Due to the induction of carotenoid synthesis, *Haematococcus pluvialis* has a direct association between the cultivation process, the astaxanthin content of cells, and overall astaxanthin output. Environmental elements such as light, pH, temperature, salt content, and nutritional stressors as well as a variety of plant hormones and their derivatives have an impact on the astaxanthin accumulation. Attempts have been made to manufacture astaxanthin using both conventional mutagenesis and more recently genetic engineering in order to genetically improve the capability of *Haematococcus pluvialis*. There are numerous techniques for extracting astaxanthin, including those that use pressurized liquids, supercritical carbon dioxide (SC-CO_2), acids, edible oils, enzymes, and liquefied dimethyl ether (DME). Although numerous studies have examined different astaxanthin extraction and downstream processing techniques from *H. pluvialis* biomass, more thorough research is necessary to fully exploit the biological potential of this microalga and its very valuable byproduct (Shah et al. 2016a, b).

7.4.4.2 Biotechnology Applied to *Haematococcus pluvialis*

In terms of the ability to produce recombinant proteins, some microalgae have tremendous potential for genetic engineering in the production of recombinant proteins. They also produce valuable biochemicals like carotenoids or Poly Unsaturated Fatty acids (PUFA)s. Despite the breakthrough in genetic engineering with the model microalga *Chlamydomonas reinhardtii*, there have been very few attempts to work with other microalgae. In the past 10 years, genetic engineering has targeted the *H. pluvialis*, either by nuclear or chloroplast transformation (Steinbrenner and Sandmann 2006). Recently, Luo et al. (2019) published the first draft genome of *H. pluvialis*, a valuable resource for understanding the genetic basis of astaxanthin production. The genome analysis of *H. pluvialis* will help us to elucidate the molecular mechanism of astaxanthin production and it can be used in the industrial sector to increase the production of astaxanthin.

References

Abdollahi S, Tofighi Z, Babaee T, Shamsi M, Rahimzadeh G, Rezvanifar H, Saeidi E, Mohajeri Amiri M, SaffariAshtiani Y, Samadi N (2020) Evaluation of anti-oxidant and anti-biofilm activities of biogenic surfactants derived from *bacillus amyloliquefaciens* and *Pseudomonas aeruginosa*. Iran J Pharm Res 19(2):115–126. https://doi.org/10.22037/IJPR.2020.1101033

Bellou S, Baeshen MN, Elazzazy AM, Aggeli D, Sayegh F, Aggelis G (2014) Microalgal lipids biochemistry and biotechnological perspectives. Biotechnol Adv 32(8):1476–1493. https://doi.org/10.1016/j.biotechadv.2014.10.003; Epub 2014 Oct 14

Cyanotech (2015) BioAstin - Natural Astaxanthin. http://www.cyanotech.com/bioastin.html. Retrieved 21 Oct 2015

de Cássia F S Silva R, Almeida DG, Rufino RD, Luna JM, Santos VA, Sarubbo LA (2014) Applications of biosurfactants in the petroleum industry and the remediation of oil spills. Int J Mol Sci 15(7):12523–12542. https://doi.org/10.3390/ijms150712523

Desai JD, Banat IM (1997) Microbial production of surfactants and their commercial potential. Microbiol Mol Biol Rev 61(1):47–64. https://doi.org/10.1128/mmbr.61.1.47-64.1997

Ekroos K (ed) (2012) Lipidomics: technologies and applications. Wiley, Weinheim

Elisngela MR, Etiele GDM, Cibele FDO, Kellen Z, Jorge AVC (2015) Microalgae cultivation for biosurfactant production. Afr J Microbiol Res 9(47):2283–2289

Floris R, Rizzo C, Giudice AL (2018) Biosurfactants from marine microorganisms. In Wael N. Hozzein (Ed.), Metabolomics - new insights into biology and medicine. IntechOpen. https://doi.org/10.5772/intechopen.80493

Gayathiri E, Prakash P, Karmegam N, Varjani S, Awasthi MK, Ravindran B (2022) Biosurfactants: potential and eco-friendly material for sustainable agriculture and environmental safety—a review. Agronomy 12(3):662

Gürkök S, Özdal M (2021) Microbial biosurfactants: properties, types, and production. Anatol J Biol 2(2):7–12

Honeycutt ME, McFarland VA, McCant DD (1995) Comparison of three lipid extraction methods for fish. Bull Environ Contam Toxicol 55(3):469–472

Kumar A, Singh SK, Kant C, Verma H, Kumar D, Singh PP, Modi A, Droby S, Kesawat MS, Alavilli H, Bhatia SK, Saratale GD, Saratale RG, Chung S-M, Kumar M (2021) Microbial biosurfactant: a new frontier for sustainable agriculture and pharmaceutical industries. Antioxidants 10(9):1472. https://doi.org/10.3390/antiox10091472

Le-Feuvre R, Moraga-Suazo P, Gonzalez J et al (2020) Biotechnology applied to *Haematococcuspluvialis* Fotow: challenges and prospects for the enhancement of astaxanthin accumulation. J Appl Phycol 32:3831–3852. https://doi.org/10.1007/s10811-020-02231-z

Liao C, Xu W, Lu G, Deng F, Liang X, Guo C, Dang Z (2016) Biosurfactant-enhanced phytoremediation of soils contaminated by crude oil using maize (Zea mays. L). Ecol Eng 92: 10–17

Luo Q, Bian C, Tao M, Huang Y, Zheng Y, Lv Y, Li J, Wang C, You X, Jia B, Xu J (2019) Genome and transcriptome sequencing of the astaxanthin-producing green microalga, Haematococcus pluvialis. Genome Biol Evol 11(1):166–173

Pérez-López P, González-García S, Jeffryes C, Agathos SN, McHugh E, Walsh D et al (2014) Life-cycle assessment of the production of the red antioxidant carotenoid astaxanthin by microalgae: from lab to pilot scale. J Clean Prod 64:332–344. https://doi.org/10.1016/j.jclepro.2013.07.011

Rahman PK, Gakpe E (2008) Production, characterisation and applications of biosurfactants-review. Biotechnology 7:360–370

Raman V, Lakshmanan V, Ravi S, Gokare AR (2008) Regulation of carotenoid biosynthetic genes expression and carotenoid accumulation in the green alga Haematococcuspluvialis under nutrient stress conditions. J Exp Bot 59:1409–1418. https://doi.org/10.1093/jxb/ern048

Rao AR, Reddy A, Aradhya SM (2010) Antibacterial properties of Spirulina platensis, Haematococcus pluvialis, Botryococcus braunii micro algal extracts. Curr Trends Biotechnol Pharm 4:809–819

Sargent JR, Tocher DR, Bell JG (2003) The lipids. In: Hardy RW (ed) Fish nutrition. Elsevier, pp 181–257

Shah N, Nikam R, Gaikwad S, Sapre V, Kaur J (2016a) Biosurfactant: types, detection methods, importance and applications. Indian J Microbiol Res 3(1):5–10

Shah MM, Liang Y, Cheng JJ, Daroch M (2016b) Astaxanthin-producing green microalga Haematococcuspluvialis: from single cell to high value commercial products. Front Plant Sci 7:531. https://doi.org/10.3389/fpls.2016.00531

Soccol CR, Pandey A, Larroche C (eds) (2013) Fermentation processes engineering in the food industry, 1st edn. CRC Press, Boca Raton. https://doi.org/10.1201/b14070

Steinbrenner J, Sandmann G (2006) Transformation of the green alga Haematococcus pluvialis with a phytoene desaturase for accelerated astaxanthin biosynthesis. Appl Environ Microbiol 72: 7477–7484

Vijayakumar S, Saravanan V (2015) Biosurfactants-types, sources and applications. Res J Microbiol 10:181–192

Wenk MR (2010) Lipidomics: new tools and applications. Cell 143(6):888–895

Wu Z, Shon JC, Liu KH (2014) Mass spectrometry-based lipidomics and its application to biomedical research. J Lifestyle Med 4(1):17

Yang K, Han X (2016) Lipidomics: techniques, applications, and outcomes related to biomedical sciences. Trends Biochem Sci 41(11):954–969

Zhang W, Wang J, Wang J, Liu T (2014) Bioresource technology. Attached cultivation of *Haematococcus pluvial* for astaxanthin production. Bioresour Technol 158:329–335. https://doi.org/10.1016/j.biortech.2014.02.044

Chapter 8
Genetic Engineering Approaches for *Haematococcus*: Recent Developments in Genome Sequencing and Strain Improvement

Bahar Aslanbay Guler and Esra Imamoglu

8.1 Introduction

The *Haematococcus* species, mainly *Haematococcus pluvialis* have attracted a great interest of scientists due to their unique life cycle and rich content of bioactive compounds. The complex life cycle of *H. pluvialis* is divided into two stages involving green vegetative stage and red encysted stage (Hoys et al. 2021; Saini et al. 2021). In the latter stage, a hematocyst cell is able to accumulate large amounts of secondary metabolites including, violoxanthin, zeaxanthin, lutein, cantaxanthin, and especially astaxanthin (Grujić et al. 2022). Astaxanthin is a red valuable keto-carotenoid pigment that has a wide range of applications in the pharmaceutical, cosmetics, nutraceuticals, and aquaculture industries because of its strong bioactive properties (Guler et al. 2022; Zhao et al. 2019). Currently, a large fraction of industrially used astaxanthin is produced synthetically, however, it has some challenges including negative impacts on environment, safety concerns, and low antioxidant capacity (Zhang et al. 2020). Therefore, natural astaxanthin has become more popular in recent years, and researchers have focused on the improvement of astaxanthin availability in *H. pluvialis* by using different genetic engineering approaches.

Although metabolic pathways of microalgae have complex mechanisms, numerous genetic engineering technologies have been carried out mainly in *Chlamydomonas reinhardtii*, *Phaeodactylum tricornutum*, and *Thalassiosira pseudonana* species because most part of their genome sequences are available (Sharon-Gojman et al. 2015). Recent progresses have largely shifted from conventional processes towards systematic and technological regulations. Some of the

B. Aslanbay Guler · E. Imamoglu (✉)
Department of Bioengineering, Faculty of Engineering, Ege University, Izmir, Turkey
e-mail: esra.imamoglu@ege.edu.tr

R. Raja et al. (eds.), *Haematococcus*, https://doi.org/10.1007/978-981-99-2901-6_8

innovative applications performed in microalgal biotechnology are synthetic biology (involving the design of expression vectors with desired genetic elements), system biology (including bioinformatics, genomics, metabolomics, proteomics, and computational models), and metabolic engineering (aiming at enhancement of the production of target products through genome editing and metabolic pathway manipulation) (Grama et al. 2022). Despite the striking breakthroughs reported in the genetic engineering of microalgae, there is quite a few research studies done with *H. pluvialis* cells because of limited genome information. Given the complexity of the metabolic pathway of *H. pluvialis* and the importance of astaxanthin in commercial applications, recent intensive efforts have been made to evaluate the whole genome sequence of this strain and to provide new insights into genetic variations.

To further improve genetic engineering studies, there is currently a need for a comprehensive review presenting concise and precise updates on the latest progress. In this chapter, a detailed overview is presented to highlight recent advances and challenges in the genetic engineering of *H. pluvialis* microalgae. Firstly, a brief introduction is given to summarize the physiological and biochemical properties of the strain along with the growth profile and metabolic pathway. Then, the recent and significant advancements for the sequencing and mapping of the *H. pluvialis* genome are reviewed with special attention to the latest technologies. Moreover, genetic transformation techniques and tools are described to improve specific characteristics of microalgae cells, and also a literature survey covering the application of genetic manipulation to *H. pluvialis* is represented.

8.2 An Overview to *H. pluvialis* and Its Metabolic Pathway for Carotenoid Production

The unicellular green microalga *H. pluvialis* is mainly distributed in freshwater sources such as natural or artificial ponds, bird baths, and rain pools located in temperate regions. As mentioned before, it has a unique life cycle involving two different stages as vegetative and encysted, in which cells appear in the four different morphologies. In the green vegetative stage, cells are observed in the forms of macrozooid, microzooid, and palmella. Under optimal conditions, macrozooids are the major morphologies, which cells have spherical or irregular pear shapes with two flagella and dimensions of 8–20 μm. They grow rapidly, resulting the formation of daughter cells by the asexual reproduction. When conditions become unfavorable, the size of cells increases, both flagella are lost, and macrozooids become non-motile forms called palmella. The prolonged stress factors lead to turn the cells into the red hematocyst phase, which is characterized by high astaxanthin accumulation. At this stage, hematocysts form a rigid thick cell wall consisting of three layers and deposit large amounts of astaxanthin in its lipid droplets. It is important to note that hematocysts cells are able to turn into green vegetative stage, which called microzooids, once the conditions become optimal (Khoo et al. 2019; Ma et al. 2018).

From the biochemical perspective, cellular composition of *H. pluvialis* cells shows differences between green and red stages. According to the latest reports, green vegetative cells include a high proportion of protein up to 45% of their dry weight. When cells turn into red encysted morphology, protein content decreases to 17%. In green stage, cells consist of 15–17% dry weight of carbohydrates and 20–25% dry weight of lipid involving neutral lipids, phospholipids, and glycolipids. In the red stage, stress conditions cause the increase in the content of carbohydrates and lipids to the amounts of 60–70% and 32–40%, respectively (Mularczyk et al. 2020; Nishshanka et al. 2022; Oslan et al. 2021). Most importantly, carotenoid content significantly varies during the life cycle of *H. pluvialis*. The main carotenoids produced in vegetative stage are lutein, β-carotene, chlorophylls, violoxanthin, neoxanthin, and zeaxanthin. These pigments are not found in the red stage or are only present in trace amounts. The primary carotenoid pattern of cells observed during red stage comprises of astaxanthin, which can reach up to 80–90% of total carotenoids (Gong et al. 2019; Mularczyk et al. 2020). The variations in carotenoid contents are associated with the alterations in the carotenoid pathway of *H. pluvialis* cells under unfavorable conditions.

A lot of research has been recently established to clarify the carotenoid synthesis pathway of *H. pluvialis* during its growth stages. This process is highly complicated, which is regulated by external factors such as high light, nutrient deprivation, UV radiation, etc. The metabolic pathway of carotenoids, which is presented in (Fig. 8.1), is mainly derived from the isopentenyl pyrophosphate (IPP) precursor. There are two different pathways for the production of IPP; (a) the mevalonate (MVA) pathway, and (b) the non-mevalonate (MEP) pathway, which is also called 1-deoxy-D-xylulose 5-phosphate pathway (DOXP). It is well known that IPP synthesis in *H. pluvialis* is carried out through MEP pathway, since these cells do not have major enzymes of MVA pathway (Le-Feuvre et al. 2020; Velmurugan and Kodiveri Muthukaliannan 2022). In beginning, pyruvate combined with glyceraldehyde 3-phosphateis transformed to IPP, and then it undergoes isomerization to an allelic isomer dimethyl allyl diphosphate (DAMPP) in the plastids. This reversible reaction is known to be catalyzed by the enzyme of isopentenyl pyrophosphate isomerase (IPI), which is coded by two genes as ipi1 and ipi2. However, latest transcriptomic studies have shown that another enzyme, called 4-hydroxy-3-methylbut-2-enyl diphosphate reductase (HDR), is more associated with the catalyzing the conversion between IPP and DMAPP (Ren et al. 2021a, b, c; Wan et al. 2021).

Later, the DAMPP molecule is elongated with the linear additions of three IPP molecules, resulting in the formation of geranylgeranyl pyrophosphate (GGPP). GGPP synthase (GGPS) is the catalyzation enzyme during this process. The next step, which plays an important role for the transition to carotenoid biosynthesis, is the condensation of two GGPP molecules to form a colorless carotene phytoene by using phytoene synthase (PSY) enzyme. According to many published studies, accumulation of high amounts of carotenoids is related to the over expression of the PSY, which shows changes as a response to stress factors (Chen et al. 2020; Hoys et al. 2021). The generated phytoene carotene undergoes steps of desaturation

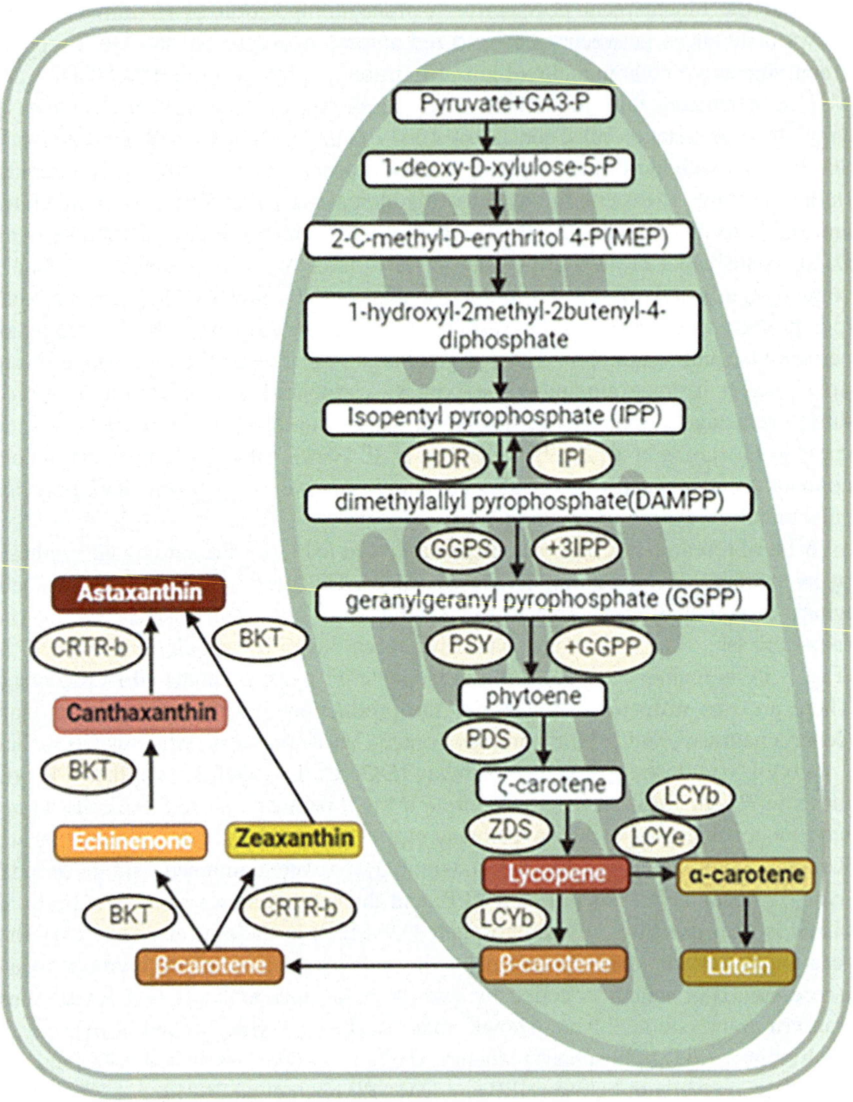

Fig. 8.1 Carotenoid synthesis pathway of *H. pluvialis*, *HDR* 4-hydroxy-3-methylbut- 2-enyl diphosphate reductase, *IPI* isopentenyl pyrophosphate isomerase, *GGPS* geranylgeranyl pyrophosphate synthase, *PSY* phytoene synthase, *PDS* phytoene desaturase, *ZDS* ζ-carotene desaturase, *LCYe* lycopene α-cyclase, *LCYb* lycopene β-cyclase, *CRTR-b* β-carotene hydroxylase, *BKT* β-carotene ketolase

reactions by the enzymes of phytoene desaturase (PDS) and ζ-carotene desaturase (ZDS). It should be noted that plastid terminal oxidase (PTOX) and plastoquinones (PQ) are assumed to be cofactors for the electron transport chain in these desaturation steps. The potential relationship between carotenoid synthesis and

PTOX has been reported in *H. pluvialis*, and two copies of PTOX1 and PTOX2 were found to be upregulated under unfavorable conditions, resulting in carotenoid accumulation (Ding et al. 2019). Desaturation of phytoene produce pink-colored lycopene, which is a highly unstable carotenoid, via the intermediate products of phytofluene, ζ-carotene, and neurosporene. From this point, the carotenoid pathway proceeds in two different routes, starting with the cyclization of lycopene. One of the pathways leads to the production of α-carotene by the catalyzations of lycopene α-cyclase (LCYe) and lycopene β-cyclase (LCYb) enzymes. After that, the subsequent step involves the hydroxylation of α-carotene to form lutein (Liang et al. 2018; Srivastava et al. 2022).

In *H. pluvialis*, the second pathway is more dominant during stress conditions, in which lycopene is cyclized at both ends by LCYb, yielding β-carotene. It has been reported that this carotenoid is transferred from the chloroplast to the cytoplasm by creating cytoplasmic lipid droplets (Pick et al. 2018). At this level, two oxygenation processes catalyzed by β-carotene hydroxylase (CRTR-b) and β-carotene ketolase (BKT) can take place for astaxanthin production. One of them is based on the addition of a keto group to β-carotene ring by the BKT enzyme, resulting in the formation of echinenone and cantaxanthin. Then, cantaxanthin can be hydroxylated by the CRTR-b to form astaxanthin. In the alternative pathway, β-carotene produces zeaxanthin with the addition of hydroxyl groups, and zeaxanthin is converted into astaxanthin by BKT (Jannel et al. 2020). The detailed information about the biosynthetic pathway of carotenoids coupled with the genome sequencing efforts offer promising strategies to further improve the *H. pluvialis* strain via genome editing technologies.

8.3 Genome Sequencing Approaches of *H. pluvialis*

H. pluvialis encounters some constraints during the cultivation and astaxanthin accumulation because of the slow growth rate, low production yield, contamination risk, and high operation cost. Given these limitations and the above-mentioned complex metabolism, it is highly important to investigate and improve the genetic mechanisms and gene expression profiles of *H. pluvialis* cells through genetic engineering applications. In the current microalgae research-to-genetic engineering value chain, the first step is the identification of genome sequences that are responsible for the regulation of metabolic pathways. With current advances in molecular biotechnology, the experimental annotation of microalgae genome has become much practical and quicker (Kumar et al. 2020; Teng et al. 2020).

Whole genome sequencing (WGS) is a key tool, allowing the identification of entire genome sequences of interested organisms. It determines the order of nucleotide bases involving coding and noncoding regions, and thus provides highly detailed information about the genetic structure (Fuentes-Pardo and Ruzzante 2017). It is well known that the whole genomes of three microalgae species as *C. reinhardtii*, *P. tricornutum*, and *T. pseudonana* were sequenced in the early

2000s. With representative insights about these organisms and advances in WGS technologies, a number of sequencing projects have been carried out focusing on different microalgae species (Fu et al. 2019). As a part of genomic survey of *Haematococcus* genus, Luo et al. (2019) performed the first whole genome sequencing of *H. pluvialis* SAG-192.80. They reported one of the biggest genomes with the size of 669 Mb and 18.545 genes among the sequenced green algae. Also, they identified six BKT copies, which are thought to contribute to high astaxanthin accumulation. These findings have remarkable potential for the development of novel strains with enriched astaxanthin content. In a more recent study, researchers have focused on the chloroplast genome of *H. pluvialis* to explore its possible relation with astaxanthin production. The sequenced chloroplast had an extraordinary large genome size of 1,351,631 Kbp compared to the other green algae strains. Such a high genome size has been associated with the powerful antioxidant capacity of astaxanthin, which is generated under photo oxidative stress. In addition to special attention to *H. pluvialis*, the chloroplast genome of *H. lacustris* strain, a member of *Haematococcus* genus, was also evaluated for a comparative analysis, and their chloroplast genomes were found highly similar. Despite the reported data and detailed comparisons with other closely related species, they revealed that more chloroplast genomes and molecular research are required to clarify the connection between chloroplast and astaxanthin (Ren et al. 2021b).

Genome mapping is another technology used for scanning the relative organization of genes on a DNA. A "genome map" can be considered a visual map, which presents the locations and relative distances between genes along chromosomes. Up to date, there have been two mapping approaches, involving genetic mapping and physical maps. In order to determine locations and distances, genetic mapping uses the frequency of recombination, defined as the ratio of meiosis in which a chromosomal crossover takes place between two genes. In comparison, physical mapping is based on the measurement of actual physical distance between genetic markers by performing molecular biology techniques (Sahu et al. 2020; Seyum et al. 2021). With the increasing availability of genome-based technologies, a number of methods, including optical, consensus, linkage, association, and quantitative trait locus (QTL) mapping are presented for obtaining crucial information about genome structure. Genome mapping approaches have been extensively applied in plants, animals, and even macroalgae species. However, much fewer genome maps have ever been reported in microalgae, so far only in *Chromochloris zofingiensis* (Roth et al. 2017), *Chlorella vulgaris* (Cecchin et al. 2019), *C. reinhardtii* (Schierenbeck et al. 2015), *Chlorococcum* sp.(Maeda et al. 2019), and *Trebouxia* sp. (Martínez-Alberola et al. 2020) to our knowledge. As mentioned above, Ren and co-workers studied the genome of *H. pluvialis*, and they constructed the map of circular chloroplast genome after sequencing analysis (Ren et al. 2021b). Unfortunately, there is not much research concerning the genome map of *H. pluvialis,* so more study is needed to clarify the genetic structure of this strain.

8.4 Recent Advances in the Improvement of *H. pluvialis* Via Genetic Engineering Applications

Molecular genetic studies of *H. pluvialis* strain are mainly focused on the enhancement of astaxanthin accumulation, because of the limitations during astaxanthin production and competitiveness with synthetic alternatives. There are multiple genetic engineering approaches, which aim at the overexpression or knocking out of specific genes that are responsible for the regulation of metabolic pathways. One of the applied methods for *H. pluvialis* is random mutagenesis, which is the simplest way to enhance particular characteristics. It is based on the exposure of cells to different mutagenic agents such as chemical compounds, UV radiation, or low-temperature plasma (Le-Feuvre et al. 2020). Chen et al. (2020) investigated the effect of high light irradiation on the astaxanthin productivity of mutant *H. pluvialis* strain, which was obtained by the utilization of low-temperature plasma. They reported that the mutant cells showed higher photosynthetic activity, and increasing content of some specific pigments (lutein, β-carotene, and astaxanthin) because of random modifications in metabolic pathways. Another study focused on the improvement of astaxanthin accumulation in *H. pluvialis* by using UV radiation. After exposure to radiation, mutant cells were screened with phototaxis technology by developing a microfluidic device in which a specific light source was used for irradiation. The phototaxis-based screening showed that the astaxanthin concentration of mutant strain was 1.26-fold higher than wild type, and thus this method offered a novel screening strategy to obtain new strains with enhanced astaxanthin accumulation (Lee et al. 2022). Despite the reported achievements, random mutagenesis has critical disadvantages, including low efficiency and risk of undesirable mutations. Currently, targeted mutagenesis is becoming popular to genetic engineering applications, thus more technological approaches are presented for strain improvement (Ren et al. 2021c).

Genetic transformation of cells by using different recombinant DNA technologies has gained attention to improve specific properties, especially carotenoid productivity. This technique allows the insertion of foreign genes into the algal genome in a strategic way. Since the carotenogenic pathway of *H. pluvialis* occurs in the chloroplast and nucleus of cells, both nuclear and chloroplast transformation methods have been developed (Le-Feuvre et al. 2020; Ng et al. 2017). Nuclear transformation refers to random integration of modified DNA into the nuclear genome, while chloroplast transformation allows control of exogenous gene insertion into the plastid genome. Moreover, chloroplast transformation provides high-level expression of transgene resulting in the over-accumulation of interested metabolites. However, the lack of post-translational glycosylation in chloroplast hinders the use of this method.

Some of the common transformation methods are electroporation, biolistic particle bombardment, agitation with glass bead, and *Agrobacterium*-mediated transformation. These methods have their own prospects and challenges in terms of mechanism, efficiency, stability, and cost-effectiveness (Fajardo et al. 2020;

Sreenikethanam et al. 2022). It is worth noting that the choice of the appropriate transformation method affects the overall efficiency of genetic engineering strategy. There have been several reports focusing on the improvement of *H. pluvialis* strain by using different transformation methods, which are summarized in Table 8.1. Galarza et al. (2018) aimed at overproduction of astaxanthin through chloroplast transformation by using biolistic bombardment method. In order to modify *H. pluvialis* cells, the endogenous PDS coding sequence was overexpressed under the control of psbA promoter/UTR. After transformation and selection, mutant strains were exposed to high light intensity under nutrient-depleted conditions, and their astaxanthin content increased up to 67% compared to the wild-strain. This is one of the pioneer studies that achieved the expression of nuclear PDS gene in an algal chloroplast.

In another study, a transgenic *H. pluvialis* strain was developed by nuclear transformation of the hexose uptake protein gene from the *Parachlorella kessleri*. Following the selection of transgene integrated cells, experiments showed that mutant strains had the ability of grown on glucose-supplemented nutrient media under dark conditions. Besides that, a significant level of astaxanthin productivity was achieved in developed cells under heterotrophic conditions within a relatively short time (Waissman-Levy et al. 2019). As mentioned in the section of carotenoid synthesis pathway (Fig. 8.1), BKT is one of the major enzymes responsible for the synthesis of astaxanthin; thus Kathiresan et al. (2015) have focused on the cloning and overexpression of this gene. They isolated the genomic BKT from *H. pluvialis* strain, cloned it in pRT100 vector and then transformed the cloned BKT to the same homologous host through *Agrobacterium* mediation. The transcription efficiencies of carotenogenesis genes including BKT, PDS, PSY, and LCYb were found higher in mutant cells, resulting in over-accumulation of total carotenoids and astaxanthin. Although astaxanthin productivity is generally the main goal of the genetic engineering applications, some studies have been carried out to improve other particular characteristics. For example, antimicrobial peptide was expressed in the chloroplast of *H. pluvialis* via biolistic technique in order to provide resistance to pathogens (Wang et al. 2020). Moreover, formation of thick cell wall was down-regulated by the inhibiting the expression of α-1,6-mannosyltransferase (HpOCH1) gene through *Agrobacterium*-mediated transformation (Xu et al. 2022).

Despite their potential for genetic engineering applications, microalgae cells represent some challenges during transformation because of the presence of thick cell wall. In order to overcome this drawback, recent efforts have been made in the formation of protoplasts and their fusion for genome transformation (Ortiz-Matamoros et al. 2018). Polyethylene glycol (PEG) is widely used for the insertion of DNA or plasmids into the protoplast cells. A successful application was reported by Guo et al. (2022), wherein they obtained protoplast from *H. pluvialis* by using cellulase and macerozyme enzymes. Then, PEG-mediated transformation was performed by incubating the protoplast with the pH124-EGFP-Ble vector. After incubation, transformants were selected by the resistance to zeomycin, and the expression level of exogenous genes was successfully achieved in these cells. Besides transformation of foreign genes, protoplasts of different organisms can be

Table 8.1 Application of genetic transformation approaches in *H. pluvialis*

Transformation type	Transformation method	Promoter	Selection strategy	Results	References
Chloroplast	Biolistic bombardment	Nuclear PDS gene	Growth on 200 μg/mL spectinomycin	67% higher astaxanthin accumulation in mutant strain	Galarza et al. (2018)
Nuclear	Agrobacterium mediation	CaMV 35S	Growth on 10 mg/L hygromycin	2–3 fold higher astaxanthin and total carotenoid production	Kathiresan et al. (2015)
Nuclear	Biolistic particle bombardment	Endogenous PDS	Resistance to 2 μM norflurazon	Ability of grown on glucose-supplemented nutrient media in very low light or darkness, astaxanthin productivity under heterotrophic conditions within a short time	Waissman-Levy et al. (2019)
Chloroplast	Biolistic bombardment	psbA, rbcL	Growth on 15 μg/mL phosphinothricin	Expression of antimicrobial peptide piscidin-4 in *H. pluvialis*	Wang et al. (2020)
–	Agrobacterium-mediated	CaMV 35S	Growth on spectinomycin 100 mg/L	Downregulation of cell wall formation-related enzyme, HpOCH1, and simplification of astaxanthin extraction	Xu et al. (2022)
Protoplast	PEG-mediated	RBCS2	Growth on zeomycin 5 μg/mL	Successful integration of foreign gene, and thus resistance to zeomycin	Guo et al. (2022)
Protoplast fusion between *O. danica* and *H. pluvialis*	PEG-mediated	–	Growth of green colonies in BG11 medium	Hybrid fusants with the characteristics of both cells, involving photoautotrophic growth on BG11 and a broad fatty acid profile	Abomohra et al. (2016)
Protoplast	Magnetic nanoparticle integration by using electroporation	–	Magnetism and Prussian blue staining	Maximum transformation efficiency with magnetic particles was 30%	Savvidou et al. (2020)

fused to obtain new strains that have properties of both cells. Researchers have reported an interphylum protoplast fusion by using PEG between a fatty acid-secreting microalga, *Ochromonas danica*, and astaxanthin producer *H. pluvialis*. The obtained hybrid fusants showed characteristics of both cells, involving photo-autotrophic growth on BG11 and a broad fatty acid profile from two microalgae (Abomohra et al. 2016). As an alternative to PEG mediation, Savvidou et al. (2020) reported the use of iron oxide magnetic nanoparticles to transform *H. pluvialis* via electroporation. The integration of magnetic nanoparticles was confirmed by magnetism, microscopy observations, chemical distribution analysis, and Prussian blue test. Also, transformation efficiency was found to be a maximum value of 30% after regeneration of protoplasts. The results obtained from protoplast studies offer a simple and efficient technique for further research at the intersection of protoplast and genetic transformation of microalgae cells.

To date, the rapid and technological developments in genetic engineering applications provide advanced gene manipulation strategies for most microalgae and especially *Haematococcus* strain. However, there are a number of emerging novel technologies that should be implemented into the transformation process of these cells. These technologies are mostly related to in site-specific gene editing techniques, which aim at the transformation of microorganisms via a more controlled way. Some of the reported genome editing methods are Clustered Regularly Inter-Spaced Palindromic Repeats (CRISPR), Zinc Finger Nuclease (ZFN), Transcription Activator-Like Effect or Nuclease (TALEN), and RNA interference (RNAi) (Fayyaz et al. 2020; Zhang et al. 2019). In comparison with traditional methods, these novel technologies offer attractive superiorities, such as high accuracy, simplicity, low cost and, reproducibility. Despite reported technologies are excellent tools for *H. pluvialis* improvement; further study is required to integrate the above-mentioned strategies into the genetic engineering applications.

8.5 Conclusion

This chapter presents an insight into the latest genomic and technological advancements in the field of genetic engineering applications of *H. pluvialis* microalga. Genome sequencing and mapping techniques are highly crucial and helpful to understand the molecular basis of metabolic pathways in living systems. The availability of whole genome sequences from *H. pluvialis* provides important information about the metabolism of algal metabolites, most importantly the carotenoid pathway. The rapid developments in genome sequencing technologies coupled with multi-omics approaches have accelerated genetic transformation studies of *H. pluvialis* to obtain mutant strains with specific characteristics. These studies have mainly focused on the improvement of astaxanthin accumulation by applying classical or genetic transformation techniques. Considering the latest progress and challenges in genetic engineering applications, this review will serve as a powerful guide in the field of *H. pluvialis* research.

Up to date, reported approaches have shown promising results for successfully improving *H. pluvialis* microalgae. However, there is still a long way before the potential offered by these microalgae as source of high-value metabolites. Recent progress in multi-omics approaches (genomics, transcriptomics, metabolomics, and proteomics) and in-site specific gene editing technologies (CRISPR/Cas, TALEN, ZFN, and RNAi) should be coupled with the traditional methods to obtain more detailed information and to increase transformation efficiency. These novel technologies are surely promising strategies towards the genetic engineering applications in microalgal biotechnology.

References

Abomohra AE-F, El-Sheekh M, Hanelt D (2016) Protoplast fusion and genetic recombination between *Ochromonas danica* (Chrysophyta) and *Haematococcus pluvialis* (Chlorophyta). Phycologia 55:65–71. https://doi.org/10.2216/15-88.1

Cecchin M, Marcolungo L, Rossato M, Girolomoni L, Cosentino E, Cuine S, Li-Beisson Y, Delledonne M, Ballottari M (2019) Chlorella vulgaris genome assembly and annotation reveals the molecular basis for metabolic acclimation to high light conditions. Plant J 100:1289–1305. https://doi.org/10.1111/TPJ.14508

Chen Z, Chen J, Liu J, Li L, Qin S, Huang Q (2020) Transcriptomic and metabolic analysis of an astaxanthin-hyperproducing *Haematococcus pluvialis* mutant obtained by low-temperature plasma (LTP) mutagenesis under high light irradiation. Algal Res 45:101746. https://doi.org/10.1016/J.ALGAL.2019.101746

Ding W, Peng J, Zhao Y, Zhao P, Xu JW, Li T, Yu X (2019) A strategy for boosting astaxanthin accumulation in green microalga *Haematococcus pluvialis* by using combined diethyl aminoethyl hexanoate and high light. J Appl Phycol 31:171–181. https://doi.org/10.1007/S10811-018-1561-8/FIGURES/4

Fajardo C, De Donato M, Carrasco R, Martínez-Rodríguez G, Mancera JM, Fernández-Acero FJ (2020) Advances and challenges in genetic engineering of microalgae. Rev Aquac 12:365–381. https://doi.org/10.1111/RAQ.12322

Fayyaz M, Chew KW, Show PL, Ling TC, Ng IS, Chang JS (2020) Genetic engineering of microalgae for enhanced biorefinery capabilities. Biotechnol Adv 43:107554. https://doi.org/10.1016/J.BIOTECHADV.2020.107554

Fu W, Nelson DR, Mystikou A, Daakour S, Salehi-Ashtiani K (2019) Advances in microalgal research and engineering development. Curr Opin Biotechnol 59:157–164. https://doi.org/10.1016/J.COPBIO.2019.05.013

Fuentes-Pardo AP, Ruzzante DE (2017) Whole-genome sequencing approaches for conservation biology: advantages, limitations and practical recommendations. Mol Ecol 26:5369–5406. https://doi.org/10.1111/MEC.14264

Galarza JI, Gimpel JA, Rojas V, Arredondo-Vega BO, Henríquez V (2018) Over-accumulation of astaxanthin in *Haematococcus pluvialis* through chloroplast genetic engineering. Algal Res 31: 291–297. https://doi.org/10.1016/J.ALGAL.2018.02.024

Gong F, Zhang C, Zhang L, Liu J (2019) Changes of carotenoids contents and analysis of astaxanthin geometrical isomerization in *Haematococcus pluvialis* under outdoor high light conditions. Aquac Res 51:770–778. https://doi.org/10.1111/are.14427

Grama SB, Liu Z, Li J (2022) Emerging trends in genetic engineering of microalgae for commercial applications. Mar Drugs 20:285. https://doi.org/10.3390/MD20050285

Grujić VJ, Todorović B, Ambrožič-Dolinšek J, Kranvogl R, Ciringer T (2022) Diversity and content of carotenoids and other pigments in the transition from the green to the red stage of

Haematococcus pluvialis microalgae identified by HPLC-DAD and LC-QTOF-MS. Plants 11: 1026. https://doi.org/10.3390/PLANTS11081026

Guler BA, Saglam-Metiner P, Deniz I, Demirel Z, Yesil-Celiktas O, Imamoglu E (2022) Aligned with sustainable development goals: microwave extraction of astaxanthin from wet algae and selective cytotoxic effect of the extract on lung cancer cells. Prep Biochem Biotechnol 1–7. https://doi.org/10.1080/10826068.2022.2116455

Guo C, Anwar M, Mei R, Li X, Zhao D, Jiang Y, Zhuang J, Liu C, Wang C, Hu Z (2022) Establishment and optimization of PEG-mediated protoplast transformation in the microalga *Haematococcus pluvialis*. J Appl Phycol 34:1595–1605. https://doi.org/10.1007/s10811-022-02718-x

Hoys C, Romero-Losada AB, del Río E, Guerrero MG, Romero-Campero FJ, García-González M (2021) Unveiling the underlying molecular basis of astaxanthin accumulation in *Haematococcus* through integrative metabolomic-transcriptomic analysis. Bioresour Technol 332:125150. https://doi.org/10.1016/J.BIORTECH.2021.125150

Jannel S, Caro Y, Bermudes M, Petit T (2020) Novel insights into the biotechnological production of *Haematococcus pluvialis*-derived astaxanthin: advances and key challenges to allow its industrial use as novel food ingredient. J Mar Sci Eng 8:789. https://doi.org/10.3390/JMSE8100789

Kathiresan S, Chandrashekar A, Ravishankar GA, Sarada R (2015) Regulation of astaxanthin and its intermediates through cloning and genetic transformation of β-carotene ketolase in *Haematococcus pluvialis*. J Biotechnol 196–197:33–41. https://doi.org/10.1016/J.JBIOTEC.2015.01.006

Khoo KS, Lee SY, Ooi CW, Fu X, Miao X, Ling TC, Show PL (2019) Recent advances in biorefinery of astaxanthin from *Haematococcus pluvialis*. Bioresour Technol 288:121606. https://doi.org/10.1016/J.BIORTECH.2019.121606

Kumar G, Shekh A, Jakhu S, Sharma Y, Kapoor R, Sharma TR (2020) Bioengineering of microalgae: recent advances, perspectives, and regulatory challenges for industrial application. Front Bioeng Biotechnol 8:914. https://doi.org/10.3389/FBIOE.2020.00914/BIBTEX

Lee KH, Chun Y, Lee JH, Park C, Yoo HY, Kwak HS (2022) Improved productivity of astaxanthin from photosensitive *Haematococcus pluvialis* using Phototaxis technology. Mar Drugs 20:220. https://doi.org/10.3390/MD20040220

Le-Feuvre R, Moraga-Suazo P, Gonzalez J, Martin SS, Henríquez V, Donoso A, Agurto-Muñoz C (2020) Biotechnology applied to *Haematococcus pluvialis* Fotow: challenges and prospects for the enhancement of astaxanthin accumulation. J Appl Phycol 32:3831–3852. https://doi.org/10.1007/S10811-020-02231-Z/TABLES/4

Liang MH, Wang L, Wang Q, Zhu J, Jiang JG (2018) High-value bioproducts from microalgae: strategies and progress. Crit Rev Food Sci Nutr 59:2423–2441. https://doi.org/10.1080/10408398.2018.1455030

Luo Q, Bian C, Tao M, Huang Y, Zheng Y, Lv Y, Li J, Wang C, You X, Jia B, Xu J, Li J, Li Z, Shi Q, Hu Z (2019) Genome and transcriptome sequencing of the astaxanthin-producing green microalga, Haematococcus pluvialis. Genome Biol Evol 11:166–173. https://doi.org/10.1093/GBE/EVY263

Ma R, Thomas-Hall SR, Chua ET, Eltanahy E, Netzel ME, Netzel G, Lu Y, Schenk PM (2018) Blue light enhances astaxanthin biosynthesis metabolism and extraction efficiency in *Haematococcus pluvialis* by inducing haematocyst germination. Algal Res 35:215–222. https://doi.org/10.1016/J.ALGAL.2018.08.023

Maeda Y, Nojima D, Sakurai M, Nomaguchi T, Ichikawa M, Ishizuka Y, Tanaka T (2019) Genome analysis and genetic transformation of a water surface-floating microalga *Chlorococcum* sp. FFG039. Sci Rep 9:11200. https://doi.org/10.1038/s41598-019-47612-8

Martínez-Alberola F, Barreno E, Casano LM, Gasulla F, Molins A, Moya P, González-Hourcade M, del Campo EM (2020) The chloroplast genome of the lichen-symbiont microalga *Trebouxia* sp. Tr9 (Trebouxiophyceae, Chlorophyta) shows short inverted repeats with a single gene and loss of the rps4 gene, which is encoded by the nucleus. J Phycol 56:170–184. https://doi.org/10.1111/JPY.12928

Mularczyk M, Michalak I, Marycz K (2020) Astaxanthin and other nutrients from *Haematococcus pluvialis*-multifunctional applications. Mar Drugs 18:459. https://doi.org/10.3390/MD18090459

Ng IS, Tan SI, Kao PH, Chang YK, Chang JS (2017) Recent developments on genetic engineering of microalgae for biofuels and bio-based chemicals. Biotechnol J 12:1600644. https://doi.org/10.1002/BIOT.201600644

Nishshanka GKSH, Liyanaarachchi VC, Nimarshana PHV, Ariyadasa TU, Chang JS (2022) Haematococcus pluvialis: a potential feedstock for multiple-product biorefining. J Clean Prod 344:131103. https://doi.org/10.1016/J.JCLEPRO.2022.131103

Ortiz-Matamoros MF, Villanueva MA, Islas-Flores T (2018) Genetic transformation of cell-walled plant and algae cells: delivering DNA through the cell wall. Brief Funct Genomics 17:26–33. https://doi.org/10.1093/bfgp/elx014

Oslan SNH, Shoparwe NF, Yusoff AH, Rahim AA, Chang CS, Tan JS, Oslan SN, Arumugam K, Bin Ariff A, Sulaiman AZ, Mohamed MS (2021) A review on *Haematococcus pluvialis* bioprocess optimization of green and red stage culture conditions for the production of natural astaxanthin. Biomolecules 11:256. https://doi.org/10.3390/BIOM11020256

Pick U, Zarka A, Boussiba S, Davidi L (2018) A hypothesis about the origin of carotenoid lipid droplets in the green algae *Dunaliella* and *Haematococcus*. Planta 249:31–47. https://doi.org/10.1007/S00425-018-3050-3

Ren Y, Sun H, Deng J, Huang J, Chen F (2021a) Carotenoid production from microalgae: biosynthesis, salinity responses and novel biotechnologies. Mar Drugs 19:713. https://doi.org/10.3390/md19120713

Ren Q, Wang Y-C, Lin Y, Zhen Z, Cui Y, Qin S (2021b) The extremely large chloroplast genome of the green alga *Haematococcus pluvialis*: genome structure, and comparative analysis. Algal Res 56:102308. https://doi.org/10.1016/J.ALGAL.2021.102308

Ren Y, Deng J, Huang J, Wu Z, Yi L, Bi Y, Chen F (2021c) Using green alga *Haematococcus pluvialis* for astaxanthin and lipid co-production: advances and outlook. Bioresour Technol 340:125736. https://doi.org/10.1016/J.BIORTECH.2021.125736

Roth MS, Cokus SJ, Gallaher SD, Walter A, Lopez D, Erickson E, Endelman B, Westcott D, Larabell CA, Merchant SS, Pellegrini M, Niyogi KK (2017) Chromosome-level genome assembly and transcriptome of the green alga *Chromochloris zofingiensis* illuminates astaxanthin production. Proc Natl Acad Sci U S A 114:E4296–E4305. https://doi.org/10.1073/PNAS.1619928114/SUPPL_FILE/PNAS.1619928114.SM05.MPG

Sahu PK, Sao R, Mondal S, Vishwakarma G, Gupta SK, Kumar V, Singh S, Sharma D, Das BK (2020) Next generation sequencing based forward genetic approaches for identification and mapping of causal mutations in crop plants: a comprehensive review. Plants 9:1355. https://doi.org/10.3390/PLANTS9101355

Saini KC, Yadav DS, Mehariya S, Rathore P, Kumar B, Marino T, Leone GP, Verma P, Musmarra D, Molino A (2021) Overview of extraction of astaxanthin from *Haematococcus pluvialis* using CO_2 supercritical fluid extraction technology *vis-a-vis* quality demands. In: Ravishankar GA, Ambati RR (eds) Global perspectives on astaxanthin: from industrial production to food, health, and pharmaceutical applications. Academic Press, London, pp 341–354. https://doi.org/10.1016/B978-0-12-823304-7.00032-5

Savvidou MG, Ferraro A, Hristoforou E, Mamma D, Kekos D, Kolisis FN (2020) Incorporation of magnetic nanoparticles into protoplasts of microalgae *Haematococcus pluvialis*: a tool for biotechnological applications. Molecules 25:5068. https://doi.org/10.3390/MOLECULES25215068

Schierenbeck L, Ries D, Rogge K, Grewe S, Weisshaar B, Kruse O (2015) Fast forward genetics to identify mutations causing a high light tolerant phenotype in *Chlamydomonas reinhardtii* by whole-genome-sequencing. BMC Genomics 16:1–15. https://doi.org/10.1186/S12864-015-1232-Y/FIGURES/7

Seyum EG, Bille NH, Abtew WG, Ntsomboh-Ntsefong G, Bell JM (2021) Genome mapping to enhance efficient marker-assisted selection and breeding of the oil palm (Elaeis guineensis Jacq.). Adv Biosci Biotechnol 12:407–425. https://doi.org/10.4236/abb.2021.1212026

Sharon-Gojman R, Maimon E, Leu S, Zarka A, Boussiba S (2015) Advanced methods for genetic engineering of *Haematococcus pluvialis* (Chlorophyceae, Volvocales). Algal Res 10:8–15. https://doi.org/10.1016/J.ALGAL.2015.03.022

Sreenikethanam A, Raj S, Rajesh Banu J, Gugulothu P, Bajhaiya AK (2022) Genetic engineering of microalgae for secondary metabolite production: recent developments, challenges, and future prospects. Front Bioeng Biotechnol 10:420. https://doi.org/10.3389/FBIOE.2022.836056/BIBTEX

Srivastava A, Kalwani M, Chakdar H, Pabbi S, Shukla P (2022) Biosynthesis and biotechnological interventions for commercial production of microalgal pigments: a review. Bioresour Technol 352:127071. https://doi.org/10.1016/J.BIORTECH.2022.127071

Teng SY, Yew GY, Sukačová K, Show PL, Máša V, Chang JS (2020) Microalgae with artificial intelligence: a digitalized perspective on genetics, systems and products. Biotechnol Adv 44: 107631. https://doi.org/10.1016/J.BIOTECHADV.2020.107631

Velmurugan A, Kodiveri Muthukaliannan G (2022) Genetic manipulation for carotenoid production in microalgae an overview. Curr Res Biotechnol 4:221–228. https://doi.org/10.1016/J.CRBIOT.2022.03.005

Waissman-Levy N, Leu S, Khozin-Goldberg I, Boussiba S (2019) Manipulation of trophic capacities in *Haematococcus pluvialis* enables low-light mediated growth on glucose and astaxanthin formation in the dark. Algal Res 40:101497. https://doi.org/10.1016/j.algal.2019.101497

Wan M, Zhao H, Guo J, Yan L, Zhang D, Bai W, Li Y (2021) Comparison of C-phycocyanin from extremophilic *Galdieria sulphuraria* and *Spirulina platensis* on stability and antioxidant capacity. Algal Res 58:102391. https://doi.org/10.1016/j.algal.2021.102391

Wang K, Cui Y, Wang Y, Gao Z, Liu T, Meng C, Qin S (2020) Chloroplast genetic engineering of a unicellular green alga *Haematococcus pluvialis* with expression of an antimicrobial peptide. Mar Biotechnol 22:572–580. https://doi.org/10.1007/s10126-020-09978-z

Xu R, Zhang L, Yu W, Liu J (2022) A strategy for interfering with the formation of thick cell walls in *Haematococcus pluvialis* by down-regulating the mannan synthesis pathway. Bioresour Technol 362:127783. https://doi.org/10.1016/j.biortech.2022.127783

Zhang YT, Jiang JY, Shi TQ, Sun XM, Zhao QY, Huang H, Ren LJ (2019) Application of the CRISPR/Cas system for genome editing in microalgae. Appl Microbiol Biotechnol 103:3239–3248. https://doi.org/10.1007/S00253-019-09726-X/FIGURES/1

Zhang C, Chen X, Too HP (2020) Microbial astaxanthin biosynthesis: recent achievements, challenges, and commercialization outlook. Appl Microbiol Biotechnol 104:5725–5737. https://doi.org/10.1007/S00253-020-10648-2/FIGURES/4

Zhao T, Yan X, Sun L, Yang T, Hu X, He Z, Liu F, Liu X (2019) Research progress on extraction, biological activities and delivery systems of natural astaxanthin. Trends Food Sci Technol 91: 354–361. https://doi.org/10.1016/J.TIFS.2019.07.014

Part III
Food and Environmental Applications

Chapter 9
Food and Food Packaging Technology

Chalani Akmeemana, Sankha Karunarathna, and Indira Wickramasinghe

9.1 Introduction

The generation of astaxanthin from *Haematococcus pluvialis* is one of the most recently developed microalgae technologies. A high-value carotenoid pigment called astaxanthin finds use in the nutraceutical, cosmetic, and food and feed sectors. Due to public awareness of the health benefits of natural ingredients, consumers are currently favoring natural food colors over synthetic ones. As a result, there is a genuine need for natural colorants in society. Therefore, the demand for astaxanthin increases with time. The creation of food products containing green algal components, especially astaxanthin, has received significant interest, according to a recent study on functional food ingredients. The richest source of natural astaxanthin is *Haematococcus pluvialis*, grown on an industrial scale. Astaxanthin is often used as a nutritional supplement in the food and feed sectors. The European Food and Safety Panel on Nutrition, Novel Foods, and Food Allergens concludes that 8 mg of astaxanthin from food supplements is safe for adults to consume (Oslan et al. 2021).

Astaxanthin, a carotenoid pigment, finds significant use in the dietary supplement, cosmetics, food, and feed sectors. Natural astaxanthin is recognized as a food color by the European Commission. *Haematococcus pluvialis* develops significant levels of astaxanthin under stress circumstances such as high salinity, nitrogen shortage, and high temperature. The primary source of astaxanthin for human consumption is *H. pluvialis*. Salmon, trout, and shrimp are fed on it as a source of color. Astaxanthin is isolated from *H. pluvialis* or acquired from seafood as a dietary supplement in humans and animals (Ambati et al. 2014). Oils, dried aplanospores, and dietary supplements are the major commercially accessible forms of astaxanthin.

C. Akmeemana · S. Karunarathna · I. Wickramasinghe (✉)
Department of Food Science and Technology, Faculty of Applied Sciences, University of Sri Jayewardenepura, Nugegoda, Sri Lanka
e-mail: indiraw@sjp.ac.lk

R. Raja et al. (eds.), *Haematococcus*, https://doi.org/10.1007/978-981-99-2901-6_9

Due to the flavor and aroma of algae, oils containing astaxanthin are not esthetically pleasing (Mularczyk et al. 2020). Thermal stability is an important parameter when considering the application of astaxanthin in food products.

9.2 Applications in the Food Industry

Microalgae contain well-balanced chemical composition and can be utilized to enhance the nutritional value of functional foods. These microalgae can effectively convert inorganic substances to organic matter. Microalga biomass can be added to foods to incorporate biologically active substances such as PUFA-3 and antioxidants (Batista et al. 2013). *Haematococcus pluvialis*, a unicellular microalga, is an excellent natural source of astaxanthin, which is classified as a Chlorophyceae in the microalga category. When considering sensorial characteristics, the odor of any food product is essential for consumer acceptance and attractiveness. However, due to the off-flavor of *H. pluvialis* powder, its direct application in food products has been limited (Liu et al. 2022). The authors used roasting treatments to enhance the flavor and investigated the effects of temperature increment on nutritional components and antioxidant activity. The study measures the effect of different roasting temperatures such as 100, 120, 140, and 160 °C for 30 min. According to the study's findings, most volatiles were increased, except esters, while exhibiting favorable odors such as violet-like, fruity, and fresh odors. The recommended roasting temperature for *H. pluvialis* to enhance flavor without affecting the nutrients is between 100 and 120 °C (Liu et al. 2022).

Astaxanthin is classified into two types. One foam is created naturally, while the other is created chemically. Petrochemical products and an industrial dye can be used to create chemically synthetic astaxanthin. Microalgae, plants, microorganisms, and aquatic animals can be used to extract natural astaxanthin. *Haematococcus pluvia*lis and *Chlamydomonas nivalis* are two microalgae that can be used to extract astaxanthin. Similarly, the petals of *Adonis annua* and *Adonis amurensis* can be used to extract astaxanthin. Moreover, Antarctic krill, Crustacyanine (lobster), *Pandalus borealis* (shrimp), Salmon salar, Red crab langostilla-like animals, and *Phaffiar hodozyma*, *Paracoccus carotinifaciens*-like microorganisms can be used to extract astaxanthin (Sun et al. 2022). Thermal stability is an important parameter when considering the application of astaxanthin in food products. (Casella et al. 2020) analysis of the stabilizing effect of the 40–80 °C temperature range for 16 h of storage period and room temperature (20 °C) for 7 days. Thus, astaxanthin is categorized as a thermal liable molecule and the use of high temperatures during cooking time may affect the structure of the molecule. The temperatures used for the extraction of astaxanthin also can affect its antioxidant capacities (Casella et al. 2020). *Haematococcus pluvialis* has few direct applications in the food industry as a result of the aforementioned factors. The direct applications of *Haematococcus* in the fields of food supplements, nutraceuticals, bakery products such as cookies, meat industry, and as a sweetener are discussed here.

9.2.1 Food Supplement/Nutraceuticals

The definition of a nutraceutical product is a food that contains natural components that are beneficial for disease prevention and the improvement of the organism's physiological functions. A chemical or biological material can be used as a component. A pharmaceutical product and a nutraceutical product are fundamentally different from one another since the former is used as medications to treat illnesses, whereas the latter is frequently ingested as dietary supplements to avoid illnesses (López and Morales 2018). In the USA, Japan, and numerous European nations, the use of *H. pluvialis* has been permitted as a color additive and nutritional supplement ingredient for human consumption (Pan-utai et al. 2021).

H. pluvialis has several health advantages and is regarded as one of the greatest food supplements for people who are at risk of heart attacks. It can also be used to improve cognitive function in Alzheimer's patients (El-Baz et al. 2018). The primary carotenoid form of *H. pluvialis*is astaxanthin, which has greater antioxidant capabilities than lutein, lycopene, alpha- and beta-carotene combined (Hwang et al. 2020). Astaxanthin has antioxidant properties that are 11 times stronger than beta-carotene, 2.75 times stronger than lutein, and surpass the antioxidant capacity of vitamins A and E. Casella et al. (2020) described that age-related macular degeneration can be slowed down, the immune system can function better, and it can shield the skin from damage brought on by toxins, chemicals, or radiation. As a result, it has found widespread application in a variety of fields, including foods, dietary supplements, pharmaceuticals, and cosmetics. Astaxanthin's unique molecular structure, which includes hydroxyl and keto moieties on each ionone ring, is the basis for these advantageous properties (Hwang et al. 2020).

Hwang et al. (2020) stated that the maximum level of Ast-mE obtained the highest level of antioxidant and anti-inflammatory activities accordingly increasing the expression of genes interrelated to antioxidant and anti-inflammatory activities. Thus these results give a basic awareness for using the natural extract of astaxanthin in a more productive, cost-effective way to apply in dietary supplements (Hwang et al. 2020). Astaxanthin supplements showed potential in lowering oxidative stress and several blood-based biochemicals, including low-density lipoprotein, total cholesterol, fasting glucose, triglycerides, glycated hemoglobin, and others. The bioactive compound is also related to lower blood pressure, higher blood levels of HDL and adiponectin, and increased blood levels of adiponectin, supporting the idea that astaxanthin holds promise as a tool for the prevention and treatment of chronic diseases and metabolic syndrome. Nevertheless, no supplement will have a "marvelous" effect and will not be able to replace consistent changes in lifestyle, such as a good diet. The antioxidant effects of astaxanthin have been confirmed and related to a vast range of metabolic and clinical outcomes (Altoé et al. 2022).

El-Baz et al. (2022) studied the acute and chronic toxicity effect of astaxanthin enrich fraction of microalgae *H. pluvialis*. The experiment results retrieved that there is no toxicological effect encountered with supplementation series conducted with 500 up to 5000 mg/kg body weight. Similar research about the toxicity effect of supplements enriching astaxanthin fraction of microalgae *H. pluvialis* was done by (Stewart et al. 2008). These researchers also obtained the same observation as there was not any toxicological effect of supplements enriching astaxanthin fraction of microalgae *H. pluvialis* (El-Baz et al. 2022). For instance, despite having a minimal dietary intake, astaxanthin benefits mitochondrial redox status and functional integrity. For instance, natural astaxanthin from *H. pluvialis* has been demonstrated to be safe as a supplement with no negative consequences for human health as well as a feed additive in the aquaculture and animal feed industries, and no documented adverse effects have been detected over its 20 years of use as a dietary supplement. According to the European Commission Implementing Regulation (EU) 2017/2470, dietary supplements containing astaxanthin-rich oleoresin made from *H. pluvialis* can increase absorption by up to 40–80 mg/day. However, the recommended daily intake of total carotenoids is between 5 and 10 mg. This includes exposure to ß-carotene at a safe level of fewer than 15 mg daily (Oslan et al. 2021)

9.2.2 *Bakery Industry*

Fats and oils were used in large quantities during the bakery product manufacturing process. As a result, they oxidize and become rancid during storage, resulting in the generation of off-flavors and a reduction in sensory attributes. To reduce the rancidity of the lipids, antioxidants or oxidation inhibitors can be used. To control lipid rancidity, synthetic antioxidants such as tertiary butyl hydroxyquinone (TBHQ), butylated hydroxy anisole (BHA), and butylated hydroxytoluene (BHT) are commonly used in bakery products. Because of the health risks associated with artificial antioxidants, customers are moving their attention to natural antioxidants. Natural antioxidants such as β-carotene, ascorbic acid, and α-tocopherol can be used instead of artificial antioxidants (San et al. 2022). Furthermore, astaxanthin extracted from *H. pluvialis* has a higher antioxidant capacity than natural antioxidants like lycopene, alpha- and beta-carotene etc. (Hwang et al. 2020). As a result, it is possible to incorporate *H. pluvialis* microalgae into bakery products Hossain et al. (2017). Castillo et al. (2021) conducted similar research, which is described in the following sections.

9.2.2.1 *Haematococcus pluvialis* Enriched Cookies

Hossain et al. (2017) developed cookies enriched with *Haematococcus pluvialis* and analysis of the effect of cookies' physical and functional properties after enrichment. Wheat, barley, and oat flour were used to formulate astaxanthin-enriched cookies. The cookies demonstrated a considerable decrease in the rate of glucose released during in vitro digestion, while an increment of the food's total phenolic and antioxidant capacity was observed. The rise in astaxanthin content corresponds to an increase in TPC, and antioxidant capability (Hossain et al. 2017). *Haematococcus pluvialis* powder addition has a considerable impact on the cookies' physical characteristics, including height, diameter, weight loss, moisture content, and hardness. When astaxanthin powder increases, weight loss, diameter, and height decrease correspondingly. The cookies prepared with whole meal oat flour experienced the highest diameter changes, whereas the cookies made with whole meal wheat flour encountered the most significant height changes. At all levels of astaxanthin addition, the moisture content of wheat, oat, and barley biscuits increased significantly. Astaxanthin's addition also led to a decrease in the cookies' hardness. According to the study, barley cookies were harder than control cookies when astaxanthin was added, whereas wheat and oat cookies were softer (Hossain et al. 2017).

Three different flour-based cookies' lightness (L*), and yellowness (b*) were dramatically decreased once astaxanthin was added. The pigment of astaxanthin powder is the main cause of the color change in the cookies; as the degree of substitution increased, the lightness of the cookies reduced and the greenness enhanced. When compared to the control cookies, the addition of astaxanthin powder decreases the amount of reduced sugar released. This could be due to the high antioxidant activity found in astaxanthin powder, which can inhibit enzyme activity during digestion. The interaction of phenolic compounds with digestive enzymes may affect non-covalent starch-phenolic interactions, restricting starch degradation (Hossain et al. 2017).

9.2.2.2 *H. pluvialis* Incorporated Filloas

The authors utilized *H. pluvialis* antioxidant and colorant potential to develop filloas, a type of traditional Peninsula dessert. When compared to the commercial colorant, the analysis revealed an 8% increase in redness stability. Besides that, when *H. pluvialis* was incorporated, the texture, carotenoids, phenolic compounds, and fatty acids improved (Castillo et al. 2021).

9.2.3 *Meat Industry*

Meat is a key component of the human diet because it is high in protein, and contains essential macronutrients like amino acids, as well as micronutrients like vitamins and minerals. Saturated fat acid, monounsaturated fatty acid, and polyunsaturated fatty acid are the three types of fat found in meat. The main cause of meat quality degradation is oxidative rancidity. It occurs as a result of the breakdown of unsaturated fatty acids and the formation of oxidized cholesterol products. The shelf life of meat products can be extended with the help of antioxidants, which can prevent the oxidation of meat fat. Neither the meat's sensory qualities nor its nutritional value are impacted by antioxidants. Natural antioxidants are a safer foam that can be used to extend the shelf life of meat products without negatively impacting the health of consumers. Astaxanthin is a type of carotenoid derived from the microalga

H. pluvialis that can be used as a natural antioxidant to prevent meat fat oxidation (AL-Ghanimi and Saleh Al-Rubeii 2020). The effect of raw ground pork meat enriched with *H. pluvialis* extract and used as meat analogues are described here.

9.2.3.1 Meat Analogues

Vegetarian or vegan ingredients known as "meat analogues" can be used as a replacement for animal meat. Unlike conventional plant proteins, proteins from microalgae are easy to produce in large quantities, occupy less space, are highly flexible, and are nutrient-rich. A meat analogue can be created using the *H. pluvialis* residue that remains after astaxanthin extraction. As a result, less waste is produced during the production of astaxanthin, and the limitation of plant proteins used as a substitute for meat is also resolved. The authors developed meat analogues by a combination of *H. pluvialis* residue (HPR) and pea protein. In ratios of 0 g/100 g, 10 g/100 g, 20 g/100 g, 30 g/100 g, and 40 g/100 g, HPR is combined with pea protein. Extrusion was carried out with a high moisture content of 50 g/100 g. The final developed sample was reddish, similar to dried red meat (Xia et al. 2022).

One of the most important characteristics of meat substitutes is texture. The textural differences occupied when incorporating HPR were measured using texture characteristics such as hardness, springiness, chewiness, crosswise shear force, lengthwise shear force, and fibrous degree of samples. The hardness of meat analogues increased at first and then decreased as the HPR ratio increased. The addition of HPR to meat analogues reduced their springiness, chewiness, crosswise shear force, and lengthwise shear force. While the addition of HPR of meat analogues improved the fibrous degree (Xia et al. 2022).

Color has a significant impact on consumer acceptance of food. The astaxanthin that remains in the HPR after the extraction process has a noticeable effect on meat analogues. As a result, HPR could be utilized as a natural colorant, providing meat analogues with a distinctive red color. Because of the residual astaxanthin in HPR, the brightness (L^*) of meat analogues significantly decreased with the addition of HPR. All samples containing HPR showed a significant increase in redness (a^*) when compared to the control. The yellowness (b^*) of samples increased initially and then decreased as HPR content increased from 0 to 40 g/100 g (dry basis) (Xia et al. 2022). SEM images were used to examine the microstructure of the meat analogues. Structure formation within extruded samples can be identified using these images. The presence of noticeable layered and fibrous structures was observed in both transverse and longitudinal SEM images of meat analogues with the addition of HPR. The round-shaped *H. pluvialis* cell was visible at 1.5 k in the layered and fibrous structures of the meat analogues that had been treated with HPR. Additionally, as *H. pluvialis* has thick cell walls that are resistant to mechanical damage, the layered structures of meat analogues became looser with increasing HPR content. The addition of HPR reduced the apparent viscosity of the meat analogues. This could be because the presence of HPR interferes with the cross-linking of protein molecular chains during extrusion (Xia et al. 2022).

9.2.3.2 Raw Ground Pork Meat Enrich with *Haematococcus pluvialis* Extract

The most popular meat in the European Union is pork. A significant portion of sold pork (40%) is consumed as ground meat. This amount is expected to rise due to increased demand for convenience foods. Due to the health risks associated with artificial antioxidants, consumers now prefer natural antioxidants. Astaxanthin is one of the most powerful natural antioxidants. Astaxanthin-rich *H. pluvialis* extract has considerable DPPH scavenging activity. When used in higher doses (0.3 and 0.45 g per kg of meat), the *H. pluvialis* extract reduced the oxidation of lipids. The study discovered a significant color difference when astaxanthin was added to ground pork meat. When the color parameters are considered, the a* (redness) and b* (yellowness) color values increase while the L* (lightness) values decrease on the surface of the raw patties. The reason for this is astaxanthin's unique purple-red color. The C* color parameter increased during the 7-day refrigerated storage period, while the hue angle value decreased. Therefore, meat that contains astaxanthin has a more vivid color and was less close to gray throughout the refrigeration period. The amount of brown metmyoglobin in meat is an indicator of discoloration. During the experiment, the percentage of metmyoglobin remained constant. Nevertheless, there was not any antimicrobial or antioxidative activity of *H. pluvialis* extract against myoglobin oxidation. According to the findings, *H. pluvialis* extract could be utilized successfully as a natural antioxidant in the meat industry (Pogorzelska et al. 2018).

9.2.4 *Other Food Industries Sectors*

Foods enriched with astaxanthin have a wide range of applications. Huang et al. investigate to improve astaxanthin consistency and bioavailability by stacking it into nanostructured lipid carriers and afterwards combining them into alginate microgels. Yogurt, ice cream, dairy products, and other food products can be made with the modified astaxanthin foam (Huang et al. 2021). The astaxanthin found in shrimp shells was used to create Acid Curd Cheese (Tvarog) with antioxidant activity. The addition of astaxanthin changes the cheese color parameters so that they are more likely to resemble salmon color. The final product can withstand higher antioxidant activities and develop new functional qualities that will draw more customers to the products (Dmytrów et al. 2021). The astaxanthin from *Phaffia rhodozyma* was incorporated into yogurt in a recent study. In this study, microencapsulation with sodium carboxymethyl cellulose and microcrystalline cellulose was used to enhance the stability, solubility, and antioxidant activity of astaxanthin. In comparison to plain yogurt, the astaxanthin-infused variety displayed an attractive orange-red color, and higher DPPH activity (Feng et al. 2018). The impact of adding astaxanthin from *H. pluvialis* to the aforementioned food products and other areas of the food industry can be investigated by additional research. The applications of astaxanthin extracted from *Haematococcus* on ready-to-use syrup are also described in the following.

9.2.4.1 Ready-To-Use Syrup

Alternative methods should be investigated due to the high cost of conventional astaxanthin extraction. The authors investigate the possibility of extracting astaxanthin through the use of osmotic shock by exposing whole cells of *H. pluvialis* to a highly concentrated sucrose solution, or syrup, at high temperatures. The osmotic shock method was successfully used by the authors to extract astaxanthin, and a ready-to-use syrup with antioxidant properties was developed. This syrup can be used to make antioxidant-rich fruit marmalades, jams, and juices. A syrup's astaxanthin content can be controlled by adjusting the temperature or the quantity of *Haematococcus* culture added (López and Morales 2018).

9.3 Packaging Industry

Food spoilage is easily caused by microbiological contamination if the food product is not properly preserved. Various approaches have been developed to extend the storage time of food products. Incorporating active ingredients into food packaging materials is an effective technique for extending the shelf life of food products (Xu et al. 2020). As a result, active food packaging has captured the attention of food packaging researchers (Inthamat et al. 2022a). Among the active food packaging films, antioxidant enrich packages are gaining popularity because they can reduce food spoilage by slowing the release of incorporated antioxidants in the film matrix onto the food surface (Łupina et al. 2021). Carotenoid pigments provide protection against photooxidation. Regardless of the presence of 700 naturally occurring carotenoids, only a limited amount of carotenoids such as lycopene, canthaxanthin, and astaxanthin are produced on an industrial scale. Astaxanthin is gaining popularity due to its high antioxidant potential, which can be up to 100 times than that α-tocopherol. (Łupina et al. 2022) As a result, incorporating astaxanthin into packaging materials will have promising advantages over food products. Accordingly (Łupina et al. 2022) and (Inthamat et al. 2022b) created astaxanthin-enriched packaging materials, and the effects of incorporating astaxanthin into packaging films are discussed below. Furthermore, the research literature did not reveal any instances of direct application of *H. pluvialis* to packaging materials. The greater astaxanthin yield in *H. pluvialis* makes it a promising option for the application of active food packaging materials in the future.

Łupina et al. (2022) examine the physical and chemical characteristics, release kinetics, and antioxidant activity of the astaxanthin-incorporated films made from the mixture of gelatin and carboxymethyl cellulose as well as the octenyl succinic anhydride starch/gelatin. The addition of astaxanthin gives the films a distinct red color. The incorporation of astaxanthin reduces the puncture strength and oxygen permeability of CMC-based films significantly. Its water vapor permeability is unaffected by the addition of astaxanthin. As the astaxanthin content increased, solubility and swelling in the octenyl succinic anhydride starch/gelatin films decreased. Astaxanthin incorporation was found to reduce oxygen permeability, allowing the films to protect the food from oxidation. The incorporation of astaxanthin results in red-colored films with excellent oxygen barrier and free radical scavenging properties. Another study developed a genipin-crosslinked chitosan/astaxanthin film with improved moisture barrier and functional properties. Heat curing was used as the flm-producing method, with three different temperature ranges of 25 °C, 80 °C, and 105 °C. The color of the heat-cured films changed from light brown to reddish brown. As a result, the opacity of the films increased, resulting in a lower light transmittance value. By forming new chemical bonds between chitosan and genipin, heat curing at 105 °C improved the water barrier properties. As a result, the developed film can be used to protect packaged foods from moisture and light (Inthamat et al. 2022b).

9.4 Challenges and Opportunities

On the industrial scale astaxanthin commonly produced synthetic astaxanthin is followed by chemical methods to overcome the obstacles encountered in the natural extraction processes. Chemically extracted astaxanthin shows a different configuration which does not contain several biological properties as natural astaxanthin. However, synthetic astaxanthin has encountered health risks in humans and is thus not recommended as a nutraceutical for humans. Stereoisomers of synthetic astaxanthin affect the stability of the molecule and also toxic and carcinogenic phytochemicals used in this process. Due to the health risks associated with chemically synthesized astaxanthin, natural astaxanthin is preferable. (Mota et al. 2022). Though astaxanthin can be produced by diverse organisms such as bacteria, microalgae, and yeast, the highest production yield was obtained by the *H. pluvialis* microalga (Mota et al. 2022, Azli et al. 2021). Besides that, the main opportunity associated with the astaxanthin extracted by *H. pluvialis* is that FDA (US Food and Drug Administration) categorized it as GRAS (Generally Recognized as Safe) grade. Natural astaxanthin gains 20–50 times higher antioxidant capacities when compared to synthetic ones (Mota et al. 2022).

The major limitation of applying the astaxanthin pigment extract through the *Haematococcus* in the food industry is heat sensitivity, which may degrade in the cooking process, low chemical stability, low water dispersibility, and limited bioavailability other constitutions give limitations to the food industry applications. To overcome this effect encapsulation of the astaxanthin can be accounted for in the encapsulation process. During the encapsulation procedure a polymeric matrix or wall material forms as a coating around the unstable compounds to preserve their biological activity from external influences (Pan-utai et al. 2021). The research carried out by (Pan-utai et al. 2021) found that maltodextrin and gum Arabic solution with a ratio of 7.5:2.5 is most suitable for microencapsulation. The results of the study retrieved that more than 60% efficiency was reported in astaxanthin encapsulation. The water activity reported in the study was 0.41 thus more stable against microbial growth (Pan-utai et al. 2021).

9.5 Conclusion

Astaxanthin extracted from *H. pluvialis* can be used in the food industry in different food products. Due to its high antioxidant activity and health benefits, it can be used in a variety of food products. However, a limited number of research studies is carried out to incorporate astaxanthin in food products. Therefore, there is a large research gap in this area. Less amount of research has been carried out on laboratory scales to investigate the effect of the addition of astaxanthin in different food products such as the bakery industry, meat industry, and dairy industry. This chapter will help new researchers to find the research gap in the usage of astaxanthin in the food and packaging industry.

References

AL-Ghanimi GMM, Saleh Al-Rubeii AM (2020) Effect of antioxidant potential of Astaxanthin and Allylisothiocyanate in quality characteristics of raw ground beef meat during cold storage. Plant Archives 20(1):673–679

Altoé AF, de Aquino Paes C, da Costa Cirino FPS, da Silva Franco A, de Souza Lima EC, Rodrigues PRM, de Velasco PC, de Souza Cardoso F (2022) Astaxanthin supplementation extracted from Haematococcuspluvialis on clinical parameters associated with chronic diseases and metabolic syndrome. Braz J Dev 8(1):4503–4527

Ambati RR et al (2014) Astaxanthin: sources, extraction, stability, biological activities and its commercial applications - a review. Mar Drugs 12(1):128–152. https://doi.org/10.3390/md12010128

Batista AP, Gouveia L, Bandarra NM, Franco JM, Raymundo A (2013) Comparison of microalgal biomass profiles as novel functional ingredient for food products. Algal Res 2(2):164–173

Casella P, Iovine A, Mehariya S, Marino T, Musmarra D, Molino A (2020) Smart method for carotenoids characterization in Haematococcuspluvialis red phase and evaluation of astaxanthin thermal stability. Antioxidants 9(5):422

Castillo A, Finimundy TC, Heleno SA, Rodrigues P, Fernandes FA, Pereira S, Garcia-Jares C, Barros L, Lores M (2021) Novel incorporation of red-stage Haematococcuspluvialis wet paste as a Colourant and enhancer of the organoleptic and functional properties of Filloas. In: Biology and life sciences forum, vol 6, no. 1. Multidisciplinary Digital Publishing Institute, Basel, p 72

Dmytrów I, Szymczak M, Szkolnicka K, Kamiński P (2021) Development of functional acid curd cheese (Tvarog) with antioxidant activity containing astaxanthin from shrimp shells preliminary experiment. Foods 10(4):895

El-Baz FK, Hussein RA, Mahmoud K, Abdo SM (2018) Cytotoxic activity of carotenoid rich fractions from Haematococcuspluvialis and Dunaliellasalina microalgae and the identification of the phytoconstituents using LC-DAD/ESI-MS. Phytother Res 32(2):298–304

El-Baz FK, Aly HF, Ali SI, Salama A (2022) Safety of Astaxanthin-rich fraction of Haematococcuspluvialis microalgae. Egypt J Chem 65(3):479–489

Feng ZZ, Li MY, Wang YT, Zhu MJ (2018) Astaxanthin from Phaffiarhodozyma: microencapsulation with carboxymethyl cellulose sodium and microcrystalline cellulose and effects of microencapsulated astaxanthin on yogurt properties. LWT Food Sci Technol 96:152–160

Hossain AM, Brennan MA, Mason SL, Guo X, Zeng XA, Brennan CS (2017) The effect of astaxanthin-rich microalgae "Haematococcuspluvialis" and wholemeal flours incorporation in improving the physical and functional properties of cookies. Foods 6(8):57

Huang J, Xie W, Liu L, Song Y, Pan F, Bai H, Pan T, Lv Y, Chen J, Shi J, Hu X (2021) Nanostructured lipid carriers in alginate microgels for the delivery of astaxanthin. Eur J Lipid Sci Technol 123(2):2000191

Hwang SH, Kim JM, Kim S, Yoon MJ, Park KS (2020) Chemical transformation of Astaxanthin from Haematococcuspluvialis improves Itsantioxidative and anti-inflammatory activities. ACS Omega 5(30):19120–19130

Inthamat P, Boonsiriwit A, Lee YS, Siripatrawan U (2022a) Effects of genipin as natural crosslinker on barrier and mechanical properties of chitosan-astaxanthin film. J Food Process Preserv 46(8): e15707

Inthamat P, Lee YS, Boonsiriwit A, Siripatrawan U (2022b) Improving moisture barrier and functional properties of active film from genipin-crosslinked chitosan/astaxanthin film by heat curing. Int J Food Sci Technol 57(1):137–144

Liu L, Liu X, Jia J, Chen H, Zheng Z, Zhao C, Wang B (2022) Economic analysis of volatile characteristics of Haematococcuspluvialis and the effect of roasting temperature on the profile of volatiles and chemical components. Algal Res 64:102690

López, B.H. and Morales, J.M.V., 2018. Extraction of antioxidants from Haematococcuspluvialis by osmotic shock with sucrose, for use in the preparation of syrup for jams. In 2018 XIV international engineering congress (CONIIN). IEEE, pp 1–5

Łupina K, Kowalczyk D, Kazimierczak W (2021) Gum arabic/gelatin and water-soluble soy polysaccharides/gelatin blend films as carriers of astaxanthin—A comparative study of the kinetics of release and antioxidant properties. Polymers 13(7):1062

Łupina K, Kowalczyk D, Lis M, Raszkowska-Kaczor A, Drozłowska E (2022) Controlled release of water-soluble astaxanthin from carboxymethyl cellulose/gelatin and octenyl succinic anhydride starch/gelatin blend films. Food Hydrocoll 123:107179

Mota GCP, Moraes LBSD, Oliveira CYB, Oliveira DWS, Abreu JLD, Dantas DMM, Gálvez AO (2022) Astaxanthin from Haematococcuspluvialis: processes, applications, and market. Prep Biochem Biotechnol 52(5):598–609

Mularczyk M, Michalak I, Marycz K (2020) Astaxanthin and other nutrients from Haematococcus. Mar Drugs 18(459):1–22

Oslan SNH, Tan JS, Oslan SN, Matanjun P, Mokhtar RAM, Shapawi R, Huda N (2021) Haematococcuspluvialis as a potential source of astaxanthin with diverse applications in industrial sectors: current research and future directions. Molecules 26(21):6470

Pan-utai W, Boonpok S, Pornpukdeewattana S (2021) Combination of mechanical and chemical extraction of astaxanthin from Haematococcuspluvialis and its properties of microencapsulation. Biocatal Agric Biotechnol 33:101979

Pogorzelska E, Godziszewska J, Brodowska M, Wierzbicka A (2018) Antioxidant potential of Haematococcuspluvialis extract rich in astaxanthin on colour and oxidative stability of raw ground pork meat during refrigerated storage. Meat Sci 135:54–61

San H, Laorenza Y, Behzadfar E, Sonchaeng U, Wadaugsorn K, Sodsai J, Kaewpetch T, Promhuad K, Srisa A, Wongphan P, Harnkarnsujarit N (2022) Functional polymer and packaging technology for bakery products. Polymers 14(18):3793

Stewart JS, Lignell Å, Pettersson A, Elfving E, Soni MG (2008) Safety assessment of astaxanthin-rich microalgae biomass: acute and subchronic toxicity studies in rats. Food Chem Toxicol 46(9):3030–3036

Sun J, Yan J, Dong H, Gao K, Yu K, He C, Mao X (2022) Astaxanthin with different configurations: sources, activity, post-modification and application in foods. Curr Opin Food Sci 49: 100955

Xia S, Xue Y, Xue C, Jiang X, Li J (2022) Structural and rheological properties of meat analogues from Haematococcuspluvialis residue-pea protein by high moisture extrusion. LWT 154: 112756

Xu J, Wei R, Jia Z, Song R (2020) Characteristics and bioactive functions of chitosan/gelatin-based film incorporated with ε-polylysine and astaxanthin extracts derived from by-products of shrimp (Litopenaeusvannamei). Food Hydrocoll 100:105436

Chapter 10
Application of Nano-materials

Salem S. Salem

10.1 Introduction

Nanotechnology is a multidisciplinary area that includes nanoscience, nanophysics, nanochemistry, nanoelectronics, nanomaterials, nanobionics, and nanometrology, other disciplines (Salem 2023; Salem et al. 2023; Elfadel et al. 2023; Shehabeldine et al. 2022a; Salem and Fouda 2021). Nanotechnology is a relatively new scientific subject with multiple applications that cover energy production to industrial manufacturing procedures to therapeutic uses (Sharaf et al. 2022; Doghish et al. 2022; Cortez-Jugo et al. 2021). A diverse and interdisciplinary discipline, nanotechnology examines numerous areas of nanoscale science and technology (Salem et al. 2023; Salem and Fouda 2021). Nanoparticles, the fundamental units of nanotechnology, have a size between 1 and 100 nm. Due to their fundamental and technological relevance, metal nanoparticles including gold (AuNPs), zinc (ZnNPs), silver (AgNPs), selenium (SeNPs), and copper (CuNPs) have recently attracted a lot of attention (Abdelghany et al. 2022; Soliman et al. 2022a; Salem et al. 2020, 2021; Shehabeldine et al. 2022b). These nanoparticles are distinct from metallic particles in terms of their catalytic, electrical, and optical properties. Usually, physical, chemical, and biological processes may be used to create nanoparticles (Salem et al. 2023). In the physical and chemical processes, sodium borohydride and other strong-chemical reducing agent are utilized, whereas sodium citrate, alcohols, gamma- and UV radiation, among others, are weak-chemical reducing agent (Elakraa et al. 2022; Hashem et al. 2022a; Abdelmoneim et al. 2022). According to studies, producing nanoparticles via biological techniques is a cheap and environmentally benign process. To date, biological agents which including bacteria, fungi, yeast, actinomycetes, and plants have been used to show the production of nanoparticles (Salem et al. 2023; Salem and Fouda 2021). A new and

S. S. Salem (✉)
Botany and Microbiology Department, Faculty of Science, Al-Azhar University, Cairo, Egypt
e-mail: salemsalahsalem@azhar.edu.eg

R. Raja et al. (eds.), *Haematococcus*, https://doi.org/10.1007/978-981-99-2901-6_10

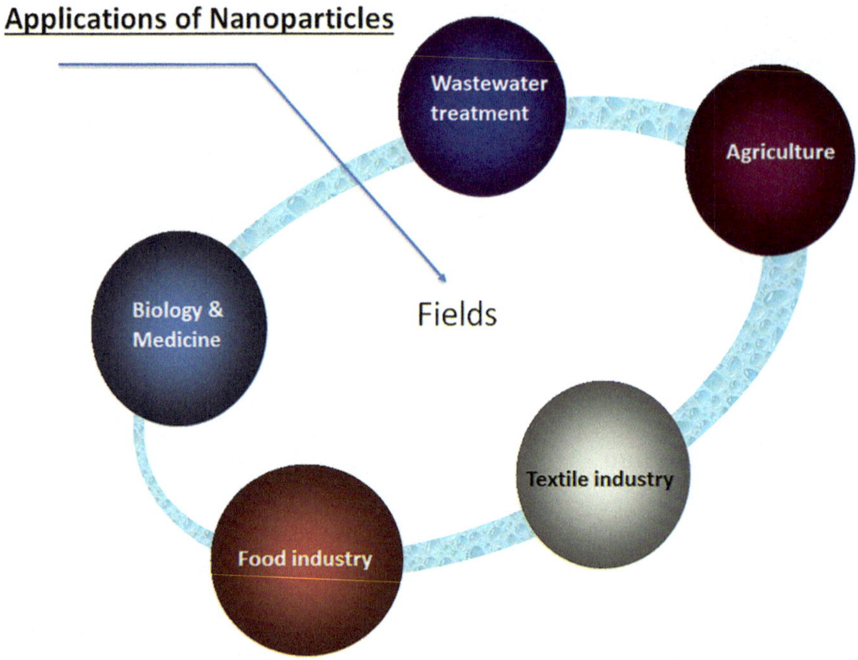

Fig. 10.1 Applications of nanotechnology (Salem 2023)

developing environmentally friendly science is the synthesis of nano-materials utilizing bacteria or plants. Many researchers have been synthesizing nano-materials using biological models (Soliman et al. 2022b; Al-Rajhi et al. 2022; Al-Zahrani et al. 2022a, b). The use of science in nanotechnology to manipulate matter at the atomic level. The characteristics of matter are very different from bulk materials at this level. The production, use, and assessment of structures, devices, and system where the shape and size are regulated at the tiny size regime are also included by this term (Dong et al. 2021). In other phrases, the fields of nano-science and nanotechnology concentrate on (a) developing synthetic approaches and surface observational tools for creating materials and structures, (b) observing the changes in physical and chemical attributes brought about by miniaturization, and (c) utilizing such features to create new and useful devices and materials. In the fields of solar energy conversion, catalysis, medicine, and treatment of water, nano-materials may offer answers to technical and environmental problems (Ismael 2020; Aref and Salem 2020; Saied et al. 2021, 2022). Fields of application of nanoparticles are represented (Fig. 10.1).

10.2 Methods of Synthesis of Nanoparticles

Numerous procedures have been developed for the synthesis of nano-materials due to the large variety of uses afforded by these materials in various branches of research and industry (Jamkhande et al. 2019). To create nanoparticles of various sizes and forms, diverse synthesis techniques employ a variety of chemical, physical, and biological factors (Salem et al. 2023). The chemical reduction process, which involves reducing metal particles to nanoparticles using chemical reducing agents like sodium borohydride or sodium citrate, is the technique that produces nanoparticles most often.

10.3 Green Synthesis of NPs

Utilizing a variety of species as long-lasting, ecologically secure precursors, biological synthesis can create durable, biofunctional nanoparticles. There have been reports of the greener synthesis of NPs using biomass filtrate derived from a variety of biological systems, including yeast, actinomycetes, plant extract, fungi, algae, and bacteria (El-Refaey and Salem 2023; Hashem et al. 2021; Salem and Fouda 2021). Recently, different organisms including unicellular and multicellular are used for biosynthesis of nanoparticles as represented (Fig. 10.2). The synthesis process of NPs may be viewed as a bottom-up process in which biomolecules released by the organism, including such enzymes, proteins, polysaccharides, and carbohydrates, oxidize or reduce metallic ions to generate NPs (Salem et al. 2023). Because different types of microbes interact with metallic ions in different ways, a thorough understanding of the process behind the synthesis of microbial NPs has not yet been achieved. The size, morphology, and form of the biosynthesized NPs are ultimately influenced by the biochemical metabolism, interactions, pH, and temperature effects of a particular microorganism. Nanoparticles are formed either by intra-cellular or extra-cellular depending on the microorganisms type. Researchers have made use of cell extracts for the biological production of NPs (Eid et al. 2020; Saied et al. 2021; Salem et al. 2022a, b; Suba et al. 2021).

10.4 Mechanism of Biosynthesis

The mechanism for intra- and extra-cellular formation of nano-materials is different in numerous biological models. Microorganisms' cell walls play a significant influence in the intra-cellular creation of nanoparticles. The positively (+) charged metal ions interact electrostatically with the negatively (−) charged cell wall. The metal ions are bioreduced to nanoparticles by the enzymes that are present in the organism cell wall, and then the smaller nanoparticles are dispersed out through the cell wall. The exact mechanism for creating nanoparticles utilizing biological models has not

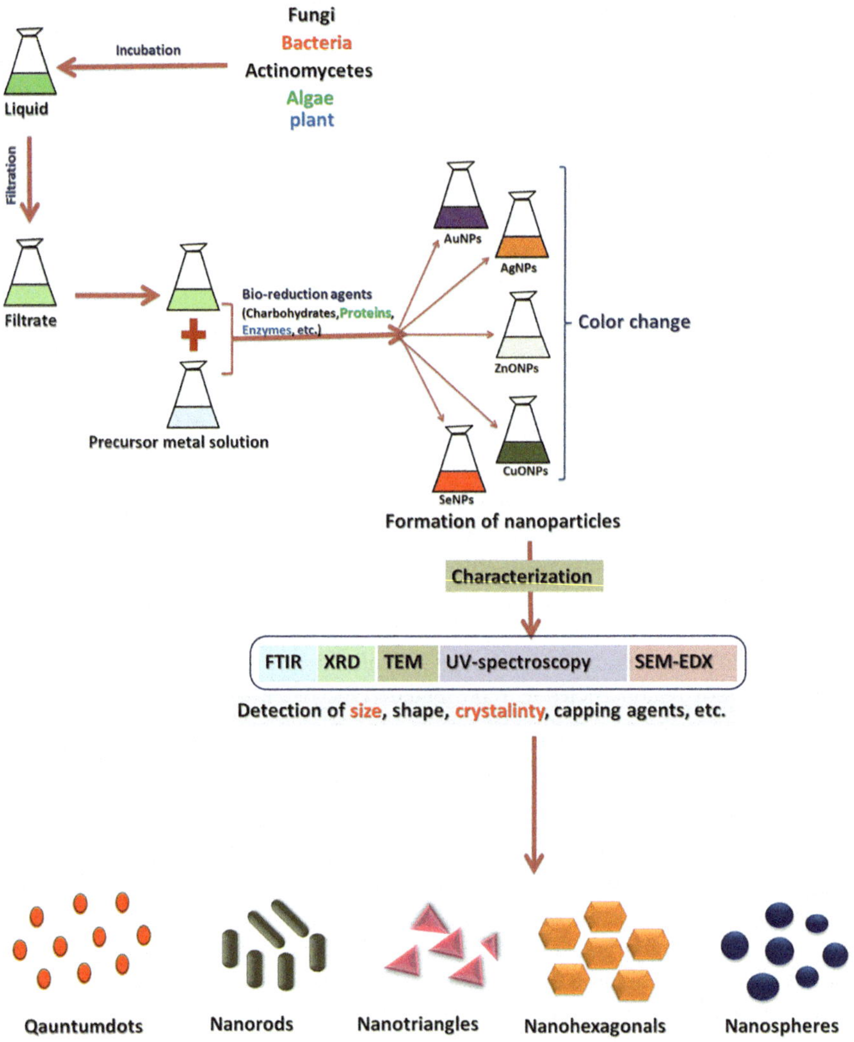

Fig. 10.2 Schematic outlines of the various biological approaches for the formation of nanoparticles

yet been thought of. This is due to the fact that various biological agents interact with metal ions and metal oxide in different ways, resulting in the creation of nanoparticles. Metal ions and metal oxide are converted to nanoparticles by the enzymes found within the cell wall, which then diffuse away from the cell membrane. A stepwise mechanism for intra-cellular formation of nanoparticles using *Verticillium* sp. (Mukherjee et al. 2001) explains the mechanism of the biosynthesis of nanoparticles involving bio-reduction capping, and trapping. When the cell surface contact with metal ions, it inter-acts electro statically and traps the ions.

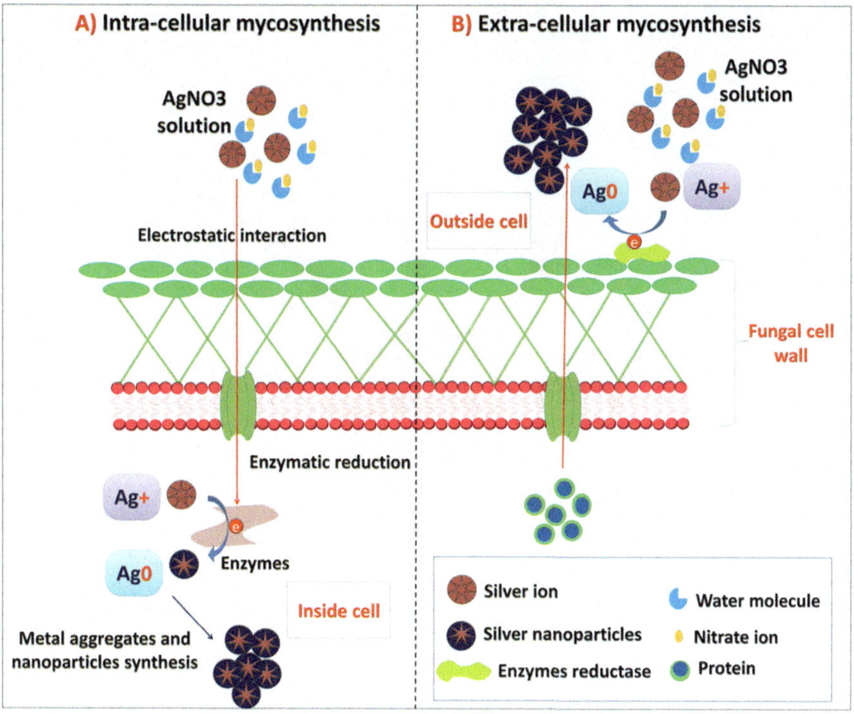

Fig. 10.3 Hypothetical mechanism for NPs biosynthesis

The cell wall's enzymes convert metal ions into metal and/or metal oxide nanoparticles. The mechanism involves the reduction of Ag^+ to Ag^0 via intracellular redox value, and proteins (enzymes) which formed by the cell. The mechanism of nanoparticle formation may occur either inside or outside the cell (Fig. 10.3) (Shaheen et al. 2021a).

10.5 Factors Affecting NPs Biosynthesis

There are numerous factors that influence the biosynthesis process for NPs as biomass, temperature, precursor conc. and time contact, pH and the being of certain enzyme (Fig. 10.4) (Shaheen et al. 2021a). It has become possible to change the size and structure of metal NMs by either forcing their environmental development or moving the functional molecules. Aeration, pH, incubation time, redox conditions, temperature, mixing ratio, salt content, and irradiation have all been looked into as ways to improve the reaction conditions for the biogenesis of NPs. The physical and chemical parameters affect the size and form of NPs. The ability to identify the size and structure of nanoparticles depends heavily on the reaction mixture's optimal

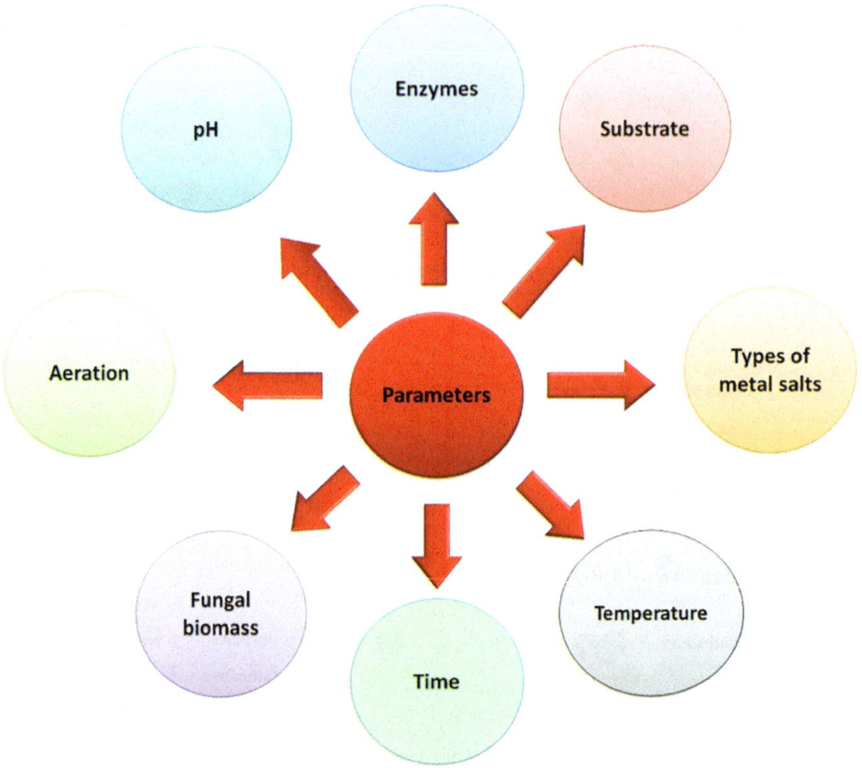

Fig. 10.4 Parameters affecting biosynthesis of nanoparticles

metal ions concentration, temperature, and pH. Planning variables including substrate concentration, pH, temperature, and exposure time to substrate may, to some extent, impact the pace of intra-cellular nanoparticle formation and later the size of the NPs. Researchers investigated the optimization of numerous parameters as temperature, inoculum of biomass, pH, and precursor conc. in the mycosynthesis of AgNPs, ZnONPs, and AuNPs (Al-Kordy et al. 2021; Thakral et al. 2021; Soleimani et al. 2022; Koçer and Özçimen 2022; Desai et al. 2021; Shirzadi-Ahodashti et al. 2022).

10.6 Characterization of Nanoparticles

The physico-chemical characterization of the produced nanoparticles is a crucial step in the biosynthesis of nanoparticles. Understanding properties such as size, shape, surface area, homogeneity, and others can help one better understand how to regulate the synthesis of nanoparticles for use in industry. Different methodologies may be used to characterize the nano-materials, which is crucial for understanding their diverse physio-chemical properties (Salem et al. 2022c; Ibrahim et al. 2021). The following categories cover the most often used tools:

1. X-Ray diffraction (XRD).
2. Scanning electron microscopic (SEM).
3. Transmission electron microscopic (TEM).
4. Ultraviolet-visible (UV-Vis) spectroscopic.
5. Fourier transform-infrared (FTIR) spectroscopic.
6. X-ray photoelectron spectroscopic (XPS).
7. Atomic absorption spectroscopic (AAS).

10.7 Application of Nanoparticles

Nanotechnology's benefits are rapidly expanding in a variety of sectors. Today, nanotechnology is considered an important factor that influences science and industry. Experts claim that nanotechnology has an impact on nations' economics and future prospects (Fajardo et al. 2022). Through the application of nanotechnology and the bio-production of nanoparticles, there is now renewed optimism for finding solutions to human issues (Said et al. 2023; Dong et al. 2021; Hagde et al. 2022). The usage of NMs has grown quickly in a variety of industries, including the pharmaceutical and medical sectors, which has prompted the release of new and useful goods on the market (Mughal et al. 2021; Rabiee et al. 2021). Due to the rise and spread of medication resistance among pathogenic microorganisms, various issues have been encountered nowadays when treating them (Shen et al. 2021). The capacity of humans to cure major infectious illnesses is threatened by bacteria's rising rates of antibiotic resistance (Mubeen et al. 2021).

The biological procedures have triggered a great amount of interest in the combination of metal and/or metal oxide NPs appropriate to the resolution of less toxic substance, ecofriendly landscape, and energy dexterous. The biological approaches of ZnONPs and CuONPs are synthesized by using naturally energetic products from microbes including bacteria and fungi (Suba et al. 2021; Ganesan et al. 2020; El-Batal et al. 2020; Waris et al. 2021). This method is adept due to its efficacy, ecofriendly techniques, simple and mass productivity (Hammad et al. 2022a; Dhillon et al. 2012). Mainly, the biological fusions of metal and/or metal oxide using florae have been broadly studied (Das et al. 2017; Hulkoti and Taranath 2014). Also, bacteria and fungi take part in an imperative role in the biological mixture of metal and/or metal oxide NPs (Zikalala et al. 2018). In the last 10 years, the use of microbes has improved superior anxiety in which there have been several studies exposed using various microbes' models (Salem 2023; Hammad et al. 2022b; Alsharif et al. 2020). Nanomaterials are formed into a variety of sizes and forms using several biological models that depend on the fabrication of nanomaterials (Saravanan et al. 2021; Salem 2023). In addition, the protein masked after bacteria might act as a coating agent that confers constancy of NPs making. Mostly, the cultures are tolerable to advance in the medium. The advancement of nanoformulations and their diverse uses make the green synthesis of NMs a promising and attractive topic for various investigations (El Shafey 2020; Singh et al. 2018; Jeevanandam et al. 2016). A prior work used *Penicillium expansum* to produce ZnONPs, which presented the possible for the extra-cellular fusion of ZnONPs with a size range of 3.5–67.3 nm. In this study, ZnONPs were utilized in place of synthetic fungicides to suppress *F. oxysporum*, a harmful phytopathogenic fungus that causes the wilt-disease in *S. melongena* L (Abdelaziz et al. 2022). Abdelghany et al. (Abdelghany et al. 2022) recognized the synthesis of ZnONPs using *Salix tetrasperma*, which yielded a pure crystal and spherical shape of ZnONPs with a size range of 5–43 nm. SEM-EDX, TEM, and DLS investigations were utilized to describe the obtained ZnONPs. The disc diffusion technique was used to evaluate ZnONPs' antibacterial effectiveness. The MTT technique was used to assess the cytotoxicity of ZnONPs against prostate cancer cells (Pc3). ZnONPs showed good antibacterial action against *S. typhimurium, E. coli, K pneumonia, B. subtilis, and S aureus*, with inhibition zones of 28.1, 23.83, 28.33, 23.83, and 33.83 mm, respectively. Inhibition zone of 23.67 mm also inhibited *C. albicans*, however, ZnONPs had no effect on *M. circinelloide* or *A. fumigatus*. ZnONPs showed good antioxidant activity, with an IC_{50} of 8.73 μg/mL as opposed to the plant extract's IC_{50} of 15.91 μg/mL. The results of a cytotoxicity test showed that ZnONPs were active against Pc3.

The biological fusion of ZnONPs utilizing fungi is a hopeful slant due to their high serenity to higher metal consideration, high binding volume and their aptitude in metal bio-accumulation over bacteria (El-Saadony et al. 2021). Also, the fungi exposed the capacity to hide a large amount of extra-cellular redox protein and/or enzyme. As a result, this made it easier for the metal ions to be reduced into NPs in greater quantities, which is appropriate for the large-scale building. The greater amount of proteins concealed in the average by the fungi swapped as coating proteins that further attach and captured the NPs surface and consulted to the constancy. For illustration, Mohamed et al. (Mohamed et al. 2019) established the fusion of ZnONPs by using *A. niger* and *F. keratoplasticum* that ensued in the development of ZnONPs with the steady diameter size of 8–38 nm and 10–42 nm for hexagonal and nanorod ZnONPs, respectively, and high mono-dispersity elements (uniformly distributed) deprived of any agglomeration. In addition, the authors recommended that the protein buried by the fungi was bound and reduced the orbicular ZnONPs and prohibited the NPs from agglomerate. In addition, filtrate-cell free (FCF) of *Aspergillus terreus* substitute was utilized in the amalgamation of ZnONPs. On that version, the FCF was patent to create NPs after the snow of zinc acetate elucidation with the size of 10–45 nm. In addition, the FTIR-spectra study validated the existence of proteins and surplus biological compounds in the ZnONPs formed. The results show how the produced ZnONPs were analyzed using various tools including TEM, FTIR, TGA, and XRD analyses (Fouda et al. 2018). A recent research found that fungi can produce a large amount of extra-cellular protein, which aids in the formation of nanoparticles and their stability in nanosolution (Shaheen et al. 2021a). Therefore, the utilization of fungi for the amalgamation of ZnONPs is favorable as the fungi are competent at concealing extra-cellular enzymes (proteins)

(Akpomie et al. 2021; Abdelhakim et al. 2020). Because of their greater tolerance to toxic metals, bacteria and fungi have been shown to produce metal oxide NP. In addition, Mohamed et al. (Mohamed et al. 2021) proven an extra-cellular formation of ZnONPs and CuONPs using *P. chrysogenum*. Also, the UV-vis spectra results detailed strong peaks at 380 nm and 335 nm, respectively, implying the positive realization of ZnONPs and CuONPs, individually. The produced ZnO and CuONPs were analyzed using various methods including TEM, SEM, FTIR, and XRD analyses. The effect of ZnONPs and CuONPs on pathogenic bacteria was also investigated in this study since it demonstrated the potency of ZnONPs and CuONPs against these microbes. We also investigated how well particles prevented the formation of biofilms; the outcome was dependent on the particle concentration, and encouraging results were demonstrated as potential medicinal uses.

Since no hazardous or toxic chemicals are used in the mixing approach, the microbe-mediated synthesis of ZnONPs and CuONPs appears to be both safe and environmentally benign (Gebre and Sendeku 2019; Agarwal et al. 2017; Akintelu et al. 2020; Fouda et al. 2020). The physiologically active complexes that the microorganisms produced served as falling and covering agents when added. Accessible the spherical CuONPs in the size range of 78–80 nm was produced by a rapid gradual fusion technique by two *Streptomyces* sp. (Hassan et al. 2019). Attainable CuONPs synthetic from *Penicillium chrysogenum* can be used for the expansion of antimicrobial for hospital use to inhibit or to minimize impurity with pathogenic bacteria (Shehabeldine et al. 2022b). Singh et al. (Singh et al. 2010) described biological synthesis of CuONPs using *Escherichia coli* with adaptable size and shapes. *Aspergillus niger* present in the soil was used to produce cubical CuONPs sized 14–47.4 nm (Badawy et al. 2021). *Aspergillus terreus* at ambient temperature was used for the amalgamation of CuONPs, which were partitioned for eliminating copper from combined circuits and allowed to exist in nano form. CuONPs of an average size of 11–47 nm were assembled extracellularly by biomass of fungal cell isolated from a soil sample in Egypt, and infrared spectroscopy [IR] analysis revealed that amide groups within proteins were responsible for the consistency and coating agents surrounding the CuONPs (Shaheen et al. 2021b). Nanotechnology offers novel solutions in making elemental Se safe, absorbable, and bioavailable for aquatic organisms in the form of Se nanoparticles (Ferro et al. 2021; Hashem et al. 2022b). This is because of the tiny size and large surface area of the element. Selenium nanoparticles (SeNPs) have shown outstanding bioavailability, with a low toxicity level, and can exert an envisioned activity at much lower concentrations than organic or in-organic Se compounds (Menon et al. 2018). Furthermore, SeNPs exhibit biological activities that promote optimum substitutions for various seleno compounds (Yanhua et al. 2016). Extra-cellular biogenesis of SeNPs by *Penicillium corylophilum* has been characterized to be of polydispersed nature (Salem et al. 2021). It has been discovered that a cell-free extract of *Penicillium expansum* has high potential for producing SeNPs, and that the process can be regulated to alter SeNP's structure. SeNPs with a size range of 4–12.7 nm have only been found to be spherical (Abu-Elghait et al. 2021). The extra-cellular bio-formation of SeNPs was performed by using yeast (*Saccharomyces cerevisiae*) extract. The production of

SeNPs is confirmed by the absorption peak at 300 nm in UV-Vis spectroscopy due to the surface Plasmon resonance of SeNPs. It is also characterized by FT-IR and XRD. The SeNPs around 5–51 nm were formed. The effectiveness of biosynthesized SeNPs as antimicrobial agents against food-borne pathogens was evaluated. With minimum inhibitory concentrations (MIC) of 62.5, 125, 250, and 500 μg/mL against *S. aureus*, *E. coli*, *A. fumigatus*, and *A. niger*, respectively, SeNPs demonstrated potential antibacterial efficacy against food-borne pathogens. Finally, a mechanism was designed for the effect of nanoparticles on microbes (Salem 2022a, b).

Among metallic nanoparticles, SeNPs have attracted the attention of researchers and have special properties such as stability in environmental conditions and synthesis at low temperatures (Wadhwani et al. 2016). In general, there are numerous approaches for the biosynthesis of SeNPs, one of these methods is the plant-mediated method, that is, the synthesis of nanoparticles using plant (Hashem and Salem 2022). Hashem and Salem (Hashem and Salem 2022) recognized the synthesis of Se NPs using *Urticadioica* that formed a pure crystalline and spherical shape of SeNPs with the size ranged from 5 to 43 nm. The production of SeNPs is confirmed by the absorption peak at 300 nm in UV-Vis spectroscopy. In addition, the obtained SeNPs was characterized by different tools as SEM-EDX, TGA, FTIR, TEM, and DLS analysis. SeNPs created by biosynthesis were tested for their antimicrobial and anticancer properties. SeNPs shown promising antibacterial efficacy against unicellular and multicellular fungus, as well as Gram-positive [*B. subtilis*, and *S. aureus*] and Gram-negative [*E. coli*, *P. aeruginosa*] bacteria. The Vero normal cell line (CCL-81) was used to test the cytotoxicity of SeNPs, and the IC_{50} value was 173.2 μg/mL. Abu-Elghait et al. (Abu-Elghait et al. 2021) synthesized selenium nano-composite and investigated its anti-biofilm effects against *P. aeruginosa* and *Saureus* strains. The results showed that this myco-synthesized novel selenium nano-composite can inhibit biofilm and have anti-biofilm effects. Nanotechnology enables the creation of advanced and multi-functional textiles with numerous creative applications in health, medicine, design, athletics, the military, and transportation. Textiles augmented with any of these NMs have several advantages in wound healing, pharmaceutical delivery, air purification, cosmetics, and other applications. Active NMs such as AgNPs, ZnONPs, SiO_2/TiO_2NPs, and CuONPs can be physically or chemically incorporated into textiles to develop antibacterial textiles (Eid et al. 2020; Fouda et al. 2018; Shaheen et al. 2019; Shaheen et al. 2021b). In the near future, nanotechnology will surely add special qualities to two major domains of food manufacturing: packaged food and food ingredient functions. To improve the antibacterial properties of polymeric compounds used in packaging sheets, certain nano-metal oxides, like ZnONPs, were added (Smaoui et al. 2023). The NMs were used in packaging operations to prevent food contamination. As an outcome, the use of packing containers connected by nanoparticles is critical and an acceptable technique for keeping food fresh for a more prolonged period of time and preventing food-borne contamination. Suitable for treating wastewater, metal oxide dependent nano-adsorbents have a significant affinity for metal ions in contaminated wastewater. Several studies have investigated the potential of metal oxide nanoparticles including MgONPs, Fe_2O_3NPs, and

CuONPs in water treatment and dye removal (Saied et al. 2021, 2022; Hammad et al. 2022a, b) The usage of NMs in agriculture is expected to grow in the next years. The use of nano fertilizers enhances crop output and quality by increasing the utilization of nutrients while decreasing spillage in the ecosystem and production costs, therefore contributing to environmentally friendly agriculture (Hashem et al. 2022c; Salem and Husen 2023).

10.8 Conclusion

Nanotechnology has attracted a lot of interest because of its future potential to advance biotechnology and health care research. The primary current objective is to focus on the design of nanomaterials via engineering and production, as well as to develop standards for novel materials and products in terms of both human and safety for the environment. Nanotechnology, being a rapidly emerging sector of technology, has found its way into a variety of businesses due to its favorable properties. Biotechnology, on the other hand, is a fundamental component of most modern enterprises as an environmentally benign green method for biosynthesis of bioactive constituents. The domains of nano-biotechnology and nanotechnology are currently undergoing intense investigation for different treatment applications. In current years, metal nanoparticles have been measured widely for numerous biomedical, bio-remediation, and bio-sensor needs because of their strange antibacterial, antioxidant, and optical possessions, and large superficial area to volume ratio and higher efficacy.

References

Abdelaziz AM et al (2022) Potential of biosynthesized zinc oxide nanoparticles to control Fusarium wilt disease in eggplant (Solanum melongena) and promote plant growth. Biometals 35(3): 601–616

Abdelghany TM et al (2022) Phytofabrication of zinc oxide nanoparticles with advanced characterization and its antioxidant, anticancer, and antimicrobial activity against pathogenic microorganisms. Biomass Convers Biorefin 13:417

Abdelhakim HK, El-Sayed ER, Rashidi FB (2020) Biosynthesis of zinc oxide nanoparticles with antimicrobial, anticancer, antioxidant and photocatalytic activities by the endophytic Alternaria tenuissima. J Appl Microbiol 128(6):1634–1646

Abdelmoneim HEM et al (2022) Multiple applications of CdS/TiO2 nanocomposites synthesized via microwave-assisted Sol–Gel. J Clust Sci 33(3):1119–1128

Abu-Elghait M et al (2021) Ecofriendly novel synthesis of tertiary composite based on cellulose and myco-synthesized selenium nanoparticles: characterization, antibiofilm and biocompatibility. Int J Biol Macromol 175:294–303

Agarwal H, Venkat Kumar S, Rajeshkumar S (2017) A review on green synthesis of zinc oxide nanoparticles – an eco-friendly approach. Resource-Efficient Technol 3(4):406–413

Akintelu SA et al (2020) Green synthesis of copper oxide nanoparticles for biomedical application and environmental remediation. Heliyon 6(7):e04508

Akpomie KG et al (2021) One-pot synthesis of zinc oxide nanoparticles via chemical precipitation for bromophenol blue adsorption and the antifungal activity against filamentous fungi. Sci Rep 11(1):8305

Al-Kordy HM, Sabry SA, Mabrouk ME (2021) Statistical optimization of experimental parameters for extracellular synthesis of zinc oxide nanoparticles by a novel haloalaliphilic Alkalibacillus sp.W7. Sci Rep 11(1):10924

Al-Rajhi AMH et al (2022) Ecofriendly synthesis of silver nanoparticles using Kei-apple (Dovyalis caffra) fruit and their efficacy against cancer cells and clinical pathogenic microorganisms. Arab J Chem 15(7):103927

Alsharif SM et al (2020) Multifunctional properties of spherical silver nanoparticles fabricated by different microbial taxa. Heliyon 6(5):e03943

Al-Zahrani FAM et al (2022a) Green synthesis and antibacterial activity of Ag/Fe2O3 nanocomposite using Buddleja lindleyana extract. Bioeng (Basel) 9(9):452

Al-Zahrani FAM et al (2022b) Synthesis of Ag/Fe2O3 nanocomposite from essential oil of ginger via green method and its bactericidal activity. Biomass Conv Bioref. https://doi.org/10.1007/s13399-022-03248-9

Aref MS, Salem SS (2020) Bio-callus synthesis of silver nanoparticles, characterization, and antibacterial activities via Cinnamomum camphora callus culture. Biocatal Agric Biotechnol 27:101689

Badawy AA et al (2021) Efficacy assessment of biosynthesized copper oxide nanoparticles (CuO-NPs) on stored grain insects and their impacts on morphological and physiological traits of wheat (Triticum aestivum L.). Plant Biol 10:233. https://doi.org/10.3390/biology10030233

Cortez-Jugo C et al (2021) A focus on "bio" in bio–nanoscience: the impact of biological factors on nanomaterial interactions. Adv Healthc Mater 10(16):2100574

Das RK et al (2017) Biological synthesis of metallic nanoparticles: plants, animals and microbial aspects. Nanotechnol Environ Eng 2(1):1–21

Desai MP et al (2021) Bacterium mediated facile and green method for optimized biosynthesis of gold nanoparticles for simple and visual detection of two metal ions. J Clust Sci 32(2):341–350

Dhillon GS et al (2012) Green approach for nanoparticle biosynthesis by fungi: current trends and applications. Crit Rev Biotechnol 32(1):49–73

Doghish AS et al (2022) Nanocomposite based on gold nanoparticles and carboxymethyl cellulose: synthesis characterization antimicrobial and anticancer activities. J Drug Deliv Sci Technol 77:103874. https://doi.org/10.1016/j.jddst.2022.103874

Dong Y et al (2021) Nanotechnology shaping stem cell therapy: recent advances, application, challenges, and future outlook. Biomed Pharmacother 137:111236

Eid AM et al (2020) Endophytic streptomyces laurentii mediated green synthesis of Ag-NPs with antibacterial and anticancer properties for developing functional textile fabric properties. Antibiotics 9(10):641. https://doi.org/10.3390/antibiotics9100641

El Shafey AM (2020) Green synthesis of metal and metal oxide nanoparticles from plant leaf extracts and their applications: a review. Green Process Synth 9(1):304–339

Elakraa AA et al (2022) Cefotaxime incorporated bimetallic silver-selenium nanoparticles: promising antimicrobial synergism, antibiofilm activity, and bacterial membrane leakage reaction mechanism. RSC Adv 12(41):26603–26619

El-Batal AI et al (2020) Penicillium chrysogenum-mediated mycogenic synthesis of copper oxide nanoparticles using gamma rays for in vitro antimicrobial activity against some plant pathogens. J Clust Sci 31(1):79–90

Elfadel RG et al (2023) Preparation of new surface coating based on modified oil-based polymers blended with ZnO and CuZnO NPs for steel protection. Sci Rep 13:7268. https://doi.org/10.1038/s41598-023-34085-z

El-Refaey AA, Salem SS (2023) Algae materials for bionanopesticides: nanoparticles and composites. Algae materials. Academic Press, Elsevier, pp 219–230. https://doi.org/10.1016/B978-0-443-18816-9.00004-6

El-Saadony MT et al (2021) Impact of mycogenic zinc nanoparticles on performance, behavior, immune response, and microbial load in Oreochromis niloticus. Saudi J Biol Sci 28(8): 4592–4604

Fajardo C et al (2022) Nanotechnology in aquaculture: applications, perspectives and regulatory challenges. Aquacult Fish 7(2):185–200

Ferro C, Florindo HF, Santos HA (2021) Selenium nanoparticles for biomedical applications: from development and characterization to therapeutics. Adv Healthc Mater 10(16):2100598

Fouda A et al (2018) In-vitro cytotoxicity, antibacterial, and UV protection properties of the biosynthesized zinc oxide nanoparticles for medical textile applications. Microb Pathog 125: 252–261

Fouda A et al (2020) Optimization of green biosynthesized visible light active CuO/ZnO nano-photocatalysts for the degradation of organic methylene blue dye. Heliyon 6(9):e04896

Ganesan V et al (2020) Periconium sp.(endophytic fungi) extract mediated sol-gel synthesis of ZnO nanoparticles for antimicrobial and antioxidant applications. Mater Sci Semicond Process 105: 104739

Gebre SH, Sendeku MG (2019) New frontiers in the biosynthesis of metal oxide nanoparticles and their environmental applications: an overview. SN Appl Sci 1(8):928

Hagde P et al (2022) Therapeutic potential of quercetin in diabetic foot ulcer: mechanistic insight, challenges, nanotechnology driven strategies and future prospects. J Drug Deliv Sci Technol 74: 103575

Hammad EN et al (2022a) Environmental impacts of ecofriendly iron oxide nanoparticles on dyes removal and antibacterial activity. Appl Biochem Biotechnol 194:6053

Hammad EN et al (2022b) Purpureocillium lilacinum mediated biosynthesis copper oxide nanoparticles with promising removal of dyes. Biointerface Res Appl Chem 12(2):1397–1404

Hashem AH, Salem SS (2022) Green and ecofriendly biosynthesis of selenium nanoparticles using Urtica dioica (stinging nettle) leaf extract: antimicrobial and anticancer activity. Biotechnol J 17(2):2100432

Hashem AH et al (2021) Biomedical applications of mycosynthesized selenium nanoparticles using Penicillium expansum ATTC 36200. Biol Trace Elem Res 199:3998–4008. https://doi.org/10.1007/s12011-020-02506-z

Hashem AH et al (2022a) Synthesis of chitosan-based gold nanoparticles: antimicrobial and wound-healing activities. Polymers 14(11):2293. https://doi.org/10.3390/polym14112293

Hashem AH et al (2022b) Unveiling antimicrobial and insecticidal activities of biosynthesized selenium nanoparticles using prickly pear Peel waste. J Funct Biomater 13(3):112

Hashem AH et al (2022c) Selenium and nano-selenium-mediated biotic stress tolerance in plants. In: Hossain MA, Ahammed GJ, Kolbert Z, El-Ramady H, Islam T, Schiavon M (eds) Selenium and nano-selenium in environmental stress management and crop quality improvement. Sustainable plant nutrition in a changing world. Springer, Cham. https://doi.org/10.1007/978-3-031-07063-1_11

Hassan SE-D et al (2019) Endophytic actinomycetes Streptomyces spp mediated biosynthesis of copper oxide nanoparticles as a promising tool for biotechnological applications. J Biol Inorg Chem 24(3):377–393

Hulkoti NI, Taranath T (2014) Biosynthesis of nanoparticles using microbes—a review. Colloids Surf B: Biointerfaces 121:474–483

Ibrahim S et al (2021) Optimization for biogenic microbial synthesis of silver nanoparticles through response surface methodology, characterization, their antimicrobial, antioxidant, and catalytic potential. Sci Rep 11(1):1–18

Ismael M (2020) A review and recent advances in solar-to-hydrogen energy conversion based on photocatalytic water splitting over doped-TiO2 nanoparticles. Sol Energy 211:522–546

Jamkhande PG et al (2019) Metal nanoparticles synthesis: an overview on methods of preparation, advantages and disadvantages, and applications. J Drug Deliv Sci Technol 53:101174

Jeevanandam J, Chan YS, Danquah MK (2016) Biosynthesis of metal and metal oxide nanoparticles. ChemBioEng Reviews 3(2):55–67

Koçer AT, Özçimen D (2022) Eco-friendly synthesis of silver nanoparticles from macroalgae: optimization, characterization and antimicrobial activity. Biomass Convers Biorefin 1–12

Menon S et al (2018) Selenium nanoparticles: a potent chemotherapeutic agent and an elucidation of its mechanism. Colloids Surf B: Biointerfaces 170:280–292

Mohamed AA et al (2019) Fungal strain impacts the shape, bioactivity and multifunctional properties of green synthesized zinc oxide nanoparticles. Biocatal Agric Biotechnol 19:101103

Mohamed AA et al (2021) Correction to: eco-friendly mycogenic synthesis of ZnO and CuO nanoparticles for in vitro antibacterial, antibiofilm and antifungal applications (Biological Trace Element Research, (2021), 199, 7, (2788-2799), 10.1007/s12011-020-02369-4). Biol Trace Elem Res 199(7):2800–2801

Mubeen B et al (2021) Nanotechnology as a novel approach in combating microbes providing an alternative to antibiotics. Antibiotics (Basel) 10(12):1473

Mughal B et al (2021) Biogenic nanoparticles: synthesis, characterisation and applications. Appl Sci 11(6):2598

Mukherjee P et al (2001) Fungus-mediated synthesis of silver nanoparticles and their immobilization in the mycelial matrix: a novel biological approach to nanoparticle synthesis. Nano Lett 1(10):515–519

Rabiee N et al (2021) Diatoms with invaluable applications in nanotechnology, biotechnology, and biomedicine: recent advances. ACS Biomater Sci Eng 7(7):3053–3068

Saied E et al (2021) The catalytic activity of biosynthesized magnesium oxide nanoparticles (Mgo-nps) for inhibiting the growth of pathogenic microbes, tanning effluent treatment, and chromium ion removal. Catalysts 11(7):821

Saied E et al (2022) Mycosynthesis of hematite (α-Fe2O3) nanoparticles using aspergillus Niger and their antimicrobial and photocatalytic activities. Bioengineering (Basel) 9(8):397

Said A et al (2023) Antibacterial activity of green synthesized silver nanoparticles using Lawsonia inermis against common pathogens from urinary tract infection. Appl Biochem Biotechnol. https://doi.org/10.1007/s12010-023-04482-1

Salem SS (2022a) Bio-fabrication of selenium nanoparticles using Baker's yeast extract and its antimicrobial efficacy on food borne pathogens. Appl Biochem Biotechnol 194(5):1898–1910

Salem SS (2022b) Baker's yeast-mediated silver nanoparticles: characterisation and antimicrobial biogenic tool for suppressing pathogenic microbes. BioNanoSci 12:1220–1229. https://doi.org/10.1007/s12668-022-01026-5

Salem SS (2023) A mini review on green nanotechnology and its development in biological effects. Arch Microbiol 205:128. https://doi.org/10.1007/s00203-023-03467-2

Salem SS, Fouda A (2021) Green synthesis of metallic nanoparticles and their prospective biotechnological applications: an overview. Biol Trace Elem Res 199(1):344–370

Salem SS, Husen A (2023) Chapter 14 - Effect of engineered nanomaterials on soil microbiomes and their association with crop growth and production. In: Husen A (ed) Engineered nanomaterials for sustainable agricultural production, soil improvement and stress management. Academic Press, pp 311–336. https://doi.org/10.1016/B978-0-323-91933-3.00010-6

Salem SS et al (2020) Bactericidal and in-vitro cytotoxic efficacy of silver nanoparticles (Ag-NPs) fabricated by endophytic actinomycetes and their use as coating for the textile fabrics. Nanomaterials (Basel) 10:2082. https://doi.org/10.3390/nano10102082

Salem SS et al (2021) Antibacterial, cytotoxicity and larvicidal activity of green synthesized selenium nanoparticles using Penicillium corylophilum. J Clust Sci 32(2):351–361

Salem SS et al (2022a) Pseudomonas indica-mediated silver nanoparticles: antifungal and antioxidant biogenic tool for suppressing Mucormycosis fungi. J Fungi 8(2):126

Salem SS et al (2022b) Green biosynthesis of selenium nanoparticles using orange peel waste: characterization, antibacterial and antibiofilm activities against multidrug-resistant bacteria. Life 12(6):893. https://doi.org/10.3390/life12060893

Salem SS et al (2022c) Synthesis of silver nanocomposite based on carboxymethyl cellulose: antibacterial, antifungal and anticancer activities. Polymers 14(16):3352. https://doi.org/10.3390/polym14163352

Salem SS et al (2023) A comprehensive review of nanomaterials: types, synthesis, characterization, and applications. Biointerface Res Appl Chem 13(1):41

Saravanan A et al (2021) A review on biosynthesis of metal nanoparticles and its environmental applications. Chemosphere 264:128580

Shaheen TI et al (2019) A new facile strategy for multifunctional textiles development through in situ deposition of SiO2/TiO2 nanosols hybrid. Ind Eng Chem Res 58(44):20203–20212. https://doi.org/10.1021/acs.iecr.9b04655

Shaheen TI, Salem SS, Fouda A (2021a) Current advances in fungal Nanobiotechnology: mycofabrication and applications. In: Lateef A et al (eds) Microbial nanobiotechnology: principles and applications. Springer Singapore, Singapore, pp 113–143

Shaheen TI, Fouda A, Salem SS (2021b) Integration of cotton fabrics with biosynthesized CuO nanoparticles for bactericidal activity in the terms of their cytotoxicity assessment. Ind Eng Chem Res 60(4):1553–1563

Sharaf MH et al (2022) A new strategy to integrate silver nanowires with waterborne coating to improve their antimicrobial and antiviral properties. Pigment Resin Technol. https://doi.org/10.1108/PRT-12-2021-0146

Shehabeldine AM et al (2022a) Multifunctional silver nanoparticles based on chitosan: antibacterial, antibiofilm, antifungal, antioxidant, and wound-healing activities. J Fungi 8 (6):612. https://doi.org/10.3390/jof8060612

Shehabeldine AM et al (2022b) Potential antimicrobial and antibiofilm properties of copper oxide nanoparticles: time-kill kinetic essay and ultrastructure of pathogenic bacterial cells. Appl Biochem Biotechnol 195:467

Shen Y et al (2021) Recent advances in nanotechnology for simultaneous detection of multiple pathogenic bacteria. Nano Today 38:101121

Shirzadi-Ahodashti M et al (2022) Optimization and evaluation of anticancer, antifungal, catalytic, and antibacterial activities: biosynthesis of spherical-shaped gold nanoparticles using Pistacia vera hull extract (AuNPs@ PV). Arab J Chem 16:104423

Singh VA et al (2010) Biological synthesis of copper oxide nano particles using Escherichia coli. Curr Nanosci 6(4):365–369

Singh J et al (2018) 'Green' synthesis of metals and their oxide nanoparticles: applications for environmental remediation. J Nanobiotechnol 16(1):1–24

Smaoui S et al (2023) Zinc oxide nanoparticles in meat packaging: a systematic review of recent literature. Food Packag Shelf Life 36:101045

Soleimani P et al (2022) Optimization of silver nanoparticle biosynthesis by entomopathogenic fungi and assays of their antimicrobial and antifungal properties. J Invertebr Pathol 190:107749

Soliman MKY et al (2022a) Multifunctional properties of silver and gold nanoparticles synthesis by Fusarium pseudonygamai. Biomass Convers Biorefin

Soliman MKY et al (2022b) Biosynthesis of silver and gold nanoparticles and their efficacy towards antibacterial, antibiofilm, cytotoxicity, and antioxidant activities. Appl Biochem Biotechnol 195:1158

Suba S et al (2021) Microbial mediated synthesis of ZnO nanoparticles derived from Lactobacillus spp: characterizations, antimicrobial and biocompatibility efficiencies. Sensors Int 2:100104

Thakral F et al (2021) Zinc oxide nanoparticles: from biosynthesis, characterization, and optimization to synergistic antibacterial potential. Curr Pharmacol Rep 7(1):15–25

Wadhwani SA et al (2016) Biogenic selenium nanoparticles: current status and future prospects. Appl Microbiol Biotechnol 100(6):2555–2566

Waris A et al (2021) A comprehensive review of green synthesis of copper oxide nanoparticles and their diverse biomedical applications. Inorg Chem Commun 123:108369

Yanhua W et al (2016) Selenium-substituted hydroxyapatite nanoparticles and their in vivo antitumor effect on hepatocellular carcinoma. Colloids Surf B: Biointerfaces 140:297–306

Zikalala N et al (2018) Biosynthesis protocols for colloidal metal oxide nanoparticles. Nanostruct Nanoobjects 16:288–299

Chapter 11
Environmental Impacts Related to Upstream and Downstream Processing of *Haematococcus pluvialis*

Adriane Terezinha Schneider, Rosangela Rodrigues Dias, Mariany Costa Deprá, Darissa Alves Dutra, Leila Queiroz Zepka, and Eduardo Jacob-Lopes

11.1 Introduction

Astaxanthin is a bright red pigment of the carotenoid class produced by chemical synthesis or by biosynthesis by microalgae cultivation, plants, yeasts, and bacteria. The main applications of this pigment are in human and animal nutrition. Given its claims of benefits to human health, astaxanthin is marketed as a food supplement, but the main application remains in aquaculture. Furthermore, as the cost of natural astaxanthin is comparatively higher, most aquaculture feed applications utilize synthetic astaxanthin. However, it is of note that synthetic astaxanthin is not approved for direct human consumption due to concerns related to its pharmacological activity and food safety (Li et al. 2011; Oslan et al. 2021). Notably, although chemical synthesis can provide astaxanthin in large quantities and at a lower cost, these human health concerns, make natural astaxanthin a better choice. In light of this, several biotechnological processes are being developed to replace the synthetic production. However, in the state of the art, large-scale biological production of astaxanthin is only carried out through *Haematococcus pluvialis* cultivation (Li et al. 2020). The current production process of *Haematococcus pluvialis* follows a two-stage approach, first producing biomass under favorable conditions and then exposing the microalgae to stressful environmental conditions that promote the accumulation of astaxanthin (Maroneze et al. 2020). In nature, *Haematococcus pluvialis*is known as the richest source of this pigment, being able to produce up to 5% of the dry weight of astaxanthin. Of note, the antioxidant activity of the pigment astaxanthinis 65 times stronger than that of vitamin C, 10 times stronger

A. T. Schneider · R. R. Dias · M. C. Deprá · D. A. Dutra · L. Q. Zepka · E. Jacob-Lopes (✉)
Bioprocess Intensification Group, Federal University of Santa Maria, UFSM, Santa Maria, RS, Brazil

R. Raja et al. (eds.), *Haematococcus*, https://doi.org/10.1007/978-981-99-2901-6_11

than that of β-carotene, canthaxanthin, zeaxanthin, and lutein, and 100 times stronger than that of α-tocopherol (Shah et al. 2016; Mularczyk et al. 2020).

The internal structure gives astaxanthin from *Haematococcus pluvialis* a series of biological functions, such as antioxidant, anti-aging, anti-inflammatory, antihypertensive, antitumor, and anticancer, which makes it possible to expand its use in pharmaceutical and cosmetic products (Ren et al. 2021). About the global *Haematococcus pluvialis* market, it is expected to reach approximately UDS 148.1 million by 2027, at a compound annual growth rate of 13.2% during the forecast period from 2020 to 2027. The increasing demand for natural astaxanthin for pharmaceuticals and cosmetics as well as the growing demand for natural food coloring, are some of the factors that keep this market solidly anchored. Furthermore, it is expected to the increasing use of natural astaxanthin in animal nutrition create profitable opportunities (Meticulous Research® 2021). However, regardless of this, the high cost associated with upstream and downstream processing can hamper, to some extent, the growth of the *Haematococcus pluvialis* market.

Besides that, what may also limit the growth of this market is the environmental footprint associated with the process of obtaining astaxanthin. This is because the large-scale cultivation of astaxanthin from *Haematococcus pluvialis* raises environmental concerns due to high energy consumption, especially because the overexploitation of fossil energy resources has contributed to global warming and climate change (Deprá et al. 2020b). About this, sustainability, in the environmental sense, is now an imperative for competitiveness and perhaps even survival of microalgae-based processes and products (Dias et al. 2022a). Therefore, in this chapter, the environmental sustainability indicators and metrics of astaxanthin production processes from *Haematococcus pluvialis* are critically addressed. We provide an overview of the commercial production process and discuss prospects for a more sustainable operation that meets the demand for sustainability.

11.2 State of the Art

Microalgal biotechnology has shown to be a promising route for astaxanthin production. Currently, *Haematococcus pluvialis* leads the commercial production of this natural pigment. However, the current state of the art of astaxanthin production by the microalgal *Haematococcus pluvialis* is surrounded by challenges that still need to be overcome (Mussagy et al. 2022). To begin with, the conditions suitable for cell growth and astaxanthin accumulation are substantially far-off (Yu et al. 2021). This is because the biochemical synthesis of microalgal astaxanthin is generally induced as a defense mechanism against environmental oxidative stress. Therefore, for more significant accumulation, it is necessary to subject microalgal cells to stress-inducing conditions, such as nutrient shortages, exposure to intense light, high temperatures, and high salinity (Fitriana et al. 2021). On the other hand, cellular biomass production has high nutritional requirements, in addition to strict cultivation control, to obtain high rates of cellular productivity (Marinho et al. 2022).

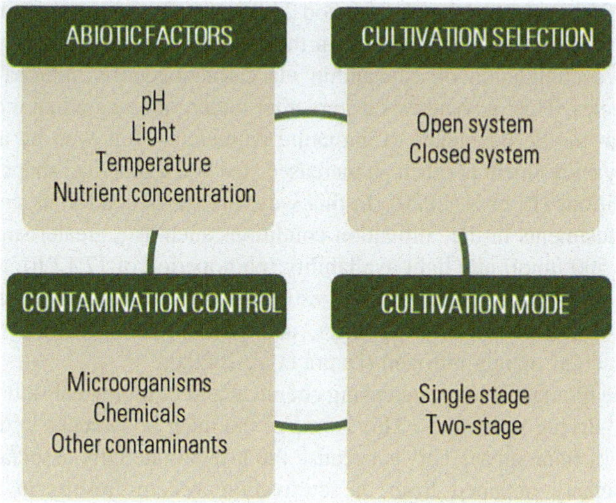

Fig. 11.1 Parameters tested to obtain a high-performance culture for astaxanthin production by *Haematococcus pluvialis*

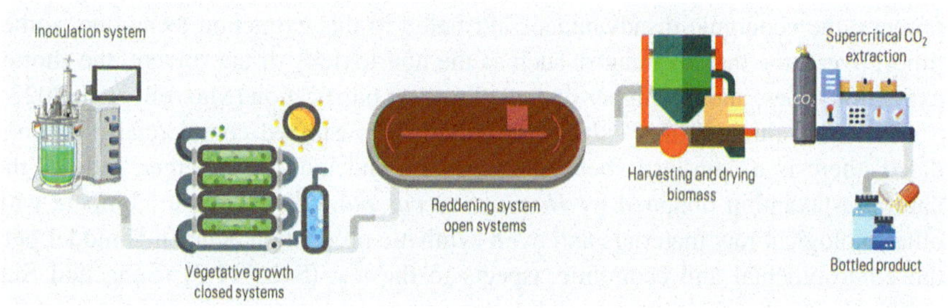

Fig. 11.2 Flowchart of industrial astaxanthin production by *Haematococcus pluvialis*

In this way, trade-offs from the cultivation process to commercial facilities were scientifically tested to obtain a consolidated technological route (Fig. 11.1).

Currently, commercial astaxanthin production facilities adopt two-stage cultivation models: (a) green vegetative and (b) red hematocyst (Fig. 11.2). At first, motile flagellates proliferate under low-stress conditions. That is, microalgae cells are grown under optimal growth conditions for a high accumulation of biomass. Technically, this step takes place in closed cultivation systems, more specifically in tubular photobioreactors, in order to have greater control of the process. In addition, closed systems prevent cellular contamination, since *Haematococcus pluvialis* strains are easily contaminated with other fast-growing strains such as *Scenedesmus* or *Chlorella* genus and, consequently, can interfere with product purity. Therefore, the ideal cultivation conditions in commercial photobioreactor systems are established at a pH of 8.0, the controlled temperature of ~20 °C, light intensity adjusted to 250 μmol, and the photoperiod 24:0 hours. It is important to note that this phase lasts on average 2 days. Later, in the red hematocyst stage, vegetative cells initiate the accumulation of astaxanthin via carotenogenesis induced by reactive oxygen species. This biosynthesis occurs after undergoing a gradual morphological change from motile flagellates to immature palmeloids, followed by aplanospores maturing by encystment to reach an increased size and a spherical shape under high-stress conditions (Li et al. 2020). In this way, the stress conditions are modulated through adjustments in the cultivation conditions such as a greater supply of light intensity (~500 μmol) and light availability (photoperiod of 12:12 h), as well as an increase in temperature (~27 °C). It is estimated that the carotenogenesis phase is carried out for a period of 5–10 days, and is associated with the depletion of nutrients, such as mainly nitrogen (Deprá et al. 2020a).

Then, the biomass is harvested using chemical and/or biological settlers and dried using spray-drying techniques. This is one of the main bottlenecks in the industrial production of microalgae. This is because the low productivity associated with the low cell density obtained from the cultivation systems results in downstream processing with high energy demand, representing about 60% of the total costs (Deprá et al. 2020b). Subsequently, when dried whole biomass is obtained, the technique applied to extract the pigments is attributed to extraction by supercritical fluid, using carbon dioxide as a solvent (Aravena et al. 2022). In the industrial process, the economic disadvantages attributed to this extraction technique, sometimes, overcome the advantages, such as the non-toxicity of the solvent, the shorter extraction times, and the higher degree of extract purification (Mazzelli et al. 2022).

However, although this cultivation model is considered commercially consolidated, there is a consensus between academia and industry whether, indeed, the natural astaxanthin obtained by *Haematococcus pluvialis* is able to compete with other biological raw materials and even synthetic production, since this model puts the environmental and economic aspects to the test (Sect. 11.3) (Sung and Sim 2022). Thus, the most recent studies have focused on strategies to increase the productivity of this pigment, in order to alleviate the burden of commercial production in the face of process requirements. Therefore, the main niches of investigation have focused on the application of data modeling, such as multivariate statistical analysis, which has been used to assess the astaxanthin accumulation capacity of *Haematococcus pluvialis*. This small and important scientific advance provides the possibility to determine the optimal cellular phase of transfer from the green to the red stage, beyond allowing a direction for the optimization of cultivation in the green stage (Fei et al. 2022a). At the same time, under this culture optimization bias, the base criterion for cleaning and disinfecting culture systems was investigated in order to avoid harmful contaminants. By establishing a model of degradation kinetics of chemical disinfectants, it was possible to reduce contaminants and increase astaxanthin volume by ~3 times more than non-optimized strategies (Fei et al. 2022b).

Additionally, the selection and improvement of strains may be the most pertinent step for the development of the industrial production process of astaxanthin by *Haematococcus pluvialis* (Basiony et al. 2022). Studies are emerging on the genetic modification of microalgae to optimize the production of isoprenoids such as astaxanthin (Sproles et al. 2021). Excessive accumulation of astaxanthin in *Haematococcus pluvialis* through genetic engineering of chloroplasts was confirmed by research carried out by Galarza et al. (2018), resulting in up to 90% greater accumulation of astaxanthin per culture volume. As for advances in the modulations of environmental cultivation conditions, the latest findings indicate that not only the availability and intensity of light interfere with astaxanthin productivity rates, but also the wavelength used (Lee et al. 2018; Ma et al. 2018).

Likewise, the chemical stability of inorganic nitrogen offered to microalgal microorganisms also interferes in the production of astaxanthin, as demonstrated by Ma et al. (2022). Finally, cell disruption and extraction of the target compound in algae-based processes are pointed out as notable technological and economic obstacles due to their high energy requirement, cost-intensive nature, and direct impacts on the quality, stability, and purity of the final pigment (Kim et al. 2022). To this end, the use of nanomaterials technology has been proposed, as it can provide promising options for the high-efficiency recovery of high-value-added biocompounds from cells that have thick and rigid cell walls (Moon et al. 2021). Throughout what has been exposed, it is clear that the efficient and effective cultivation of the microalgae *Haematococcus pluvialis* for the production of biological astaxanthin requires the alignment and balance of several biotechnological parameters, concomitant with the processes control. However, to date, the high-added value of these pure and very specific metabolites stands out against the risks of unstable industrial production. Still, it is important to emphasize that this industrial model of production was only established in favor of commercial demand needs. Therefore, we cannot consider it a plastered production model, because the scientific community is constantly evolving and elucidating the gaps in knowledge that shortly may bring these systems closer to an ideal standard process.

11.3 Potential Environmental Impacts

11.3.1 Ecosystem Quality Damage Category

The science is clear; we are living in a pathological system that threatens every ecosystem on which our lives depend. Furthermore, there is no doubt that human activity has been and continues to be responsible for increased toxicity in soil, water, and air, as well as eutrophication and acidification. The consequences of changes in the ecosystems that support life on Earth are already being felt and if nothing is done the result will be predictable. Evidently, what needs to be done is to go beyond states of denial, making a firm commitment to maintaining the health of the planet (Gills and Morgan 2020). In this sense, measuring the ecological footprint is necessary so that countries, local leaders, and individuals can understand its impact on the ecosystem and to what extent it can be changed. In the status quo, it is highlighted that humanity's demand for ecological goods and services has surpassed the Earth's regenerative capacity. We are consuming the equivalent of 1.7 lands and operating clearly outside ecological limits. In light of this, the challenge is to maintain a resource extraction rate that does not deplete the amount produced by the ecosystem (Hayden 2019).

Currently, companies and associations have increasingly invested in ecological metrics and indicators as guidelines for environmental management, thus seeking to build a more sustainable relationship with nature. With regard to indicators are concerned, there is no single eco-indicator that captures all aspects of the terrestrial and aquatic ecological system. There is a set of them that individually represent a specific operational attribute of the system, such as terrestrial ecotoxicity, freshwater ecotoxicity, marine ecotoxicity, terrestrial acidification, aquatic acidification, aquatic eutrophication, and photochemical ozone formation (Jacob-Lopes et al. 2021). In the ecosystem quality damage endpoint category, there are several individual midpoint indicators, such as those mentioned above, which together provide more significant results on environmental sustainability. However, the combination of these individual ecological performance indicators used depends, for example, on the product or service evaluated and reflects the relevant parts linked to the life cycle (Hardaker et al. 2022).

With regard to the life cycle of *Haematococcus pluvialis*, the focus of this chapter, its potential impacts on ecosystem quality, is also being investigated. This is because the premise that biotechnological pathways of *Haematococcus pluvialis*-based products provide greener alternatives when compared to chemical synthesis pathways has become questionable amid growing recognition of the high demand for natural capital. That said, although the assessment of the life cycle of *Haematococcus pluvialis* products strives to be from the cradle to the grave, that is, from the extraction of resources to the final destination, it is important to mention that most studies, in the state of the art, it is based on a partial lifecycle assessment from resource extraction (cradle) to factory gate (completion of production) (Pérez-López et al. 2014; Peter et al. 2022).

The partial life cycle assessment studies of products based on *Haematococcus pluvialis* show that the energy demand of the upstream and downstream processes is responsible for most of the environmental impacts (about 98%). Indirect emissions from the energy generation that feeds these processes are the main villain in the life cycle of these products (Deprá et al. 2020b). In summary, according to the literature data, the natural production of astaxanthin from the microalgae *Haematococcus pluvialis* carries a greater environmental burden in the eco-indicators of acidification (kg SO_2-eq), eutrophication (kg N-eq), potential photochemical ozone creation (kg NMVOC-eq), and ecotoxicity (CTUe-eq) when compared to synthetic astaxanthin production. The emission value in these eco-indicators is about 210, 213, 5, and 209 times higher, respectively (Table 11.1) (Deprá et al. 2020a).

Clearly, the assumption that natural *Haematococcus pluvialis* pigments are a more sustainable and planet-friendly alternative is far from reality. However, both the synthetic pigment and the base of *Haematococcus pluvialis* are within planetary boundaries. In other words, they do not go beyond the environmental limits within which humanity can safely operate; however, this does not exclude that they are

Table 11.1 Environmental footprint associated with the production of 1 kg of synthetic astaxanthin and based on *Haematococcus pluvialis*

	Biological route	Synthetic route
Ecosystem quality damage category		
Acidification (kg SO_2-eq)	$1.49 \times 10^{+1}$	7.10×10^{-2}
Eutrophication (kg N-eq)	$5.34 \times 10^{+0}$	2.50×10^{-2}
Potential photochemical ozone creation (kg NMVOC-eq)	$6.27 \times 10^{+1}$	$1.30 \times 10^{+1}$
Ecotoxicity (CTUe-eq)	$1.97 \times 10^{+4}$	$9.44 \times 10^{+1}$
Resources consumption damage category		
Waterfootprint (m^3)	$8.30 \times 10^{+1}$	1.00×10^{-2}
Land use and land cover (PDF/ha)	$9.90 \times 10^{+0}$	9.60×10^{-1}
Human health damage category		
Ozone depletion (kg CFC-11-eq)	8.5×10^{-4}	4.00×10^{-6}
Climate change damage category		
Global warming potential (kg CO_2-eq)	$2.80 \times 10^{+3}$	$3.66 \times 10^{+1}$

Adapted from Deprá et al. (2020a)

energy and material-intensive processes, consequently, they are not future-proof (Steffen et al. 2015). It is also worth noting that being sustainable goes beyond environmental and economic metrics. From the perspective of social indicators that also constitute the tripod of sustainability, the consumption of natural pigments offers clear benefits, with solid clinical evidence (Acién et al. 2017).

11.3.2 Resources Consumption Damage Category

Equally, but not least, the upstream and downstream processes of *Haematococcus pluvialis* have potential impacts on the resource consumption damage category, which include midpoints such as land footprint and water footprint. These midpoints attempt to measure the sustainability of the resource consumption endpoint category as a whole, which is evidently not an easy task given the complexity of the issues this category poses (Martins et al. 2018; Chen et al. 2022). However, regardless of this, it is known that indirect emissions from power generation are also associated with resource consumption in the life cycle of *Haematococcus pluvialis* products. Not only that, the production process itself requires land use and cover and direct water use to a great extent (Onorato and Rösch 2020). In the literature, water and land footprint values of 83.00 m^3 and 9.90 PDF/ha, respectively, are reported for the production of astaxanthin from *Haematococcus pluvialis*. This is considering the production of 1 kg of product. When considering the water and land footprint of synthetic astaxanthin production, these values are around 0.01 m^3 and 0.96 PDF/ha (Table 11.1). As previously mentioned, the production of astaxanthin by *Haematococcus pluvialis* requires a two-step system, which may be linked to the higher results in these indicators compared to the synthetic production of this pigment (Deprá et al. 2020a).

Besides that, it is worth mentioning that if we consider the use of the green and gray footprint in the upstream and downstream processes of *Haematococcus pluvialis*, the impacts in this category can be mitigated or nullified. By definition, the gray footprint refers to the use of wastewater and the green footprint refers to the use of rainwater. Noteworthy, beyond reducing drinking water consumption, some studies reveal significant potential for cost reduction of up to 60% (Xie et al. 2019; Ma et al. 2020). Finally, with regard to the land footprint, it is also worth remembering that if the use of non-arable land is considered, the footprint may be null. Also, it is worth mentioning that they will avoid competition with food production. Notably, it would be pertinent that the lands elected for *Haematococcus pluvialis* installations were infertile and no-cultivable (Boruff et al. 2015; Nagappan et al. 2021).

11.3.3 Human Health Damage Category

The impact of *Haematococcus pluvialis* products on the human health damage category that includes midpoints such as ozone depletion (SOD), ionizing radiation (IR), fine particulate matter formation (FPMF), human carcinogenic toxicity (HCT), and human non-carcinogenic toxicity (HNCT) still needs an in-depth assessment in the context of the cradle-to-grave life cycle assessment. However, regardless of this, it is clear that the upstream and downstream processes of *Haematococcus pluvialis* have routine releases of radioactive material, fine particulate matter, 1,4-dichlorobenzene, chlorofluorocarbon, and other compounds which come from the energy source that feeds these processes (Bulle et al. 2019). About this, as an example, recently Dias et al. (2022b) evaluated the cradle-to-gate life cycle of the photovoltaic-powered *Haematococcus pluvialis* astaxanthin production process of four distinct solar cell technologies, namely single crystalline silicon (sc-Si), multi-crystalline silicon (mc-Si), amorphous silicon (a-Si), and copper-indium-diselenide (CIS). According to the authors' results, the production of 1 t $month^{-1}$ of astaxanthin from these four solar cell technologies had impacts from 5.55×10^{-2} to 7.7×10^{-2} kg CFC11-eq for SOD, $1.25 \times 10^{+4}$ to $1.41 \times 10^{+4}$ kBq Co-60-eq for IR, $2.98 \times 10^{+2}$ to $3.90 \times 10^{+2}$ kg $PM_{2.5}$-eq for FPMF, $1.32 \times 10^{+4}$ to $2.06 \times 10^{+4}$ kg 1,4-DBC-eq for HCT and $4.63 \times 10^{+5}$ to $6.35 \times 10^{+5}$ kg 1,4-DBC-eq for HNCT (Table 11.2). It is noteworthy that the authors show that depending on the solar cell technology used for power generation that feeds the processes, impacts in all midpoint categories can be mitigated. In addition, it is worth mentioning that the HNCT impact category was the most influential midpoint among all midpoints that make up the harm to human health category.

Through the data mentioned above, it is clear that indirect emissions from the energy source, responsible for most of the environmental impacts of these processes, can be more or less significant according to the choice of technology that generates

Table 11.2 Environmental footprint associated with the production of 1 t month^{-1} of astaxanthin from *Haematococcus pluvialis* by solar cell technology. Amorphous silicon, a-Si; Copper-indium-diselenide, CIS; Multi-crystalline silicon, mc-Si; Single crystalline silicon, sc-Si; stratospheric ozone depletion, SOD; ionizing radiation, IR; fine particulate matter formation, FPMF; human carcinogenic toxicity, HCT; human non-carcinogenic toxicity, HNCT. (Adapted from Dias et al. (2022b))

	SOD (kg CFC11-eq)	IR (kBq Co-60-eq)	FPMF (kg $PM_{2.5}$-eq)	HCT (kg1,4-DBC-eq)	HNCT (kg1,4-DBC-eq)
a-Si	5.39×10^{-2}	$1.25 \times 10^{+4}$	$3.46 \times 10^{+2}$	$2.06 \times 10^{+4}$	$4.63 \times 10^{+5}$
CIS	5.55×10^{-2}	$1.26 \times 10^{+4}$	$2.98 \times 10^{+2}$	$1.33 \times 10^{+4}$	$6.35 \times 10^{+5}$
mc-Si	6.54×10^{-2}	$1.29 \times 10^{+4}$	$3.30 \times 10^{+2}$	$1.32 \times 10^{+4}$	$5.17 \times 10^{+5}$
sc-Si	7.77×10^{-2}	$1.41 \times 10^{+4}$	$3.90 \times 10^{+2}$	$1.37 \times 10^{+4}$	$5.20 \times 10^{+5}$

them. It is also important to mention here that the environmental sustainability metrics for the midpoints of the harm to human health category for astaxanthin production from *Haematococcus pluvialis* and the derivative of the synthetic route are also, according to the literature, proportionally different. While the production of 1 kg of astaxanthin from *Haematococcus pluvialis* has an impact of 0.00085 kg CFC-11-eq for SOD, the production of synthetic astaxanthin has values of 0.0000040 kg CFC-11-eq (Table 11.1). Evidently, as already mentioned, being sustainable is going beyond environmental metrics (Deprá et al. 2020a).

11.3.4 Climate Change Damage Category

Climate change is causing tragic environmental, economic, and social consequences. The global average temperature is increasing dramatically. And on this horizon, it is extremely important that we join forces to follow letter the Paris Agreement, which aims to keep the rise in the planet's temperature well below 2 °C (Overpeck and Conde 2019; Meinshausen et al. 2022). In this regard, scaling greenhouse gas emissions is a key step toward achieving the objective of minimizing associated climate risks (Tanti Watthanaphanich et al. 2022).

In addition, in the face of environmental collapse, it cannot be neglected that the food segment is one of the pillars of this ruin and inevitably needs to undergo transformative changes so that it can contribute to a future in which we live well, within planetary boundaries. In this chapter, among the four endpoints addressed, the most relevant in the assessment of the life cycle of Haematococcus pluvialis processes and products is the climate change damage category. This one comprises the global warming potential midpoint (CO_2-eq), assessed by greenhouse gas emissions (CO_2, N_2O, and CH_4) (Herrera et al. 2021). About this indicator, it is notable that the production of astaxanthin in front of production synthetic production is not future proof. This is because, according to the literature data, CO_2 emission values for 1 kg of astaxanthin from *Haematococcus pluvialis* are 2804.52 kg CO_2-eq and 36.62 kg CO_2-eq, respectively (Table 11.1) (Deprá et al. 2020a).

However, it is worth mentioning that the data mentioned above refer to the use of high CO_2 emission fossil energy. In this sense, it is clear that the replacement of the high-emission fossil matrix for low-emission sources is the backbone for carrying out less carbon-intensive operations. Recently, Dias et al. (2022a) when evaluating the replacement of coal-based energy use by nuclear, hydro, wind, biomass, and photovoltaic found a reduction in CO_2 emissions of around 94–99%. For energy sources from biogas, natural gas, and waste, the values are around 82–84%, 49–51%, and 45–46%, in that order. In light of these values, it is evident that the production of *Haematococcus pluvialis* can be guided by the principle of providing more sustainable and ecologically correct products. Finally, it is worth noting that the replacement of energy with high emission of greenhouse gases by low-emission sources does not only impact the category of global warming potential. The impact of replacement is dizzying for all midpoints of all endpoints mentioned here.

11.4 Conclusion

The ability of *Haematococcus pluvialis* to naturally synthesize astaxanthin with biological activities relevant to health makes it a bioresource with full potential for exploitation in the bioactive compounds industry. However, it should be noted that several studies' life cycle assessments have shown the environmental weaknesses of the processes for obtaining this natural pigment in the face of chemical synthesis pathways. The use of clean and renewable energy would be one of the crucial points to overcome the vulnerability of the upstream and downstream stages of *Haematococcus pluvialis* processing. Process intensification and integration, as well as the implementation of the biorefinery concept, are also promising strategies to improve environmental sustainability metrics. Finally, as the climate clock is ticking, any and all initiatives that make the microalgae biotechnology industry more sustainable and clean are welcome.

References

Acién FG, Molina E, Fernández-Sevilla JM, Barbosa M, Gouveia L, Sepúlveda C, Arbib Z (2017) Economics of microalgae production. In: Microalgae-basedbiofuels and bioproducts. Wood head Publishing, pp 485–503

Aravena RI, del Valle JM, Juan C (2022) Supercritical CO_2 extraction of aqueous suspensions of disrupted *Haematococcus pluvialis* cysts. J Supercrit Fluids 181:105392

Basiony M, Ouyang L, Wang D, Yu J, Zhou L, Zhu M, Zhang L (2022) Optimization of microbial cell factories for astaxanthin production: biosynthesis and regulations, engineering strategies and fermentation optimization strategies. Synth Syst Biotechnol 7(2):689–704

Boruff BJ, Moheimani NR, Borowitzka MA (2015) Identifying locations for large-scale microalgae cultivation in Western Australia: a GIS approach. Appl Energy 149:379–391

Bulle C, Margni M, Patouillard L, Boulay AM, Bourgault G, De Bruille V, Jolliet O (2019) IMPACT world+: a globally regionalized life cycle impact assessment method. Int J Life Cycle Assess 24(9):1653–1674

Chen C, Tang T, Shi Q, Zhou Z, Fan J (2022) The potential and challenge of microalgae as promising future food sources. Trends Food Sci Technol 126:99

Deprá MC, Santos AMD, Jacob-Lopes E (2020a) Sustainability metrics in the microalgae-based pigments production: a life cycle assessment approach. In: Pigments from microalgae handbook. Springer, Cham, pp 363–390

Deprá MC, Severo IA, dos Santos AM, Zepka LQ, Jacob-Lopes E (2020b) Environmental impacts on commercial microalgae-based products: sustainability metrics and indicators. Algal Res 51: 102056

Dias RR, Deprá MC, Zepka LQ, Jacob-Lopes E (2022a) Roadmap to net-zero carbon emissions in commercial microalgae-based products: environmental sustainability and carbon offset costs. J Appl Phycol 34(3):1255–1268

Dias RR, Deprá MC, Sartori RB, Schneider AT, Zepka LQ, Jacob-Lopes E (2022b) Distributed-generation photovoltaic systems in microalgae facilities: the choice of solar cell technology. Sustain Energy Technol Assess

Fei Z, Liao J, Fan F, Wan M, Bai W, He M, Li Y (2022a) Improving astaxanthin production by using multivariate statistical analysis to evaluate green cells of *Haematococcus pluvialis*. Biochem Eng J 183:108456

Fei Z, Fan F, Liao J, Wan M, Bai W, Wang W, Li Y (2022b) Improving astaxanthin production of *Haematococcus pluvialis* on the outdoor large scale cultivation by optimizing the disinfection strategy of photobioreactor. Algal Res 64:102708

Fitriana HN, Lee SY, Choi SA, Lee JY, Kim BL, Lee JS, Oh YK (2021) Electric stimulation of astaxanthin biosynthesis in *Haematococcus pluvialis*. Appl Sci 11(8):3348

Galarza JI, Gimpel JA, Rojas V, Arredondo-Vega BO, Henríquez V (2018) Over-accumulation of astaxanthin in *Haematococcus pluvialis* through chloroplast genetic engineering. Algal Res 31: 291–297

Gills B, Morgan J (2020) Global climate emergency: after COP24, climate science, urgency, and the threat to humanity. Globalizations 17(6):885–902

Hardaker A, Styles D, Williams P, Chadwick D, Dandy N (2022) A framework for integrating ecosystem services as endpointimpacts in life cycle assessment. J Clean Prod 370:133450

Hayden A (2019) Ecological foot print. Encyclopedia Britannica

Herrera A, D'Imporzano G, Fernandez FGA, Adani F (2021) Sustainable production of microalgae in raceways: nutrients and water management as key factors influencing environmental impacts. J Clean Prod 287:125005

Jacob-Lopes E, Zepka LQ, Depra MC (2021) Sustainability metrics and indicators of environmental impact: industrial and agricultural life cycle assessment. Elsevier, Amsterdam

Kim B, Lee SY, Narasimhan AL, Kim S, Oh YK (2022) Cell disruption and astaxanthin extraction from *Haematococcus pluvialis*: recent advances. Bioresour Technol 343:126124

Lee C, Ahn JW, Kim JB, Kim JY, Choi YE (2018) Comparative transcriptome analysis of *Haematococcus pluvialis*on astaxanthin biosynthesis in response to irradiation with red or blue LED wavelength. World J Microbiol Biotechnol 34(7):1–14

Li J, Zhu D, Niu J, Shen S, Wang G (2011) An economic assessment of astaxanthin production by large scale cultivation of *Haematococcus pluvialis*. Biotechnol Adv 29(6):568–574

Li X, Wang X, Duan C, Yi S, Gao Z, Xiao C, Li J (2020) Biotechnological production of astaxanthin from the microalga *Haematococcus pluvialis*. Biotechnol Adv 43:107602

Ma R, Thomas-Hall SR, Chua ET, Eltanahy E, Netzel ME, Netzel G, Schenk PM (2018) Blue light enhances astaxanthin biosynthesis metabolism and extraction efficiency in *Haematococcus pluvialis* by inducing haematocyst germination. Algal Res 35:215–222

Ma W, Opp C, Yang D (2020) Past, present, and future of virtual water and waterfootprint. Water 12(11):3068

Ma R, Tao X, Chua ET, Ho SH, Shi X, Liu L, Chen J (2022) Enhancing astaxanthin production in *Haematococcus pluvialis* QLD by a pH steady $NaHCO_3$-CO_2-C/NH_4Cl-N culture system. Algal Res 64:102697

Marinho YF, Oliveira CYB, Malafaia CB, Cahú TB, Oliveira APS, Napoleão TH, Gálvez AO (2022) A circular approach for the efficient recovery of astaxanthin from *Haematococcus pluvialis* biomass harvested by flocculation and water reusability. Sci Total Environ 841:156795

Maroneze MM, Dias RR, Severo IA, Queiroz MI (2020) Microalgae-based processes for pigments production. In: Pigments from microalgae handbook. Springer, Cham, pp 241–264

Martins AA, Marques F, Cameira M, Santos E, Badenes S, Costa L, Mata TM (2018) Waterfootprint of microalgae cultivation in photobioreactor. Energy Procedia 153:426–431

Mazzelli A, Valentini M, Cicci A, Iaquaniello G, Bravi M (2022) Industrial bio-fractionation process of microalgae valuable products using supercritical CO_2. A techno-economical evaluation. Chem Eng Res Des 178:50–60

Meinshausen M, Lewis J, McGlade C, Gütschow J, Nicholls Z, Burdon R, Hackmann B (2022) Realization of Paris agreement pledges may limit warming just below 2°C. Nature 604(7905): 304–309

Meticulous Research® (2021). https://www.meticulousresearch.com/product/haematococcus-pluvialis-market-5142. Accessed 13 Sept 2022

Moon G, Lee N, Kang S, Park J, Kim YE, Lee SA, Chung S (2021) Hydrothermal synthesis of novel two-dimensional α-quartz nanoplates and their applications in energy-saving, high-efficiency, microalgal biorefineries. Chem Eng J 413:127467

Mularczyk M, Michalak I, Marycz K (2020) Astaxanthin and other nutrients from *Haematococcus pluvialis*-multifunctional applications. Mar Drugs 18(9):459

Mussagy CU, Pereira JF, Santos-Ebinuma VC, Pessoa A Jr, Raghavan V (2022) Insights into using green and unconventional technologies to recover natural astaxanthin from microbial biomass. Crit Rev Food Sci Nutr 1–15

Nagappan S, Das P, AbdulQuadir M, Thaher M, Khan S, Mahata C, Kumar G (2021) Potential of microalgae as a sustainable feed ingredient for aquaculture. J Biotechnol 341:1–20

Onorato C, Rösch C (2020) Comparative life cycle assessment of astaxanthin production with *Haematococcus pluvialis* in different photobioreactor technologies. Algal Res 50:102005

Oslan SNH, Shoparwe NF, Yusoff AH, Rahim AA, Chang CS, Tan JS, Mohamed MS (2021) A review on *Haematococcus pluvialis* bioprocess optimization of green and red stage culture conditions for the production of natural astaxanthin. Biomol Ther 11(2):256

Overpeck JT, Conde C (2019) A calltoclimate action. Science 364(6443):807–807

Pérez-López P, González-García S, Jeffryes C, Agathos SN, McHugh E, Walsh D, Moreira MT (2014) Life cycleassessment of the production of the red antioxidant carotenoid astaxanthin by microalgae: from lab to pilot scale. J Clean Prod 64:332–344

Peter AP, Tan X, Lim JY, Chew KW, Koyande AK, Show PL (2022) Environmental analysis of *Chlorella vulgaris* cultivation in largescale closed system under waste nutrient source. Chem Eng J 433:134254

Ren Y, Deng J, Huang J, Wu Z, Yi L, Bi Y, Chen F (2021) Using green alga *Haematococcus pluvialis* for astaxanthin and lipid co-production: advances and outlook. Bioresour Technol 340: 125736

Shah MMR, Liang Y, Cheng JJ, Daroch M (2016) Astaxanthin-producing green microalga *Haematococcus pluvialis*: from single cell to high value commercial products. Front Plant Sci 7:531

Sproles AE, Fields FJ, Smalley TN, Le CH, Badary A, Mayfield SP (2021) Recent advancements in the genetic engineering of microalgae. Algal Res 53:102158

Steffen W, Richardson K, Rockström J, Cornell SE, Fetzer I, Bennett EM, Sörlin S (2015) Planetaryboundaries: Guidinghum an development on a changing planet. Science 347(6223): 1259855

Sung YJ, Sim SJ (2022) Multifaceted strategies for economic production of microalgae *Haematococcus pluvialis*-derived astaxanthin *via* direct conversion of CO_2. Bioresour Technol 344:126255

Tanti Watthanaphanich T, Shao X, Huang L, Yoshida Y, Long Y (2022) Evaluating carbon footprint embodied in Japanese food consumption based on global supply chain. Struct Change Econ Dyn 63:56

Xie B, Gong W, Tang X, Bai L, Guo Y, Wang J, Liang H (2019) Blending high concentration of anaerobic digestion effluent and rainwater for cost-effective *Chlorella vulgaris* cultivation in the photobioreactor. Chem Eng J 360:861–865

Yu BS, Hong ME, Sung YJ, Choi HI, Chang WS, Kwak HS, Sim SJ (2021) A green decontamination technology through selective biomineralization of algicidal microorganisms for enhanced astaxanthin production from *Haematococcus pluvialis* at commercial scale. Bioresour Technol 332:125121

Part IV
Therapeutic and Clinical Applications

Chapter 12
Therapeutic Potential of *Haematococcus pluvialis* in the Field of Drug Delivery

T. S. Kumar, A. Josephine, and G. Dharani

12.1 Introduction

Today, microalgae are becoming more and more popular due to their high nutritional value as well as its recognition as a healthy diet (Ruiz-Domínguez et al. 2019). In the food and pharmaceutical industries, proteins including enzymes, amino acids, and polypeptides finds indispensable applications (Harasym and Bogacz-Radomska 2016). Interestingly, microalgae act as a source of active biomolecules for the pharmaceutical industry (Molino et al. 2018). Researchers are continually proving the health benefits of microalgal compounds and their biological features (Nawrocka et al. 2017). Numerous bioactive substances with potent antibacterial, antifungal, and antiviral characteristics are widely used to make vaccines, medicines, agrochemicals, and cosmetics (Specht et al. 2014). Microalgae exert a key role in the production of growth factors, hormones, antibodies, vaccines, and immunological regulators in the medical and pharmaceutical industries (Aw et al. 2012; Yan et al. 2016).

Astaxanthin (AST) is a type of xanthophyll carotenoid that is commonly found in plants, microorganisms, and natural seafood such as salmon, shrimp, lobster, and crab (Johnson and An 2008). *Haematococcus pluvialis* (*H. pluvialis)* is one of the most important microalgal species, containing AST, which is hailed as the most powerful natural antioxidant (Shah et al. 2016). It is impossible to discuss *H. pluvialis* without mentioning AST. About 95% of the AST produced worldwide comes from synthetic manufacture (Jannel et al. 2020). But of all the natural AST

T. S. Kumar (✉) · G. Dharani
Marine Biotechnology, National Institute of Ocean Technology, Ministry of Earth Sciences, Government of India, Chennai, Tamil Nadu, India

A. Josephine (✉)
Central Research Laboratory, Meenakshi Academy of Higher Education and Research (MAHER – Deemed to be University), Chennai, Tamil Nadu, India

R. Raja et al. (eds.), *Haematococcus*, https://doi.org/10.1007/978-981-99-2901-6_12

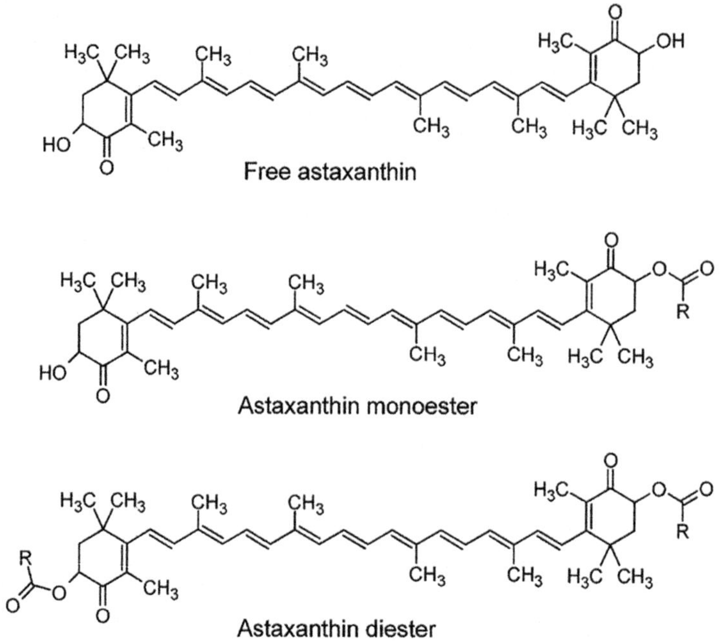

Fig. 12.1 Structure of free and esterified form of Astaxanthin (Miao et al. 2006)

derivatives that are currently available, most of them are only derived from the freshwater microalga *H. pluvialis* which has received approval from the Food and Drug Administration's for human use (Jannel et al. 2020). The Trolox equivalent antioxidant capacity (TEAC) assay confirmed the higher antioxidant capacity of natural extracts containing AST esters over free AST. The structure of free and esterified form of AST are given below, which is adopted from (Miao et al. (2006) (Fig. 12.1).

Importantly, when tested on human endothelial cells (HUVEC), *H. pluvialis* extract was found to exhibit approximately 90 times the intracellular antioxidant activity of synthetic AST (Régnier et al. 2015). As a result, one of the major goals pertaining to this microalga is the optimization of extraction techniques for maximum AST recovery (Mularczyk et al. 2020). Furthermore, its biomass contains a high proportion of proteins, essential vitamins, and essential minerals such as magnesium, iron, calcium, and zinc with lower fat content. Studies on AST as a potent therapeutic agent have yielded positive results in the last 10 years, with potential applications in the treatment of aging, cardiovascular, neurological, and cancer diseases. Majority of these illnesses are linked to inflammation, which is exerted by the interaction of free radicals with nucleic acids and proteins in the body. The ability of AST to quench singlet oxygen and scavenge reactive species is responsible for the protective action as anti-inflammatory agent (Brotosudarmo et al. 2020). According to numerous studies (Wang et al. 2015; Wang et al. 2017a; Tseng et al. 2020), AST has a variety of beneficial effects, including the reduction of oxidative stress, improvement of the immune system, treatment of cardiovascular disease, antibacterial property, improvement of macular degeneration, anti-inflammatory activity, and inhibition of certain cancers. As an interesting note, AST administration to COVID-19 patients was observed to significantly reduce cytokine storm, prevent ARDS (acute respiratory distress syndrome) and ALI (acute lung injury), according to a report by Talukdar et al. (2020).

Drug delivery mechanisms are created to direct medications or genes toward specific cell types, such as cancerous cells. Generally, delivery systems can help improve the physicochemical profile of this compound, such as stability, water solubility, antioxidant properties, drug release, and in vivo and in vitro biological activities (Lima et al. 2021). However, despite the fact that AST is a potent xanthophyll with many health benefits, its therapeutic applications have been limited by its insolubility. Due to this disadvantage, several studies have been conducted in the recent years to enhance the efficacy and bioavailability of AST on numerous paradigms (Anarjan and Ping Tan 2013; Bharathiraja et al. 2017; Wang et al. 2017b). Modern drug delivery methods have made a significant effort to increase the bioavailability, stability, and solubility of this potent antioxidant. This chapter highlights current advancements in drug delivery as well as the therapeutic potential of *H. pluvialis*.

12.2 Therapeutic Properties of *H. pluvialis*

12.2.1 Antioxidant Properties

Oxidative stress is defined as a prooxidant-antioxidant imbalance that causes macromolecular damage and disrupts redox signaling and cellular control (Sies et al. 2017). Prooxidants are substances that contribute to the formation of reactive oxygen species (ROS), which then destroy cellular macromolecules by inducing protein oxidation, lipid peroxidation, and DNA damage (Wu and Cederbaum 2003). Antioxidants, on the other hand, reduce oxidative stress by counteracting or reducing the effects of ROS (Valko et al. 2007). Antioxidants are essential components of foods and cosmetics owing to its fundamental functions in reducing radical-induced degradation and scavenging free radicals in living cells and tissues (Haung et al. 2020). Most importantly, AST due to its high antioxidant capacity has received the name, "Superstar of Antioxidant" (Shah et al. 2016). AST has exceptional antioxidant activity, and several authors have already described its positive effects on human health, including those against UV-induced skin damage, which can lead to photoaging and photo carcinogenesis (Verschooten et al. 2006; Camera et al. 2009; Suganuma et al. 2010; Zhang et al. 2014). AST is a functional food component that has been widely used in the food industry due to its antioxidant properties (Kang and Sim 2008). Although AST has the highest antioxidant activity, continuously developed reddish palmelloids cells that are rich in oleic acid and AST have the highest antioxidant capacity and appear to be the most suitable for human consumption, in nutraceuticals, or in medications (Cerón et al. 2007). Total AST was found to have the highest concentration of all carotenoids in *H. pluvialis*, followed by β -carotene, canthaxanthin, and lutein. Essentially, AST's antioxidant capacity is 38-fold greater than that of β -carotene and 500 times greater than that of vitamin E (Pan-utai et al. 2021).

Phenolic compounds are one of the most important classes of natural antioxidants, and their health benefits are gaining popularity among consumers and food producers (Machu et al. 2015). In line with this, *H. pluvialis* has been found to have higher phenolic contents than the other microalgal species such as *Tetraselmis suecica, Arthrospira platensis, Scenedesmus obliquus,* and *Phaeodactylum tricornutum* (Batista et al. 2017; Gilbert-López et al. 2017; Ruiz-Domínguez et al. 2019). The oleoresin present in *H. pluvialis* contains 74.08 mg of gallic acid equivalent (GAE)/g of oleoresin extract (OE), which is slightly higher than other direct results for the microalga reported. A few studies have reported *H. pluvialis* oleoresin's antioxidant capacity, but these studies were conducted in microalgal extracts (Ruiz-Domínguez et al. 2019).

12.2.1.1 Various Drug Delivery Modes of AST Against Oxidative Stress

Several attempts have been adopted to enhance the drug delivery efficiency of such potent antioxidant into human system. Encapsulation of *H. pluvialis* carotenoid-rich extract into biodegradable PLGA (Poly lactic-co-glycolic acid) nanocapsules has been fortified with the goal of producing an innovative pharmaceutical formulation with high antioxidant properties. Apart from their nanometric size, which may increase the intracellular capture of encapsulated substances, nanocapsules enable the transport of water-insoluble substances with high levels of entrapment efficiency, protection of the active substance against oxidation due to the presence of the polymer coating, and the control of drug release rate (Guterres et al. 2007; Mora-Huertas et al. 2010). The feasibility of encapsulating an *H. pluvialis* carotenoid-rich extract into PLGA nanocapsules was demonstrated by Vieira et al. 2021. When compared to ascorbic acid, the carotenoids in the nanocapsule formulation had a higher antioxidant potential. Technically, a thermosensitive hydrogel containing ESHp-loaded nanocapsules has been shown to have sustained carotenoid release properties on the skin as well as a prolonged antioxidant effect.

Liposomes are colloidal in nature and hence, vesicular delivery systems are made up of at least one bilayer as amphiphilic lipid membrane and a hydrophilic core (Costa and Santos 2017). Liposomes can be used as a medication delivery vehicle, partially resolving the issues with AST's poor stability, water solubility, and bioavailability (Pan et al. 2018). In a study (Dopierała et al. 2020), it was found that AST can be incorporated into the lipid monolayer to produce a stable film that regulates membrane fluidity and enhances the stability of the liposomal delivery system. It was also observed that in this study, the AST-lipid membrane interaction though decreases the membrane's fluidity, increases its micro polarity within a specific range of AST concentration, thus improving the saturated lipid bilayer's stability. This approach looks to be a viable method for enhancing lipid-based medication formulations, creating a dosage form that could potentially assist in resolving the issue of unstable bioactive chemicals.

In order to overcome the flexibility limitations of vesicles in an aqueous solution, (Lee et al. 2017) developed liposomes with solid-supported lipid bilayers as an alternative liposomal delivery system. The hydrophilic regions of lecithin vesicles were silicified using tetraethyl orthosilicate to create silicified liposomes, which were then combined with boron nitride to increase their stability. According to the 2,2-diphenyl-1-picrylhydrazyl (DPPH) free-radical scavenging assay, the resultant AST-loaded silicified-phospholipids boron nitride complex displayed a release pattern and antioxidant activity that is in line with the literature. It also offers a quick and low-cost approach of utilizing AST for drug delivery. However, because of their large particle size, regular liposomes have huge drawbacks (between 1 and 100 m). Pan et al. (2018) designed nanoliposomes with smaller particle sizes that had improved penetration and new targeting properties as an alternate method to expand the usage of this delivery system. By using nanoliposomes, encapsulation effectiveness was increased ($97.5 \pm 0.3\%$) compared to previous reports, and transmission electron microscopy verified that the average particle size was 80 ± 2 nm. AST incorporating nanoliposomes (AST-LN) was 17 times more soluble than pure AST and exhibited good water dispersibility, with no precipitation or insoluble particles. Recent studies have revealed that different AST isomers contribute to antioxidant activity both under in vitro and in vivo conditions. However, research on AST and its metabolism in biological systems is lacking. Future research should concentrate on the physicochemical properties of various AST structures, as well as their uptake mechanisms and ability to integrate into metabolic pathways. To investigate the nutraceutical and pharmaceutical applications of various AST isomers, extensive molecular studies using in vitro and in vivo models should be conducted (Brotosudarmo et al. 2020).

12.2.2 Nutrient supplements

H. pluvialis is widely used in human nutrition (food, food and beverage pigments), and health (pharmaceuticals, nutraceuticals, and dietary supplements) (Mularczyk et al. 2020). AST is well tolerated by humans, and several investigations have revealed its safety and enhanced bioavailability (Stewart et al. 2008; Ranga Rao et al. 2013). Commercial products containing AST are available in daily capsule, soft gel, energy drink, and powder forms (Ambati et al. 2014). Additionally, AST is commonly sold as dietary supplements, oils, or dried aplanospores (Zanoni et al. 2019). A significant interest in producing foods with green algal components, especially AST, has been revealed by a recent study on functional food ingredients. The food and feed industries regularly use AST as a dietary supplement. The European Food Safety Authority (EFSA) Panel on Nutrition, Novel Foods, and Food Allergens declared in 2020 that it was safe for adults to consume 8 mg of

AST from food supplements (Turck et al. 2020). A daily AST intake of 2–4 mg is considered safe by Spiller and Dewell (2003). When adults consumed up to 6 mg per day, they did not note any harmful consequences. The daily AST intake of 40 mg/day produced satisfactory AST bioavailability results. Recent studies have demonstrated that adding microalgae extract or powder to food products, including cookies, can enhance their nutritional and textural aspects (Mofasser Hossain et al. 2017; Onacik-gür et al. 2018; Sahni et al. 2019).

The *H. pluvialis* cyst cells, which are generated under unfavorable environmental conditions, contain a significant amount of AST, which has a unique and well-known antioxidant capability. However, since the bioactive substances are covered by a strong cell wall during the cyst phase, it might prevent proper release and bioavailability. As a result, cell wall disruption and the acquisition of the desired compounds typically require an extraction method (Sarada et al. 2006). *H. pluvialis* is one of the most explored microalgae in terms of encapsulation systems for food applications. Several researchers have looked on the encapsulation of its extract obtained using various techniques or pure substances, primarily the carotenoid AST (Vieira et al. 2020). The studies have revealed a wide range of encapsulating techniques for *H. pluvialis* and related compounds, including both top-down and bottom-up strategies. For the encapsulating methods, half of the authors preferred spray drying, extrusion, and emulsification (Vieira et al. 2020). Spray drying of microcapsules with the addition of maltodextrin and gelatin in a 2:1 ratio has produced satisfactory results. The microcapsules exhibited smooth surfaces, high solubility, and good spherical forms. The procedure is economically viable since it uses inexpensive carriers (Zhao et al. 2019). AST's stability and bioavailability were increased *via* liposomal encapsulation (Peng et al. 2010). Many researchers have reported similar results when using microencapsulation with b-cyclodextrin, hydroxypropyl-b-cyclodextrin, and sulfobutyl ether b-cyclodextrin (Higuera-Ciapara et al. 2004; Tachaprutinun et al. 2009; Ambati et al. 2014).

Studies on in vitro digestion demonstrated a reduction in glucose release proportionate to an increase in AST concentrations in the formulation of whole grain cakes (Mofasser Hossain et al. 2017). The antioxidant qualities and bioavailability of AST are enhanced when *H. pluvialis* biomass is consumed as a dietary supplement along with olive oil (Ranga Rao et al. 2010). AST builds up in lipid droplets in the stomach and is combined with micelles to form the plasma membranes of enterocytes. Low-density lipoprotein (LDL) and high-density lipoprotein (HDL) then carry AST in the bloodstream (Furr and Clark 1997).

AST has been used in animal feeding for a long time because of its powerful antioxidant properties, safety during usage, and capacity to improve the organoleptic aspects of animal products. Due to the fact that animals cannot produce carotenoids on their own, they must ultimately obtain this pigment from their diet, which is abundant in plants and algae (Hasan and Chakrabarti 2009; Mularczyk et al. 2020). The aquaculture industry mainly uses AST from *H. pluvialis* for the pigmentation of fish (Shah et al. 2016). In addition to the role of AST in enhancing the color of many aquatic animal species, it also improves animal survival, stress tolerance, disease resistance, and growth efficiency (Lim et al. 2018). It was observed that adding

H. pluvialis to rainbow trout diet (3 g/kg of feed) effectively increased antioxidant activity as well as some biochemical parameters (e.g., reductions in cholesterol, triglyceride, and serum glucose levels) (Ju et al. 2012). In the farming of crustaceans like the giant tiger prawn, AST is of great interest (Penaeus monodon). In farmed prawns with a pigment deficiency, this carotenoid reverses the blue color syndrome. Within 4 weeks, shrimp pigmentation was restored and harmonized by food with 50–100 g of AST per 1 kg of feed (Torrissen 2000). Additionally, it was revealed that the defatted *H. pluvialis* meal may be used as a source of protein as well as an alternate pigmentation ingredient in the feed for Pacific white shrimp (Ju et al. 2012). Furthermore, a higher growth rate was produced on by this additive (Sheikhzadeh et al. 2012).

In poultry farming, AST used as a feed ingredient strengthened muscles, improved yolk color, and prevented fat oxidation (Yang et al. 2006). AST enhanced the yellow pigmentation of chicken feet and beaks. Feeding AST-rich diet to chickens resulted in improved fecundity in addition to increased weight gain and muscle mass (Inborr 1998). AST supplementation decreased chicken mortality associated to inflammation of the yolk sac and further increased resistance to Salmonella infection (Bergstrom et al. 2009) According to Waldenstedt et al. (2003), broiler chickens' cecal *Clostridium perfringens* colonization was decreased when fed an algal meal rich in AST (7, 36, or 179 mg astaxanthin/kg feed). Compared to the control birds, the concentrations of this pigment increased in the kidney, gut, and breast muscle (Waldenstedt et al. 2003). Pork is the most popular meat in the European Union and accounts for 9% of all agricultural production, according to estimates (Eurostat 2015). The addition of AST to the pig's diet improved the meat's color, which is one of the most significant aspects of pork for the consumer. Compared to loin chops from control pigs, AST-supplemented pigs' loin chops were darker and less yellow (Carr et al. 2010). Additionally, (Bergstrom et al. 2009) reported that the deeper color of the loin muscle of pigs fed AST (5 and 10 mg/kg) may help improve customer acceptance of fresh pork. When dogs' diets contain AST, their cellular and humoral immunity is improved. Additionally, DNA damage is decreased, and inflammation is decreased (Chew et al. 2000).

12.2.2.1 Various Drug Delivery Modes of AST Against Malnutrition

In recent years, various delivery systems such as liposomes, nanoparticles, emulsion systems, and microcapsules have been developed to improve the stability and bioavailability of AST (Shuhan Dai et al. 2022). Aung et al. (2022) developed a self-micro emulsifying drug delivery system (SMEDDS) tablet containing AST and evaluated their physicochemical and biological properties. This AST SMEDDS tablets were successfully designed and tested for improved dissolution, antioxidant activity, intestinal Caco-2 cell uptake, and absorption. AST SMEDDS spray-dried powders were attempted by combining L-AST SMEDDS (rice bran oil, Kolliphor® RH 40, and Span® 20) with a solid carrier (MCC) and hydrophilic polymers (HPMC and PVA) prior to tablet compression. These tablets appeared to be a promising platform for improving AST permeation, absorption, and thus oral bioavailability. Although AST's application in nutrition has reached a moderate level, more innovations and developments are required to improve the stability of astaxanthin collected from *H. pluvialis* and its applications in the food industry.

12.2.3 Anticancer activities

Cancer has emerged as the world's most serious public health issue (Siegel et al. 2019) and it is a leading cause of death and a major impediment to increasing life expectancy in every country on the planet (Sung et al. 2021; Bray et al. 2021). Around 19.3 million new cancer cases were reported worldwide in 2020, resulting in nearly 10 million deaths (Sung et al. 2021). Malignant tumors, which are uncontrolled cell proliferations, cause significant pathologic changes (Abd El-Hack et al. 2019). This is one of the most dangerous cancers due to its ability to metastasize and invasion of cancer cells (McSherry et al. 2007). Radiotherapy, chemotherapy, and/or surgery are the three main cancer treatments. Nevertheless, due to worst impediments and jeopardy associated with chemotherapy drugs, search of various sources of novel anticancer drugs is still continuing researched. Interestingly, phytochemicals derived from microalgae have greater potential biological activity than those originating from terrestrial plants (Holst and Williamson 2008; Prabakaran et al. 2018).

It is interesting to note that carotenoids reduce the growth of various cancer cell lines, including those of the breast, prostate, melanoma, lung, and leukemia (Kotake-Nara et al. 2001; Stivala et al. 2000). Among the many types of carotenoids derived from microalgae, AST has a distinct antitumor capacity with a more potent form of carotenoids (Palozza et al. 2009). Since AST is a powerful antioxidant that protects against genotoxicity and cytotoxicity caused by oxygen species, stimulates hepatic xenotoxic-metabolizing enzymes, and boosts tumor immunity (Jyonouchi et al. 1994). An upcoming study reports that AST can reduce the side effects of doxorubicin (DOX). A document reported that AST treatment can alleviate DOX-induced cognitive impairment (chemo brain) symptoms in rats (El-Agamy et al. 2018). Liu et al. (2015) demonstrated that AST can reduce the severity of DOX-induced focal segmental glomerulosclerosis. As a result, AST has the significant potential to be used as a preventative medicine for chemotherapy patients (Ma et al. 2020). Studies showed that, in the later stages of tumor development, AST regenerates stress-induced hepatic metastasis in mice and enhances immune response via lipid peroxidation downregulation and antioxidant activation (Kurihara et al. 2002; Kang et al. 2001). A substantial body of evidences have confirmed that AST inhibits the growth of various tumor cells, including neuroblastoma, leukemia, breast cancer, colon cancer, lung cancer, bladder cancer, gastric cancer, oral cancer, and liver cancer (Naguib 2000; Salehi et al. 2013; Singh et al. 2013; Qu et al. 2013; Wakabayashi et al. 2015; Roerecke et al. 2015; Jiang et al. 2016).

In an in vitro study on breast cancer cells (McCall et al. 2018), AST was found to inhibit the proliferation and migration of breast cancer cells, phenomenally. When compared to control breast epithelial cells, AST significantly reduced proliferation rates and inhibited cell migration in ER+ MCF7 and TNBC MDA-MB-231 cells. In another study, crude extract of *H. pluvialis* was used as a medicinal agent and demonstrated anticancer activities in BC MDA-MB-231 cells. The above findings further suggested that increasing the Bax/Bcl2 ratio through inhibiting the expression of mutant p53 may be the underlying cause for causing apoptosis of cancer cells by AST (Alateyah et al. 2022).

12.2.3.1 Various Drug Delivery Modes of AST Against Cancer

Melanoma is an aggressive tumor that can spread to other organs, and its 5-year survival rate is under 15% (D'Arcangelo et al. 2018). Melanoma spreads by taking advantage of vessel wall leaking (Venza et al. 2015). High concentration and poorly soluble AST in O/W (oil in water) embedded with nanoemulsion products produce inhibitory efficacies and construct TAP-nanoemulsion to inhibit melanoma proliferation and induce apoptotic responses as anticancer drug delivery systems. AST- and peanut oil-loaded nanoemulsion with TPGS (D-tocopheryl polyethylene glycol succinate) to protect the O/W nanoemulsion from various environmental challenges. A good antioxidative ability was present in TAP-nanoemulsion in a dose-dependent way to lessen cellular ROS production. TAP-nanoemulsion revealed a high-quality oral release characteristic and adequate stability in vitro in the physiologic mimic environment. Most importantly, AST, *via* suppressing Bcl-2, cyclins D1 and E, NF-KB, ERK, MEK, and MMP-1 and MMP-9 and boosting cleaved caspase-9 and caspase-3, ATM, and p21, TAP-nanoemulsion caused human malignant melanoma in the lung to undergo apoptosis (Haung et al. 2020).

In a study on colon cancer, AST was found to suppress cancer cell development by both encouraging apoptosis through an increase in caspase 3 expressions in colon cancer cells as well as by stopping the cell cycle from progressing (Hormozi et al. 2019). By inhibiting oxidative cell damage, antioxidant compounds reduce mutagenesis and carcinogenesis. According to studies, AST not only inhibits colon cancer cell proliferation but can also cause apoptosis (Mularczyk et al. 2020). Dietary AST showed anti-malignant activity in the post-initiation phase of cell cycles in a clinical case of carcinogen-induced rat colon cancer (Tanaka et al. 1995). A study (Prabhu et al. 2009) has concluded that AST plays an efficient anti-inflammatory and anticancer role in rats that were subjected to chemical induction of colon cancer in male Wistar rats via induction of apoptotic cell death; apoptotic death was induced by NF-kB-modulated expression, as well as altered expression of matrix metalloproteinase 9 (MMP-9), cyclooxygenase 2 (COX-2), matrix metalloproteinase 2 (MMP-2), ERK-2, and AKT.

Presently, nanoparticles are used in a large number of drug delivery systems, and various materials have been used as drug stimulants or drug enhancers to improve the treatment effectiveness, durability, stability, and safety of anticancer drugs. As a biological molecule, AST can reduce metal salts to form nanoparticles that can be used in biological systems; the production of gold nanoparticles (Au NPs) using AST as a natural reducing agent has been studied. A tetrazolium-based assay was used to assess the cytotoxic effects of prepared nanostructures against human breast cancer cells (MDA-MB231); AXT-Au NPs have a powerful cytotoxic effect against cancer cells, and apoptotic morphology has been detected in the treated cells. AST reduced Au NPs, on the other hand, have the potential to be a promising agent in the field of photo-based diagnosis and therapy due to an interesting UV-vis absorption peak in the near-infrared region, which is essential in photo-based diagnosis and therapy (Jafari et al. 2022). AST-conjugated bovine serum albumin was used to synthesize polypyrrole nanoparticles for photo-based therapy and cancer detection. A study examined the effect of an AST-alpha tocopherol nanoemulsion synthesized by spontaneous and ultrasonication emulsification methods on three different types of cancer cells; wherein, it exhibited significant anticancer potential as well as antimicrobial and wound healing properties (Bharathiraja et al. 2017).

The fourth major causes of cancer-related fatalities globally are pancreatic cancer (Siegel et al. 2018). Interestingly, by upregulating the uptake membrane proteins hENTs and downregulating ribonucleotide reductase (RR) expression through a separate signaling mechanism, AST and sclareolide can both resensitize gemcitabine (GEM)-resistant pancreatic cancer cells to GEM. For the treatment of GEM-resistant pancreatic cancer, this may be a novel agent when combined with GEM (Yan et al. 2017; Chen et al. 2017). Although the mechanisms by which AST mediates anticancer action are not fully clarified, several AST molecular targets have been proposed. Advanced molecular studies on the role and efficacy of anticancer compounds, as well as the characterization of carotenoids derived from *H. pluvialis*, are still needed. Furthermore, extensive in vitro or in vivo studies on *H. pluvialis* bioactive compounds should be conducted to evaluate their potential applications against a wide range of cancers.

12.2.4 Antimicrobial Effects

Microalgae are one of the potential producers of novel antibacterial compounds among photosynthesis organisms (El-Baz et al. 2016; Little et al. 2021). Only a few research studies have looked into AST's antibacterial potential thus far. Many studies have examined the effects of microalgae extracts on cultures of pathogenic microbes. These extracts' antibacterial activity has been associated with polysaccharides. fatty acids, phenolic compounds, tannins, alkaloids, terpenoids, indole, halogenated hydrocarbons, inhibitory enzymes, acetogenin, and amides (Pane et al. 2015; Falaise et al. 2016; Smith et al. 2010). Bacteria are becoming more resistant to antibiotics, which is a major issue for public health because antibiotics are losing their effectiveness at alarming rates as a result of the advent of antibiotic-resistant microorganisms (O'Neill 2016; Mourad et al. 2020). The majority of bacterial species have created their own drug resistance mechanisms. Finding novel bioactive compounds is essential due to the widespread bacterial resistance to current antibiotics (Rossolini et al. 2014; Darwesh et al. 2019).

In the COVID-19 disease, the addition of AST can regulate and reduce the inflammations brought on by a novel species of SARS CoV-2 and high levels of cytokine storms as well as inflammatory agents including interleukins (IL-1, IL-6, IL-8), tumor necrosis factor (TNF-), etc. AST can be a promising therapeutic dietary supplement in combination with primary antiviral medications for patients with severe pneumonitis, acute lung injury (ALI), acute respiratory distress syndrome (ARDS), other lung-related infections, as well as in recovered COVID-19 patients as a post treatment care (Ahmadi and Nasrabadi 2021; Talukdar et al. 2020; Chia et al. 2021). In the COVID-19 management, AST and its preparations are becoming more well-liked and accepted globally (Patil et al. 2022).

12.2.4.1 Various drug delivery modes of AST against microbial infections

When AST was docked with various mycobacterial drug targets, it produced better interactions than the traditional anti-TB drugs, isoniazid, and rifampicin. An encapsulated AST exhibited better mycobacterial inhibition (MIC value of 500 g/mL) against MDR-TB and H37Rv (Vasudevan et al. 2021). Numerous nanotechnology-based techniques, including nanoemulsion, nanoencapsulation, nanoliposomes, carbon nanotubes, and nanomicelles, have been utilized successfully to deliver medications to the desired region without causing any negative side effects (Kaci et al. 2018; Chen and Chen 2018). Nanoemulsions are crucial in a variety of biomedical applications, including medication delivery, cosmetics, and cell culture technologies (Jonkman et al. 2014; Hu et al. 2016). Shanmugapriya et al. (2018) showed how AST can combat different bacteria when it is present in a nanoemulsion. MIC (minimum inhibition concentration) values ranged from 500 to 4000 μg/mL for both Gram-positive and Gram-negative bacteria. By using spontaneous and ultrasonic emulsification techniques, an AST-alpha-tocopherol nanoemulsion was described, and its stability with regard to particular physical features was tuned for in vitro biomedical applications. Studies on in vitro cytotoxicity, wound healing, apoptosis, and antimicrobials showed that the manufactured nanoemulsion was safe to exhibit broad antimicrobial capabilities and was appropriate for the treatment of wounds and enhanced wound repair (Shanmugapriya et al. 2018). Further in-depth experiments may highlight the usage of AST as a potent therapeutic strategy for bacterial infections.

12.2.5 Anti-Diabetic Properties

Approximately 463 million people worldwide have diabetes, and it is predicted that during the next 10 years, this figure will rise to 578 million (Landon et al. 2020).

According to a human epidemiological study, the incidence of diabetic mellitus (DM) and plasma carotenoid levels have an inverse relationship (Akbaraly et al. 2008). AST inhibited the high-glucose exposure-induced increased reactive oxygen species (ROS) generation in the mitochondria in human mesangial cells, suggesting that it may have a protective effect against diabetic neuropathy (Manabe et al. 2008). AST has been shown to increase whole-body insulin sensitivity and insulin-stimulated glucose absorption while decreasing plasma glucose and insulin levels. According to the reviewed research, it is concluded that it can increase insulin sensitivity. Many studies have demonstrated how an excessive amount of ROS contributes to the pathogenesis of type 2 diabetes and insulin resistance, as well as the onset and progression of chronic illnesses (Sztretye et al. 2019).

According to study findings, AST may lower the oxidative stress that hyperglycemia causes in the pancreatic β-cells (Uchiyama et al. 2002). Dietary AST and α-tocopherol reduces kidney oxidative damage by inhibiting the oxidative stress induced by diabetes (Nakano et al. 2008; Jannel et al. 2020). AST, with its excellent antioxidant properties, is of particular interest for application in this scenario because oxidative stress is a major contributing factor to diabetic nephropathy. AST is efficient in treating and minimizing the problems of diabetes, depending on the stage of the disease (Jafari et al. 2022). AST protects pancreatic β cells in the early stages of diabetes, boosts insulin sensitivity, and enhances glucose metabolism. As a result, blood sugar levels and insulin resistance go down (Landon et al. 2020). In experiment diabetes-affected rats supplemented with *H. pluvialis* extract exhibited healthier liver tissue. In the streptozotocin (STZ)-induced diabetic rats, an ethanolic extract of *H. pluvialis* proved successful in lowering blood glucose levels. The antioxidant activity of AST, which may be amplifying insulin release from β-cells or its insulin-like action, may be the protective mechanism of action of the algal extract (El-Baz et al. 2016).

Administration of natural AST has not been associated in studies to any unfavorable side effects (Brendler and Williamson 2019). As a result, AST has been thoroughly studied for use as an anti-diabetic agent in diabetes medications (Oslan et al. 2021). According to a study, diet-induced insulin resistance and hepatic steatosis in rats can be controlled and prevented by AST (Zhuge et al. 2021). Male Wistar rats were given a high-energy meal containing AST at doses ranging from 15 to 50 mg/kg, and then a low dose of STZ (40 mg/kg) was injected to create a diabetes model in order to study the anti-diabetic effects of AST. The study findings show that AST treatment significantly reduces STZ-induced diabetes by raising the expression of genes related with insulin sensitivity (adiponectin, adipoR1, and adipoR2).

12.2.5.1 Various Drug Delivery Modes of AST Against Diabetes

A study was conducted to evaluate the efficacy of a thin film containing AST-loaded nanoemulsion (FDT-As-NE) in lowering blood glucose levels in alloxan-induced diabetic rabbits (ADR). When AST is incorporated to a nanoemulsion system, it dissolves in the oil phase and enters the micelles system, which is thermodynamically stable and has droplet sizes of <100 nm. After 14 days of treatment, blood glucose levels in ADR treated with FDT-As-NE (51.21% $\pm$ 20.02%) were significantly lower ($P < 0.01$) than in ADR treated with pure as AST (15.75% $\pm$ 12.46%). As a result, the study suggests that a thin film containing an AST-loaded nanoemulsion administered intravenously may be useful for lowering glucose levels (Nurdianti et al. 2020). As the incidence of human diabetes continues to rise, research on AST's drug delivery is insufficient, necessitating additional human trials to determine the precise mechanism and molecule responsible for these findings.

12.2.6 Cardio Protectant Characteristics

A prominent cause of death in the worldwide is cardiovascular disease (CVD). WHO estimated that 17.9 million people died from CVDs in 2019, accounting for 32% of all deaths worldwide. Among them, heart attacks and strokes alone were responsible for 85% of deaths (WHO 2022). The pathophysiology of various cardiovascular diseases is caused by ROS; for instance, oxidative modification of low-density lipoprotein (LDL) has been linked to the pathogenesis of atherosclerosis (Epstein et al. 1989), myocardial infarction, ischemia/reperfusion damage, cardiac hypertrophy, cardiomyopathy, and heart failure (Giordano 2005). Several dietary antioxidants have been investigated in this context, with encouraging results for the treatment and prevention of CVD (Jain et al. 2015).

AST provides excellent support for CVD due to its anti-inflammatory and antioxidant properties that can greatly decrease blood transmit times (Hewlings 2019; Jinu and Chandra 2021). The effects of AST on cardiovascular health have been studied experimentally in a variety of species, but only a small number of human clinical trials have been carried out in the past two decades (Donoso et al. 2021). In an in vitro study, free AST significantly increased nitric oxide release and decreased peroxynitrite levels in human umbilical vein endothelial cells and platelets isolated from Wistar-Kyoto rats (Khan et al. 2010). It enhances cholesterol and glucose metabolism and reduces inflammation. In individuals suffering from mild hyperlipidemia, AST significantly decreased serum triglyceride (TG) levels while increasing serum adiponectin and high-density lipoprotein (HDL) cholesterol levels (Yoshida et al. 2010). Another research reveals that adults with milk hyperlipidemia who took AST had significantly lower TG levels, higher HDL cholesterol levels, and higher levels of adiponectin in their blood (Szentesi et al. 2019). AST has a lot of promise for use by people as a food ingredient to reduce the risk of coronary heart disease (CHD) since research results show that it achieves its effects, at least in part, by modifying the expression of genes involved in cholesterol and lipid metabolism as well as antioxidant mechanism (Yang et al. 2011). In a pilot study on heart failure (HF) patients with left ventricular (LV) systolic dysfunction, improvements in cardiac function and exercise tolerance were observed in relation to the suppression of oxidative stress by 3-month AST supplementation (Kato et al. 2020).

12.2.6.1 Various Drug Delivery Modes of AST Against Cardiovascular Diseases

A novel elastic hydrogel material that permits AST loading without affecting its antioxidant activity was developed using blends of PVA (polyvinyl alcohol) and Dex (dextran). This technique allows for the placement of hydrogel patches and antioxidant releases in vivo conditions in rats without affecting the surrounding tissues or causing inflammation. Furthermore, this device may be employed to decrease myofibril stress caused on a MI and deliver an antioxidant molecule to the injured heart tissue (Zuluaga et al. 2017). The main organelle in cells, the mitochondria, are essential for the healthy functioning of the heart as well as the pathogenesis and development of various heart diseases (Griffiths 2012). In both normal and clinical conditions, AST has a protective effect on the rat heart mitochondria (RHM), making it an effective drug for improving cardiac muscle function. For the treatment of pathological conditions linked to oxidative damage and mitochondrial dysfunction, AST may be a potential target in the mitochondria. Additionally, it was found that AST treatment prevented the oxidative damage to mitochondria caused by isoproterenol and increased mitochondrial efficiency (Krestinina et al. 2021), thus proving its significant beneficial role in cardio protection. Based on preclinical and clinical evidences, AST has been shown to be effective in the prevention and treatment of cardiovascular diseases with less adverse events in human clinical trials. However, more studies are required in the field of clinical trials on cardiovascular diseases using AST derived from *H. pluvialis.*

12.2.7 Anti-Aging Properties

Aging generally makes the body's antioxidant and repair systems less effective (Bjørklund et al. 2022). The two main mechanisms thought to be responsible for age-related skin changes are biological aging and exposure to UV light. Collagen and elastin are destroyed by photoaging, which leads to wrinkles, pigmentation, and degradation of skin texture (Davinelli et al. 2018). Many cuticle diseases are brought on by phototoxic damage, which is a common injury that also hastens the aging process (Chou et al. 2020). Additionally, reactive oxygen species are produced by the skin in response to UV light (Chalyk et al. 2017). A highly fascinating possible geroprotector is AST (Sorrenti et al. 2020). AST has shown potential as an anti-aging agent in an experimental oxidative stress model (Huangfu et al. 2013). According to this study, a number of environmental factors, including smoking and UV exposure, may accelerate and exacerbate the symptoms of aging skin (Ponto et al. 2021). Cognitive impairment and decreased brain mobility are also considered to be two important features of aging. Since AST can pass the blood–brain barrier, its consumption might slow down the aging process of the brain (Fu et al. 2022). The significant advantages of AST supplementation for healthy aging have been confirmed by researchers (Bjørklund et al. 2022). AST's antioxidant and anti-inflammatory characteristics have been linked to its anti-aging effects, which include halting age-related muscle degeneration and enhancing mitochondrial energy production (Sorrenti et al. 2020; Bahbah et al. 2021). It can also support increased immunity, fatigue resistance, and delay-aging (Zhao et al. 2019). Research on AST demonstrates that it counteracts reactive oxygen species (ROS) and reactive nitrogen species (RNS) that cause system overloading within the body's cells and accelerate aging (Nishida et al. 2007; Bjørklund et al. 2022). Malondialdehyde (MDA), a well-known indicator of systemic oxidative stress, was found to decrease after a human model received 4 mg of AST daily (by 11.2% on day 15 and by 21.7% on day 29), according to Chalyk et al. (2017). Furthermore, Seong-Ryeong et al. (2021) observed that AST is a nutritional supplement that reduces the phosphorylation of extracellular signal-regulated kinases (ERK) induced by ROS generation and prevents the skin fibroblastic autophagic cell death produced by bisphenol A (BPA) in normal human dermal fibroblasts (NHDF). Moreover, it inhibits UVB-induced apoptosis in vivo, thereby slowing the skin's aging process (Chou et al. 2020).

12.2.7.1 Various Drug Delivery Modes of AST Against Aging

A simple and efficient low-energy EPI method was employed to successfully create AST-loaded nanoemulsions (NEs) and carboxymethyl chitosan (CMCS)-AST-NE. The CMCS-AST-NE may be a promising delivery system for the application of AST in dermal and transdermal products due to the fact that this procedure boosts AST chemical stability and skin permeability (Hong et al. 2017). Many studies have suggested that AST has anti-inflammatory properties, and that supplementing with it may help prevent skin issues caused by environmental exposure as well as age-related skin degradation. In a study conducted by Petyaev et al. (2018) 32 healthy human subjects between the ages of 60 and 70 showing evidence of oxidative stress participated in a randomized clinical trial. This study showed that supplementing with a lysosomal dark chocolate formulation that contained 7 mg of co-crystallized AST with enhanced bioavailability had intriguing impacts on the improvement of oxidative state in aged human participants, thus pointing to the advantages of mixing dark chocolate with AST. The above findings suggested that taking AST supplements could help keep humans from suffering from dementia as they age (Nakagawa et al. 2011). This study strongly suggested that *H. pluvialis* has therapeutic potential against diseases associated with aging, such as alzheimer's disease. In another study the stability of AST and their anti-aging properties in a crude acetone extract of *H. pluvialis* from electrospun cellulose acetate (CA) nanofibers were investigated. According to this study, under accelerated conditions, the stability of released AST from the CA nanofibers loaded with crude AST extract was noticeably higher than that of free AST in the crude AST extract (Nootem et al. 2018). With the scientific evidence in hand, a new approach to studying age-related diseases may be explored in the future to elucidate the role of *H. pluvialis* in drug delivery and its bioactive components.

12.2.8 Wound Healing Applications

A complicated and crucial process of the body's response to injury is wound healing or repair. Coordination of several cells and components is necessary for this to be accomplished. Inflammation, proliferation, and remodeling are the three processes that constitute the healing process (Braiman-Wiksman et al. 2007). For effective wound healing, oxidative and antioxidative forces must be balanced. By activating the NF-kB/Rel pathway (Ojha et al. 2008), which is believed to be the primary cause of chronic unhealed wounds, prolonged creation or significant quantities of ROS, also known as oxidative stress, can cause chronic inflammation. Antioxidant enzyme systems and dietary antioxidants like carotenoids can decrease the harmful effects of oxidative stress (Auf dem Keller et al. 2006). A study showed that AST reduced the oxidative stress caused by ROS in a rat deep-burn model, thus healing the early burn wounds. By affecting xanthine oxidase (XO) and the reduced form of nicotinamide adenine dinucleotide phosphate (NADPH) oxidase (Noxs), both of which are involved in the generation of ROS, this action controls the synthesis of free radicals (Fang et al. 2017).

Tissue engineering must take into account the normal antioxidant mechanisms that promote wound healing by protecting skin and tissues from oxidative reactions. In the field of tissue engineering, Chae et al. (2022) developed a novel concept of AST exhibiting potential antioxidant and antibacterial effects with the assistance of graphene oxide (GO) nanocarriers. AST and GO can be chemically combined to enhance the dispersion qualities and retain structural stability and antioxidant activity. This AST-loaded ultrathin GO films improved the adhesion and viability of the particular cell type, which may be very helpful in tissue regeneration or wound healing. The amount of collagen in human dermal fibroblasts was increased by an enhanced AST extract from *H. pluvialis* via inhibiting matrix metalloproteinases (MMP-1) and MMP-3 expression (Chou et al. 2016). In fact, maintaining collagen turnover in a balanced way and ECM (extracellular matrix) function depend on the proper regulation of MMPs (Lu et al. 2012). When a wound heals, the ECM at the wound site dramatically reorganizes. It has been demonstrated that AST works well to quicken the healing of full-thickness dermal wounds in mice.

Wounds treated with AST had considerably higher levels of biological indicators for wound healing like collagen type I α 1 (Col1A1) and basic fibroblast growth factor. Additionally, AST increased the expression of wound healing biological markers such as collagen type I1 and basic fibroblast growth factor (bFGF) (Meephansan et al. 2017). AST is believed to be a powerful anti-inflammatory and antioxidant agent, and it is required for the incorporation of collagen in AST composite materials in order to reduce and prevent injury to mammalian cells as well as improve the speed at which injured mammalian cells recover (Kang et al. 2001). The topical wound healing properties of AST-alpha-tocopherol nanoemulsions with κ-carrageenan were verified in vitro and in vivo studies. Results showed that the treatment with AST was a promising alternative for diabetic wounds with more activity and faster healing (Shanmugapriya et al. 2018). Veeruraj et al.

(2019) developed an AST collagen film that had the highest wound healing rate among the experimental groups and healed the wound completely in just 15 days. A study demonstrated that AST at a dose of 4 mg twice per day considerably increased skin elasticity and tended to increase skin moisture content (Chiewprasit and Koonngam 2022).

12.2.8.1 Various Drug Delivery Modes of AST Against Wounds

ASTA@Lec NS was successfully developed using a simple emulsion evaporation approach (Lecithin nano-liposol). It could effectively load a large amount of AST without altering its physiochemical properties. The bioavailability of AST was significantly improved by enhanced ASTA@Lec aqueous solubility. As a result, this could eliminate ROS and improve NIH 3T3 cell wound healing more efficiently without any cytotoxicity (Oh et al. 2020). Among studies on the many health-promoting properties of *H. pluvialis* especially AST, very few studies on wound healing have been reported. According to several dermatological investigations, AST extract from *H. pluvialis* is hailed as a topical therapy and oral supplement to promote wound healing and exerts significant protective properties through well-balanced oxidative reactions. However, more extensive research on the biomolecules of *H. pluvialis* is necessary in order to treat human wounds.

12.2.9 Ocular Protectant

Eye health oxidative damage has been linked to a number of ocular diseases. The most common are age-related macular degeneration (Imamura et al. 2006), cataract (Spector 1995), retinopathy of prematurity (Niesman et al. 1997), uveitis (Gritz et al. 1991), corneal inflammation (Alio et al. 1995), and keratitis. AST has become a popular research subject in eye diseases owing to its beneficial role. Although AST is not found in the retina, unlike lutein and zeaxanthin, it can selectively penetrate the blood–retinal barrier (BRB) and shield the retinal cells from oxidative damage (Janani et al. 2021). Among other xanthophylls, AST has been shown to protect human eye epithelial cells from UVB-induced stress (Chitchumroonchokchai et al. 2004). Moreover, AST has shown to have neuroprotective properties in animal models of nitroglycerin (NTG) induced migraine. Studies have shown the efficacy of AST in the prevention and treatment of a variety of ophthalmic diseases and disorders, including retinal diseases, uveitis, asthenopia (eye fatigue), cataract, and ocular surface disorders (Giannaccare et al. 2020). Additionally, AST has been shown to improve retinal blood flow, vision acuity, and uveitis in both animal and human models (Giannaccare et al. 2020).

Decreased visual acuity results from cataract formation because it makes the eye's natural lens opaque. Surgery is necessary to restore eyesight, and it continues to be one of the leading causes of blindness worldwide (Scorcia et al. 2020).

According to a recent study, AST treatment prevented oxidative stress in diabetic mice models, which was shown by a significant decrease in the levels of advanced glycation end products. This delayed the onset and progression of metabolic cataracts (Yang et al. 2020). In chick embryo models, AST treatment reduced lens opacity and prevented the progression and growth of steroid-induced cataracts and in fact, AST also observed to decrease the glutathione levels in the lens (Ishikawa et al. 2015).

Age-related macular degeneration (ARMD) is defined by the presence of retinal pigment epithelium (RPE) cells within the damaged macula, which results in the loss of foveal photoreceptors (Votruba and Gregor 2001). AST consumption prevents and slows the progression of dry ARMD (Otsuka et al. 2013). Aldolase reductase (AR), a critical enzyme in the polyol pathway that contributes to the pathogenesis of diabetic microvascular issues, was examined both in vitro and in vivo using a mouse type 2 diabetes model. For 1 week, samples were given AST (4.8 mg/kg of body weight) once a day. Results demonstrated a significant decrease in AR activity both in vivo and ex vivo after AST intake, suggesting its involvement in both the early detection and treatment of diabetic retinopathy (Benlarbi-Ben Khedher et al. 2019).

An injury to the optic nerves known as optic neuropathy results in visual loss. It can be caused on by ischemia, inflammatory, demyelination, or traumatic causes. The second largest cause of blindness worldwide, glaucoma is one of the progressive optic neuropathies brought on by damage to the nerve fiber layer (Patil et al. 2022). The damage caused by glaucoma to a mouse's retina can heal after being treated with AST, according to research by Cort and colleagues (Cort et al. 2010). According to Nagaki et al. (2005), the AST-supplemented group (6 mg/day) showed a significantly (p 0.01) enhanced retinal capillary blood flow in both eyes without lowering intraocular pressure, whereas the intraocular pressure of the placebo group remained the same. Dry eye syndrome (DES) is a tear deficiency or excessive tear evaporation disorder that causes an unstable tear film (Schaumberg et al. 2003) A supplement of 6 mg AST twice daily improved the symptoms and signs of patients with mild-to-moderate DES (Tian et al. 2022). High affinity liposomal AST demonstrated a potential candidate for the treatment of dry eyes in an in vitro model of DES (Shimokawa et al. 2019).

12.2.9.1 Various Drug Delivery Modes of AST Against Ocular Diseases

A study (Xu et al. 2019) reported that lutein treatment was ineffective for delaying photoreceptor degeneration in MNU (N-methyl-N-nitrosourea) treated mice. However, treatment with AST can effectively alleviate MNU-induced retinal degeneration. Photoreceptor loss and visual impairments in MNU-treated mice can also be alleviated by AST nanodispersions. These protective effects are provided by AST nano dispersions, which modulate apoptosis and alleviate oxidative stress. These findings suggest that AST nanodispersions may be useful as a therapeutic strategy to ensure successful visual function recovery. Research on ocular diseases and clinical evidence promote the potential use of AST in the prevention and treatment of a number of ocular diseases. Despite numerous studies on reproducing the efficient roles of AST in ocular diseases in vivo and in vitro, additional clinical research on the drug delivery mechanism of AST in diseases such as cataract has been conducted on humans.

12.3 Conclusion

The present review is a pointer on the diverse therapeutic applications of *H. pluvialis* with thorough discussion on the key active principle behind its protective effect, AST. AST offers a wide range of health benefits and applications in nutraceuticals by its "super antioxidant" property. AST is extremely beneficial in a number of biological processes including antioxidant, anti-inflammatory, anti-apoptotic, and anticancer activities, due to its unique biological structure (Davinelli et al. 2018). Natural AST collected from *H. pluvialis* is safe to consume as a supplement (Shah et al. 2016). It is also a non-toxic healthy nutrient, and because of its strong antioxidant properties, it has been implicated in protecting cellular compounds from oxidative damage as well as cell regulating gene expression, inducing cell communication, and maintaining cell health. However, due to its slow rate of dispersion in blood vessels and low cellular absorption, AST oral intake is also limited. Therefore, extensive research has been conducted on using nanomaterials that are loaded with AST for therapeutic purposes. Wide spread efforts, such as encapsulation, designed to improve the bioavailability, stability, and solubility of this powerful antioxidant has also been highlighted in this review. Nevertheless, in-depth research is needed on the active compounds of *H. pluvialis* to assess their possibility as novel drug delivery mechanism against a wide range of diseases.

References

Abd El-Hack ME, Abdelnour S, Alagawany M, Abdo M, Sakr MA, Khafaga AF, Mahgoub SA, Elnesr SS, Gebriel MG (2019) Microalgae in modern cancer therapy: current knowledge. In: Biomedicine and pharmacotherapy, vol 111. Elsevier Masson SAS, pp 42–50. https://doi.org/10.1016/j.biopha.2018.12.069

Ahmadi AR, Nasrabadi RA (2021) Astaxanthin protective barrier and its ability to improve the health in patients with COVID-19. Iran J Microbiol 13:434–441

Akbaraly TN, Fontbonne A, Favier A, Berr C (2008) Plasma carotenoids and onset of dysglycemia in an elderly population: results of the epidemiology of vascular ageing study. Diabetes Care 31(7):1355–1359. https://doi.org/10.2337/dc07-2113

Alateyah N, Ahmad SMS, Gupta I, Fouzat A, Thaher MI, Das P, Al Moustafa AE, Ouhtit A (2022) *Haematococcus pluvialis* microalgae extract inhibits proliferation, invasion, and induces apoptosis in breast cancer cells. Front Nutr 9:882956. https://doi.org/10.3389/fnut.2022.882956

Alio JL, Ayala MJ, Mulet ME, Artola A, Ruiz JM, Bellot J (1995) Antioxidant therapy in the treatment of experimental acute corneal inflammation. Ophthalmic Res 27(3):136–143. https://doi.org/10.1159/000267648

Ambati RR, Moi PS, Ravi S, Aswathanarayana RG (2014) Astaxanthin: sources, extraction, stability, biological activities and its commercial applications-a review. Mar Drugs 12(1):128. https://doi.org/10.3390/MD12010128

Anarjan N, Ping Tan C (2013) Physico-chemical stability of astaxanthin nanodispersions prepared with polysaccharides as stabilizing agents. Int J Food Sci Nutr 64(6):744–748. https://doi.org/10.3109/09637486.2013.783001

Auf dem Keller U, Kümin A, Braun S, Werner S (2006) Reactive oxygen species and their detoxification in healing skin wounds. J Investig Dermatol Symp Proc 11(1):106–111

Aung WT, Khine HEE, Chaotham C, Boonkanokwong V (2022) Production, physicochemical investigations, antioxidant effect, and cellular uptake in Caco-2 cells of the supersaturable astaxanthin self-microemulsifying tablets. Eur J Pharm Sci 176:106263. https://doi.org/10.1016/J.EJPS.2022.106263

Aw MS, Simovic S, Yu Y, Addai-Mensah J, Losic D (2012) Porous silica microshells from diatoms as biocarrier for drug delivery applications. Powder Technol 223:52–58. https://doi.org/10.1016/J.POWTEC.2011.04.023

Bahbah EI, Ghozy S, Attia MS, Negida A, Emran TB, Mitra S, Albadrani GM, Abdel-Daim MM, Uddin MS, Simal-Gandara J (2021) Molecular mechanisms of astaxanthin as a potential neurotherapeutic agent. Mar Drugs 19:201

Batista AP, Niccolai A, Fradinho P, Fragoso S, Bursic I, Rodolfi L, Biondi N, Tredici MR, Sousa I, Raymundo A (2017) Microalgae biomass as an alternative ingredient in cookies: Sensory, physical and chemical properties, antioxidant activity and *in vitro* digestibility. Algal Res 26:161–171. https://doi.org/10.1016/J.ALGAL.2017.07.017

Benlarbi-Ben Khedher M, Hajri K, Dellaa A, Baccouche B, Hammoum I, Boudhrioua Mihoubi N, Dhifi W, Ben Chaouacha-Chekir R (2019) Astaxanthin inhibits aldose reductase activity in *Psammomysobesus*, a model of type 2 diabetes and diabetic retinopathy. Food Sci Nutr 7:3979–3985. https://doi.org/10.1002/fsn3.1259

Bergstrom JR, Houser TA, Gunderson JA, Gipe AN, Jacela J, Benz JM, Sulabo RC, Nelssen JL, Tokach MD (2009) Effects of dietary astaxanthin on the growth performance and carcass characteristics of finishing pigs. pp 239–244. https://krex.k-state.edu/dspace/handle/2097/2152

Bharathiraja S, Manivasagan P, Oh YO, Moorthy MS, Seo H, Bui NQ, Oh J (2017) Astaxanthin conjugated polypyrrole nanoparticles as a multimodal agent for photo-based therapy and imaging. Int J Pharm 517(1–2):216–225. https://doi.org/10.1016/J.IJPHARM.2016.12.020

Bjørklund G, Gasmi A, Lenchyk L, Shanaida M, Zafar S, Mujawdiya PK, Lysiuk R, Antonyak H, Noor S, Akram M, Smetanina K, Piscopo S, Upyr T, Peana M (2022) The role of astaxanthin as a nutraceutical in health and age-related conditions. Molecules 27(21):7167. https://doi.org/10.3390/molecules27217167

Braiman-Wiksman L, Solomonik I, Spira R, Tennenbaum T (2007) Novel insights into wound healing sequence of events. Toxicol Pathol 35(6):767–779. https://doi.org/10.1080/01926230701584189

Bray F, Laversanne M, Weiderpass E, Soerjomataram I (2021) The ever-increasing importance of cancer as a leading cause of premature death worldwide. Cancer 127(16):3029–3030. https://doi.org/10.1002/CNCR.33587

Brendler T, Williamson EM (2019) Astaxanthin: how much is too much? A safety review. Phytother Res 33:3090–3111

Brotosudarmo THP, Limantara L, Setiyono E, Heriyanto (2020) Structures of Astaxanthin and Their Consequences for Therapeutic Application. Int J Food Sci 2020:2156582. https://doi.org/10.1155/2020/2156582

Camera E, Mastrofrancesco A, Fabbri C, Daubrawa F, Picardo M, Sies H, Stahl W (2009) Astaxanthin, canthaxanthin and beta-carotene differently affect UVA-induced oxidative damage and expression of oxidative stress-responsive enzymes. Exp Dermatol 18(3):222–231. https://doi.org/10.1111/J.1600-0625.2008.00790.X

Carr CC, Johnson DD, Brendemuhl JH, Gonzalez JM (2010) Fresh pork quality and shelf-life characteristics of meat from pigs supplemented with natural astaxanthin in the diet. Prof Anim Sci 26(1):18–25. https://doi.org/10.15232/S1080-7446(15)30554-4

Cerón MC, García-Malea MC, Rivas J, Acien FG, Fernandez JM, Del Río E et al (2007) Antioxidant activity of *Haematococcus pluvialis* cells grown in continuous culture as a function of their carotenoid and fatty acid content. Appl Microbiol Biotechnol 74:1112–1119. https://doi.org/10.1007/s00253-006-0743-5

Chae SY, Park R, Hong SW (2022) Surface-mediated high antioxidant and anti inflammatory effects of astaxanthin-loaded ultrathin graphene oxide film that inhibits the overproduction of intracellular reactive oxygen species. Biomater Res 26:30. https://doi.org/10.1186/s40824-022-00276-4

Chalyk NE, Klochkov VA, Bandaletova TY, Kyle NH, Petyaev IM (2017) Continuous astaxanthin intake reduces oxidative stress and reverses age-related morphological changes of residual skin surface components in middle-aged volunteers. Nutr Res 48:40–48. https://doi.org/10.1016/j.nutres.2017.10.006

Chen YC, Chen BH (2018) Preparation of curcuminoid microemulsions from *Curcuma longa* L. to enhance inhibition effects on growth of colon cancer cells HT-29. RSC Adv 8(5):2323–2337. https://doi.org/10.1039/c7ra12297g

Chen YT, Kao CJ, Huang HY et al (2017) Astaxanthin reduces MMP expressions, suppresses cancer cell migrations, and triggers apoptotic caspases of in vitro and in vivo models in melanoma. J Funct Foods 31:20–31

Chew BP, Park JS, Wong TS, Kim HW, Weng BBC, Byrne KM, Hayek MG, Reinhart GA (2000) Dietary beta-carotene stimulates cell-mediated and humoral immune response in dogs. J Nutr 130(8):1910–1913. https://doi.org/10.1093/JN/130.8.1910

Chia WY, Kok H, Chew KW, Low SS, Show PL (2021) Can algae contribute to the war with COVID-19? Bioengineered 12:1226–1237

Chiewprasit P, Koonngam P (2022) Comparative effectiveness of oral astaxanthin given daily vs. every other day on skin elasticity: randomized, double-blind cohort research. J Basic Appl Pharmacol 2:26–36

Chitchumroonchokchai C, Bomser JA, Glamm JE, Failla ML (2004) Xanthophylls and alpha-tocopherol decrease UVB-induced lipid peroxidation and stress signaling in human lens epithelial cells. J Nutr 134(12):3225–3232. https://doi.org/10.1093/jn/134.12.3225

Chou HY, Lee C, Pan JL, Wen ZH, Huang SH, Lan CW, Liu WT, Hour TC, Hseu YC, Hwang BH, Cheng KC, Wang HM (2016) Enriched astaxanthin extract from *Haematococcus pluvialis* augments growth factor secretions to increase cell proliferation and induces MMP1 degradation to enhance collagen production in human dermal fibroblasts. Int J Mol Sci 17(6):955. https://doi.org/10.3390/ijms17060955

Chou HY, Ma DL, Leung CH, Chiu CC, Hour TC, Wang HD (2020) Purified astaxanthin from *Haematococcus pluvialis* promotes tissue regeneration by reducing oxidative stress and the secretion of collagen in vitro and in vivo. Oxid Med Cell Longev 2020:4946902. https://doi.org/10.1155/2020/4946902

Cort A, Ozturk N, Akpinar D et al (2010) Suppressive effect of astaxanthin on retinal injury induced by elevated intraocular pressure. Regul Toxicol Pharmacol 58(1):121–130

Costa R, Santos L (2017) Delivery systems for cosmetics - from manufacturing to the skin of natural antioxidants. Powder Technol 322:402–416. https://doi.org/10.1016/J.POWTEC.2017.07.086

Dai S, Zang J, Lv C (2022) Distribution, purification, and delivery of astaxanthin in food system. J Food Sci Technol 7(2):456–467

D'Arcangelo D, Giampietri C, Muscio M, Scatozza F, Facchiano F, Facchiano A (2018) WIPI1, BAG1, and PEX3 autophagy-related genes are relevant melanoma markers. Oxid Med Cell Longev 2018:1471682. https://doi.org/10.1155/2018/1471682

Darwesh OM, Barakat KM, Mattar MZ, Sabae SZ, Hassan SH (2019) Production of antimicrobial blue green pigment pyocyanin by marine Pseudomonas aeruginosa. Biointerface Appl Chem 9: 4334–4339

Davinelli S, Nielsen ME, Scapagnini G (2018) Astaxanthin in skin health, repair, and disease: a comprehensive review. Nutrients 10:522. https://doi.org/10.3390/nu10040522

Donoso A, González-Durán J, Muñoz AA, González PA, Agurto-Muñoz C (2021) Therapeutic uses of natural astaxanthin: an evidence-based review focused on human clinical trials. Pharmacol Res 166:105479. https://doi.org/10.1016/j.phrs.2021.105479

Dopierała K, Karwowska K, Petelska AD, Prochaska K (2020) Thermodynamic, viscoelastic and electrical properties of lipid membranes in the presence of astaxanthin. Biophys Chem 258: 106318. https://doi.org/10.1016/J.BPC.2019.106318

El-Agamy SE, Abdel-Aziz AK, Wahdan S, Esmat A, Azab SS (2018) Astaxanthin ameliorates doxorubicin-induced cognitive impairment (Chemobrain) in experimental rat model: impact on oxidative, inflammatory, and apoptotic machineries. Mol Neurobiol 55(7):5727–5740. https://doi.org/10.1007/s12035-017-0797-7

El-Baz FK, Aly HF, Abdo SM, Mahmoud R, Saad SA (2016) *Dunaliella salina* alleviates renal dysfunction and suppresses inflammatory cytokines in STZ-induced diabetic rats. Int J Pharm Sci Rev Res 38(2):210–215

Epstein FH, Steinberg D, Parthasarathy S, Carew TE, Khoo JC, Witztum JL (1989) Beyond cholesterol. N Engl J Med 320(14):915–924. http://www.ncbi.nlm.nih.gov/pubmed/2648148

Eurostat (2015) Meat production statistics. http://ec.europa.eu/eurostat/statisticsexplained/index.php/Meat_production_statistics#Pigmeat. Accessed July 2016

Falaise C, François C, Travers MA, Morga B, Haure J, Tremblay R, Turcotte F, Pasetto P, Gastineau R, Hardivillier Y, Leignel V, Mouget JL (2016) Antimicrobial compounds from eukaryotic microalgae against human pathogens and diseases in aquaculture. Mar Drugs 14(9): 159. https://doi.org/10.3390/md14090159

Fang Q, Guo S, Zhou H, Han R, Wu P, Han C (2017) Astaxanthin protects against early burn-wound progression in rats by attenuating oxidative stress-induced inflammation and mitochondria-related apoptosis. Sci Rep 7:41440

Fu M, Liang X, Zhang X, Yang M, Ye Q, Qi Y, Liu H, Zhang X (2022) Astaxanthin delays brain aging in senescence-accelerated mouse prone 10: inducing autophagy as a potential mechanism. Nutr Neurosci:1–11

Furr HC, Clark RM (1997) Intestinal absorption and tissue distribution of carotenoids. J Nutr Biochem 8(7):364–377. https://doi.org/10.1016/S0955-2863(97)00060-0

Giannaccare G, Pellegrini M, Senni C, Bernabei F, Scorcia V, Cicero AFG (2020) Clinical applications of astaxanthin in the treatment of ocular diseases: emerging insights. Mar Drugs 18(5):239. https://doi.org/10.3390/md18050239

Gilbert-López B, Mendiola JA, van den Broek LAM, Houweling-Tan B, Sijtsma L, Cifuentes A, Herrero M, Ibáñez E (2017) Green compressed fluid technologies for downstream processing of Scenedesmus obliquus in a biorefinery approach. Algal Res 24:111–121. https://doi.org/10.1016/J.ALGAL.2017.03.011

Giordano FJ (2005) Oxygen, oxidative stress, hypoxia, and heart failure. J Clin Invest 115(3): 500–508. https://doi.org/10.1172/JCI24408

Griffiths EJ (2012) Mitochondria and heart disease. Adv Exp Med Biol 942:249–267. https://doi.org/10.1007/978-94-007-2869-1_11

Gritz DC, Montes C, Atalla LR, Wu GS, Sevanian A, Rao NA (1991) Histochemical localization of superoxide production in experimental autoimmune uveitis. Curr Eye Res 10(10):927–931. https://doi.org/10.3109/02713689109020328

Guterres SS, Alves MP, Pohlmann AR (2007) Polymeric nanoparticles, nanospheres and nanocapsules, for cutaneous applications. Drug Target Insights 2:147. https://doi.org/10.1177/117739280700200002

Harasym J, Bogacz-Radomska L (2016) Colorants in foods-from past to present barwniki w żywności-od przeszłości do teraźniejszości. https://doi.org/10.15611/nit.2016.3.02

Hasan MR, Chakrabarti R (2009) Floating aquatic macrophytes -Duckweeds. Use of algae and aquatic macrophytes as feed in small-scale aquaculture – a review, pp 29–52

Haung HY, Wang YC, Cheng YC, Kang W, Hu SH, Liu D, Xiao C, Wang HD (2020) A novel oral astaxanthin nanoemulsion from *Haematococcus pluvialis* induces apoptosis in lung metastatic melanoma. Oxid Med Cell Longev 2020:2647670. https://doi.org/10.1155/2020/2647670

Hewlings SJ (2019) Human health benefits of astaxanthin derived from *Haematococcus pluvialis*: a review. EC Nutr 10:902–916

Higuera-Ciapara I, Felix-Valenzuela L, Goycoolea FM, Argüelles-Monal W (2004) Microencapsulation of astaxanthin in a chitosan matrix. Carbohydr Polym 56(1):41–45. https://doi.org/10.1016/J.CARBPOL.2003.11.012

Holst B, Williamson G (2008) Nutrients and phytochemicals: from bioavailability to bioefficacy beyond antioxidants. Curr Opin Biotechnol 19(2):73–82. https://doi.org/10.1016/J.COPBIO.2008.03.003

Hong L, Zhou CL, Chen FP, Han D, Wang CY, Li JX, Chi Z, Liu CG (2017) Development of a carboxymethyl chitosan functionalized nanoemulsion formulation for increasing aqueous solubility, stability and skin permeability of astaxanthin using low-energy method. J Microencapsul 34(8):707–721. https://doi.org/10.1080/02652048.2017.1373154

Hormozi M, Ghoreishi S, Baharvand P (2019) Astaxanthin induces apoptosis and increases activity of antioxidant enzymes in LS-180 cells. Artif Cells Nanomed Biotechnol 47:891–895

Hu Q, Gerhard H, Upadhyaya I, Venkitanarayanan K, Luo Y (2016) Antimicrobial eugenol nanoemulsion prepared by gum arabic and lecithin and evaluation of drying technologies. Int J Biol Macromol 87:130–140. https://doi.org/10.1016/j.ijbiomac.2016.02.051

Huangfu J, Liu J, Sun Z, Wang M, Jiang Y, Chen ZY, Chen F (2013) Antiaging effects of astaxanthin-rich alga *Haematococcus pluvialis* on fruit flies under oxidative stress. J Agric Food Chem 61(32):7800–7804. https://doi.org/10.1021/jf402224w

Imamura Y, Noda S, Hashizume K, Shinoda K, Yamaguchi M, Uchiyama S, Shimizu T, Mizushima Y, Shirasawa T, Tsubota K (2006) Drusen, choroidal neovascularization, and retinal pigment epithelium dysfunction in SOD1-deficient mice: a model of age-related macular degeneration. Proc Natl Acad Sci U S A 103(30):11282–11287. https://doi.org/10.1073/pnas.0602131103

Inborr J (1998) *Haematococcus*, the poultry pigmentor. Feed Mix 6:31–34

Ishikawa S, Hashizume K, Nishigori H, Tezuka Y, Sanbe A, Kurosaka D (2015) Effect of astaxanthin on cataract formation induced by glucocorticoids in the chick embryo. Curr Eye Res 40:535–540. https://doi.org/10.3109/02713683.2014.935445

Jafari Z, Bigham A, Sadeghi S, Dehdashti SM, Rabiee N, Abedivash A, Bagherzadeh M, Nasseri B, Karimi-Maleh H, Sharifi E, Varma RS, Makvandi P (2022) Nanotechnology-abetted astaxanthin formulations in multimodel therapeutic and biomedical applications. J Med Chem 65(1):2–36. https://doi.org/10.1021/acs.jmedchem.1c01144

Jain AK, Mehra NK, Swarnakar NK (2015) Role of antioxidants for the treatment of cardiovascular diseases: challenges and opportunities. Curr Pharm Des 21(30):4441–4455. https://doi.org/10.2174/1381612821666150803151758

Janani R, Anitha RE, Perumal MK, Divya P, Baskaran V (2021) Astaxanthin mediated regulation of VEGF through HIF1α and XBP1 signaling pathway: An insight from ARPE-19 cell and streptozotocin mediated diabetic rat model. Exp Eye Res 206:108555. https://doi.org/10.1016/j.exer.2021.108555

Jannel S, Caro Y, Bermudes M, Petit T (2020) Novel insights into the biotechnological production of *Haematococcus pluvialis*-derived astaxanthin: Advances and key challenges to allow its industrial use as novel food ingredient. J Mar Sci Eng 8:789

Jiang G, Li B, Liao X, Zhong C (2016) Poultry and fish intake and risk of esophageal cancer: A meta-analysis of observational studies. Asia Pac J Clin Oncol 2:e82–e91. https://doi.org/10.1111/ajco.12114

Jinu M, Chandra KM (2021) Astaxanthin: An algae-based natural compound with a potential role in human health-promoting effect: An updated comprehensive review. J Appl Biol Biotechnol 9:114–123

Johnson EA, An GH (2008) Astaxanthin from microbial sources 11(4):297–326. https://doi.org/10.3109/07388559109040622

Jonkman JE, Cathcart JA, Xu F, Bartolini M, Amon JE, Stevens KM, Colarusso P (2014) An introduction to the wound healing assay using live-cell microscopy. Cell Adh Migr 8(5): 440–451. https://doi.org/10.4161/cam.36224

Ju ZY, Deng DF, Dominy W (2012) A defatted microalgae (*Haematococcus pluvialis*) meal as a protein ingredient to partially replace fishmeal in diets of Pacific white shrimp (*Litopenaeus vannamei*, Boone, 1931). Aquaculture 354–355:50–55. https://doi.org/10.1016/J.AQUACULTURE.2012.04.028

Jyonouchi H, Zhang L, Gross M, Tomita Y (1994) Immunomodulating actions of carotenoids: enhancement of in vivo and in vitro antibody production to T dependent antigens. Nutr Cancer 21:47–58

Kaci M, Belhaffef A, Meziane S, Dostert G, Menu P, Velot É, Desobry S, Arab Tehrany E (2018) Nanoemulsions and topical creams for the safe and effective delivery of lipophilic antioxidant coenzyme Q10. Colloids Surf B Biointerfaces 167:165–175. https://doi.org/10.1016/j.colsurfb.2018.04.010

Kang CD, Sim SJ (2008) Direct extraction of astaxanthin from *Haematococcus* culture using vegetable oils. Biotechnol Lett 30(3):441–444. https://doi.org/10.1007/S10529-007-9578-0

Kang JO, Kim SJ, Kim H (2001) Effect of astaxanthin on the hepatotoxicity, lipid peroxidation and antioxidative enzymes in the liver of CCl4-treated rats. Methods Find Exp Clin Pharmacol 23(2):79–84. https://doi.org/10.1358/mf.2001.23.2.627931

Kato T, Kasai T, Sato A, Ishiwata S, Yatsu S, Matsumoto H et al (2020) Effects of 3-month astaxanthin supplementation on cardiac function in heart failure patients with left ventricular systolic dysfunction-a pilot study. Nutrients 12:1896

Khan SK, Malinski T, Mason RP, Kubant R, Jacob RF, Fujioka K, Denstaedt SJ, King TJ, Jackson HL, Hieber AD, Lockwood SF, Goodin TH, Pashkow FJ, Bodary PF (2010) Novel astaxanthin prodrug (CDX-085) attenuates thrombosis in a mouse model. Thromb Res 126:299–305

Kotake-Nara E, Kushiro M, Zhang H, Sugawara T, Miyashita K, Nagao A (2001) Carotenoids affect proliferation of human prostate cancer cells. J Nutr 131:3303–3306. https://doi.org/10.1093/jn/131.12.3303

Krestinina O, Baburina Y, Krestinin R (2021) Mitochondrion as a Target of Astaxanthin Therapy in Heart Failure. Int J Mol Sci 22(15):7964. https://doi.org/10.3390/ijms22157964

Kurihara H, Koda H, Asami S, Kiso Y, Tanaka T (2002) Contribution of the antioxidative property of astaxanthin to its protective effect on the promotion of cancer metastasis in mice treated with restraint stress. Life Sci 70(21):2509–2520. https://doi.org/10.1016/s0024-3205(02)01522-9

Landon R, Gueguen V, Petite H, Letourneur D, Pavon-Djavid G, Anagnostou F (2020) Impact of astaxanthin on diabetes pathogenesis and chronic complications. Mar Drugs 18(7):357. https://doi.org/10.3390/md18070357

Lee HS, Sung DK, Kim SH, Choi WI, Hwang ET, Choi DJ, Chang JH (2017) Controlled release of astaxanthin from nanoporous silicified-phospholipids assembled boron nitride complex for cosmetic applications. Appl Surf Sci 424:15–19. https://doi.org/10.1016/J.APSUSC.2017.03.036

Lim KC, Yusoff FM, Shariff M, Kamarudin MS (2018) Astaxanthin as feed supplement in aquatic animals. Rev Aquaculture 10(3):738–773. https://doi.org/10.1111/RAQ.12200

Lima SGM, Freire MCLC, Oliveira VDS, Solisio C, Converti A, de Lima ÁAN (2021) Astaxanthin delivery systems for skin application: a review. Mar Drugs 19(9):511. https://doi.org/10.3390/MD19090511

Little SM, Senhorinho GNA, Saleh M, Basiliko N, Scott JA (2021) Antibacterial compounds in green microalgae from extreme environments: a review. ALGAE 36(1):61–72. https://doi.org/10.4490/algae.2021.36.3.6

Liu T, Zhong S, Liao X, Chen J, He T, Lai S et al (2015) A meta-analysis of oxidative stress markers in depression. PLoS One 10(10):e0138904

Lu P, Weaver VM, Werb Z (2012) The extracellular matrix: a dynamic niche in cancer progression. J Cell Biol 196:395–406. https://doi.org/10.1083/jcb.201102147

Ma H, Chen S, Xiong H, Wang M, Hang W, Zhu X, Zheng Y, Ge B, Li R, Cui H (2020) Astaxanthin from *Haematococcus pluvialis* ameliorates the chemotherapeutic drug (doxorubicin) induced liver injury through the Keap1/Nrf2/HO-1 pathway in mice. Food Funct 11(5): 4659–4671. https://doi.org/10.1039/c9fo02429h

Machu L, Misurcova L, Ambrozova JV, Orsavova J, Mlcek J, Sochor J, Jurikova T (2015) Phenolic content and antioxidant capacity in algal food products. Molecules 20(1):1118–1133. https://doi.org/10.3390/MOLECULES20011118

Manabe E, Handa O, Naito Y, Mizushima K, Akagiri S, Adachi S, Takagi T, Kokura S, Maoka T, Yoshikawa T (2008) Astaxanthin protects mesangial cells from hyperglycemia-induced oxidative signaling. J Cell Biochem 103(6):1925–1937. https://doi.org/10.1002/jcb.21583

McCall B, Mc Partland CK, Moore R, Frank-Kamenetskii A, Booth BW (2018) Effects of astaxanthin on the proliferation and migration of breast cancer cells in vitro. Antioxidants (Basel) 7(10):135. https://doi.org/10.3390/antiox7100135

McSherry EA, Donatello S, Hopkins AM, McDonnell S (2007) Molecular basis of invasion in breast cancer. Cell Mol Life Sci 64(24):3201–3218. https://doi.org/10.1007/S00018-007-7388-0

Meephansan J, Rungjang A, Yingmema W, Deenonpoe R, Ponnikorn S (2017) Effect of astaxanthin on cutaneous wound healing. Clin Cosmet Investig Dermatol 10(259):265

Miao F, Lu D, Li Y, Zeng M (2006) Characterization of astaxanthin esters in *Haematococcus pluvialis* by liquid chromatography–atmospheric pressure chemical ionization mass spectrometry. Anal Biochem 352(2):176–181. https://doi.org/10.1016/J.AB.2006.03.006

Mofasser Hossain AKM, Brennan MA, Mason SL, Guo X, Zeng XA, Brennan CS (2017) The effect of astaxanthin-rich microalgae "*Haematococcus pluvialis*" and wholemeal flours incorporation in improving the physical and functional properties of cookies. Foods 6(8):57. https://doi.org/10.3390/FOODS6080057

Molino A, Iovine A, Casella P, Mehariya S, Chianese S, Cerbone A, Rimauro J, Musmarra D (2018) Microalgae characterization for consolidated and new application in human food, animal feed and nutraceuticals. Int J Environ Res Public Health 15(11):2436. https://doi.org/10.3390/ijerph15112436

Mora-Huertas CE, Fessi H, Elaissari A (2010) Polymer-based nanocapsules for drug delivery. Int J Pharm 385(1–2):113–142. https://doi.org/10.1016/J.IJPHARM.2009.10.018

Mourad RM, Darwesh OM, Abdel-Hakim A (2020) Enhancing physico-mechanical and antibacterial properties of natural rubber using synthesized Ag-SiO_2 nanoparticles. Int J Biol Macromol 164:3243–3249. https://doi.org/10.1016/j.ijbiomac.2020.08.063

Mularczyk M, Michalak I, Marycz K (2020) Astaxanthin and other nutrients from *Haematococcus pluvialis*-multifunctional applications. Mar Drugs 18(9):459. https://doi.org/10.3390/md18090459

Nagaki Y et al (2005) The effect of astaxanthin on retinal capillary blood flow in normal volunteers. J Clin Ther Med 21(5):73–78

Naguib YM (2000) Antioxidant activities of astaxanthin and related carotenoids. J Agric Food Chem 48:1150–1154. https://doi.org/10.1021/jf991106k

Nakagawa K, Kiko T, Miyazawa T, CarpenteroBurdeos G, Kimura F, Satoh A, Miyazawa T (2011) Antioxidant effect of astaxanthin on phospholipid peroxidation in human erythrocytes. Br J Nutr 105:1563–1571

Nakano M, Onodera A, Saito E, Tanabe M, Yajima K, Takahashi J, Chuyen NV (2008) Effect of astaxanthin in combination with α-tocopherol or ascorbic acid against oxidative damage in diabetic ODS rats. J Nutr Sci Vitaminol 54:329–334

Nawrocka D, Kornicka K, Śmieszek A, Marycz K (2017) *Spirulina platensis* improves mitochondrial function impaired by elevated oxidative stress in adipose-derived mesenchymal stromal cells (ASCs) and intestinal epithelial cells (IECs), and enhances insulin sensitivity in equine metabolic syndrome (EMS) horses. Mar Drugs 15(8):237. https://doi.org/10.3390/MD15080237

Niesman MR, Johnson K, A., & Penn, J. S. (1997) Therapeutic effect of liposomal superoxide dismutase in an animal model of retinopathy of prematurity. Neurochem Res 22(5):597–605. https://doi.org/10.1023/a:1022474120512

Nishida Y, Yamashita E, Miki W (2007) Quenching activities of common hydrophilic and lipophilic antioxidants against singlet oxygen using chemiluminescence detection system. Carotenoid Sci 11:16–20

Nootem J, Chalorak P, Meemon K, Mingvanish W, Pratumyot K, Ruckthong L, Srisuwannaket C, Niamnont N (2018) Electrospun cellulose acetate doped with astaxanthin derivatives from *Haematococcus pluvialis* for *in vivo* anti-aging activity. RSC Adv 8:37151–37158

Nurdianti L, Rusdiana T, Sopyan I (2020) Antidiabetic activity of thin film containing astaxanthin-loaded nanoemulsion using carboxymethylcellulose sodium polymer on alloxan-induced diabetic rabbit. J Adv Pharm Technol Res 11(4):189–193. https://doi.org/10.4103/japtr.JAPTR_55_20

O'Neill J (2016) Tackling drug-resistant infections globally: final report and recommendations

Oh H, Lee JS, Sung D, Lim JM, Choi WI (2020) Potential antioxidant and wound healing effect of nano-liposol with high loading amount of astaxanthin. Int J Nanomedicine 15:9231–9240. https://doi.org/10.2147/IJN.S272650

Ojha N, Roy S, He G, Biswas S, Velayutham M, Khanna S, Kuppusamy P, Zweier JL, Sen CK (2008) Assessment of wound-site redox environment and the significance of Rac2 in cutaneous healing. Free Radic Biol Med 44(4):682–691

Onacik-gür S, Zbikowska A, Majewska B (2018) Effect of spirulina (*Spirulina platensis*) addition on textural and quality properties of cookies. Italian J Food Sci 30(1):2018–2020. https://doi.org/10.14674/IJFS-702

Oslan SNH, Tan JS, Oslan SN, Matanjun P, Mokhtar RAM, Shapawi R, Huda N (2021) *Haematococcus pluvialis* as a potential source of astaxanthin with diverse applications in industrial sectors: current research and future directions. Molecules 26(21):6470. https://doi.org/10.3390/molecules26216470

Otsuka T, Shimazawa M, Nakanishi T, Ohno Y, Inoue Y, Tsuruma K, Ishibashi T, Hara H (2013) Protective effects of a dietary carotenoid, astaxanthin, against light-induced retinal damage. J Pharmacol Sci 123(3):209–218. https://doi.org/10.1254/jphs.13066fp

Palozza P, Torelli C, Boninsegna A, Simone R, Catalano A, Mele MC et al (2009) Growth-inhibitory effects of the astaxanthin-rich alga *Haematococcus pluvialis* in human colon cancer cells. Cancer Lett 283:108–117. https://doi.org/10.1016/j.canlet.2009.03.031

Pan L, Wang H, Gu K (2018) Nanoliposomes as vehicles for astaxanthin: characterization, in vitro release evaluation and structure. Molecules 23(11):2822. https://doi.org/10.3390/MOLECULES23112822

Pane G, Cacciola G, Giacco E, Mariottini GL, Coppo E (2015) Assessment of the antimicrobial activity of algae extracts on bacteria responsible of external otitis. Mar Drugs 13(10):6440–6452. https://doi.org/10.3390/md13106440

Pan-utai, W., Boonpok, S., & Pornpukdeewattana, S. (2021). Combination of mechanical and chemical extraction of astaxanthin from *Haematococcus pluvialis* and its properties of micro-encapsulation. Biocatalysis and Agricultural Biotechnology, 33, 101979. https://doi.org/https://doi.org/10.1016/J.BCAB.2021.101979

Patil AD, Kasabe PJ, Dandge PB (2022) Pharmaceutical and nutraceutical potential of natural bioactive pigment: astaxanthin. *Nat Prod Bioprospect.* 12(1):25. https://doi.org/10.1007/s13659-022-00347-y

Peng CH, Chang CH, Peng RY, Chyau CC (2010) Improved membrane transport of astaxanthine by liposomal encapsulation. Eur J Pharm Biopharm 75(2):154–161. https://doi.org/10.1016/J.EJPB.2010.03.004

Petyaev IM, Klochkov V, Chalyk N, Pristensky D, Chernyshova M, Kyle N, Bashmakov Y (2018) Markers of hypoxia and oxidative stress in aging volunteers ingesting lycosomal formulation of dark chocolate containing astaxanthin. J Nutr Health Aging 22:1092–1098. https://doi.org/10.1007/s12603-018-1063-z

Ponto T, Latter G, Luna G, Leite-Silva VR, Wright A, Benson HAE (2021) Novel self-nano-emulsifying drug delivery systems containing astaxanthin for topical skin delivery. Pharmaceutics 13(5):649. https://doi.org/10.3390/pharmaceutics13050649

Prabakaran G, Moovendhan M, Arumugam A, Matharasi A, Dineshkumar R, Sampathkumar P (2018) Quantitative analysisof phytochemical profile in marine microalgae *Chlorella vulgaris*. Int J Pharm Biol Sci 8(2):562–565

Prabhu PN, Ashokkumar P, Sudhandiran G (2009) Antioxidative and antiproliferative effects of astaxanthin during the initiation stages of 1,2-dimethyl hydrazine-induced experimental colon carcinogenesis. Fundam Clin Pharmacol 23(2):225–234. https://doi.org/10.1111/j.1472-8206.2009.00669.x

Qu X, Ben Q, Jiang Y (2013) Consumption of red and processed meat and risk for esophageal squamous cell carcinoma based on a meta-analysis. Ann Epidemiol 23:762–770. https://doi.org/10.1016/j.annepidem.2013.09.003

Ranga Rao A, Raghunath Reddy RL, Baskaran V, Sarada R, Ravishankar GA (2010) Characterization of microalgal carotenoids by mass spectrometry and their bioavailability and antioxidant properties elucidated in rat model. J Agric Food Chem 58:8553–8559

Ranga Rao A, Baskaran V, Sarada R, Ravishankar GA (2013) *In vivo* bioavailability and antioxidant activity of carotenoids from microalgal biomass - a repeated dose study. Food Res Int 54(1):711–717. https://doi.org/10.1016/J.FOODRES.2013.07.067

Régnier P, Bastias J, Rodriguez-Ruiz V, Caballero-Casero N, Caballo C, Sicilia D, Fuentes A, Maire M, Crepin M, Letourneur D, Gueguen V, Rubio S, Pavon-Djavid G (2015) Astaxanthin from *Haematococcus pluvialis* Prevents Oxidative Stress on Human Endothelial Cells without Toxicity. Mar Drugs 13(5):2857–2874. https://doi.org/10.3390/MD13052857

Roerecke M, Shield KD, Higuchi S, Yoshimura A, Larsen E, Rehm MX, Rehm J (2015) Estimates of alcohol-related oesophageal cancer burden in Japan: systematic review and meta-analyses. Bull World Health Organ 93:329–338. https://doi.org/10.2471/BLT.14.142141

Rossolini GM, Arena F, Pecile P, Pollini S (2014) Update on the antibiotic resistance crisis. Curr Opin Pharmacol 18:56–60. https://doi.org/10.1016/j.coph.2014.09.006

Ruiz-Domínguez MC, Espinosa C, Paredes A, Palma J, Jaime C, Vílchez C, Cerezal P (2019) Determining the potential of *Haematococcus pluvialis* oleoresin as a rich source of antioxidants. Molecules 24(22). https://doi.org/10.3390/molecules24224073

Sahni P, Sharma S, Singh B (2019) Evaluation and quality assessment of defatted microalgae meal of *Chlorella* as an alternative food ingredient in cookies. Nutr Food Sci 49(2):221–231. https://doi.org/10.1108/NFS-06-2018-0171

Salehi M, Moradi-Lakeh M, Salehi MH, Nojomi M, Kolahdooz F (2013) Meat, fish, and esophageal cancer risk: A systematic review and dose-response meta-analysis. Nutr Rev 71:257–267. https://doi.org/10.1111/nure.12028

Sarada R, Vidhyavathi R, Usha D, Ravishankar GA (2006) An efficient method for extraction of astaxanthin from green alga *Haematococcus pluvialis*. J Agric Food Chem 54(20):7585–7588. https://doi.org/10.1021/JF060737T

Schaumberg DA, Sullivan DA, Buring JE, Dana MR (2003) Prevalence of dry eye syndrome among US women. Am J Ophthalmol 136(2):318–326. https://doi.org/10.1016/s0002-9394(03)00218-6

Scorcia V, Soda M, Lucisano A, Lanza M, Giannaccare G (2020) A Preliminary Comparative Study of Visual Performance Between Two Newly Commercially Available Monofocal Intraocular Lenses Implanted During Cataract Surgery. Clin Ophthalmol 14:831–835. https://doi.org/10.2147/OPTH.S246319

Seong-Ryeong L, Do-Wan K, Sung J, Kim TH, Chang-Hyung C, Sei-Jung L (2021) Astaxanthin inhibits autophagic cell death induced by bisphenol A in human dermal fibroblasts. Antioxidants 10:1273. https://doi.org/10.3390/antiox10081273

Shah MMR, Liang Y, Cheng JJ, Daroch M (2016) Astaxanthin-producing green microalga *Haematococcus pluvialis*. Single cell to high value commercial products. Front Plant Sci 7: 531. https://doi.org/10.3389/fpls.2016.00531

Shanmugapriya K, Kim H, Saravana PS, Chun BS, Kang HW (2018) Astaxanthin-Alpha Tocopherol Nanoemulsion Formulation by Emulsification Methods: Investigation on Anticancer, Wound Healing, and Antibacterial Effects. Colloids Surf B Biointerfaces 172:170–179

Sheikhzadeh N, Tayefi-Nasrabadi H, Oushani AK, Enferadi MHN (2012) Effects of *Haematococcus pluvialis* supplementation on antioxidant system and metabolism in rainbow trout (*Oncorhynchus mykiss*). Fish Physiol Biochem 38(2):413–419. https://doi.org/10.1007/S10695-011-9519-7

Shimokawa T, Yoshida M, Fukuta T, Tanaka T, Inagi T, Kogure K (2019) Efficacy of high-affinity liposomal astaxanthin on up-regulation of age-related markers induced by oxidative stress in human corneal epithelial cells. J Clin Biochem Nutr 64(1):27–35. https://doi.org/10.3164/jcbn.18-27

Siegel RL, Miller KD, Jemal A (2018) Cancer statistics. CA Cancer J Clin 68(1):7–30

Siegel RL, Miller KD, Jemal A (2019) Cancer statistics, 2019. CA Cancer J Clin 69(1):7–34. https://doi.org/10.3322/CAAC.21551

Sies H, Berndt C, Jones DP (2017) Oxidative stress Annu Rev Biochem 86, 715–748. https://doi.org/10.1146/ANNUREV-BIOCHEM-061516-045037

Singh S, Sharma AN, Murad MH, Buttar NS, El-Serag HB, Katzka DA, Iyer PG (2013) Central adiposity is associated with increased risk of esophageal inflammation, metaplasia, and adenocarcinoma: a systematic review and meta-analysis. Clin Gastroenterol Hepatol 11:1399–1412. https://doi.org/10.1016/j.cgh.2013.05.009

Smith VJ, Desbois AP, Dyrynda EA (2010) Conventional and unconventional antimicrobials from fish, marine invertebrates and micro-algae. Mar Drugs 8(4):1213–1262. https://doi.org/10.3390/md8041213

Sorrenti V, Davinelli S, Scapagnini G, Willcox BJ, Allsopp RC, Willcox DC (2020) Astaxanthin as a putative geroprotector: molecular basis and focus on brain aging. Mar Drugs 18:351

Specht EA, Mayfield SP, Rosano GL, Rybicki E (2014) Algae-based oral recombinant vaccines. Front Microbiol 5:60. https://doi.org/10.3389/fmicb.2014.00060

Spector A (1995) Oxidative stress-induced cataract: mechanism of action. FASEB J 9(12): 1173–1182. https://doi.org/10.1096/fasebj.9.12.7672510

Spiller GA, Dewell A (2003) Safety of an astaxanthin-rich *Haematococcus pluvialis* algal extract: a randomized clinical trial. J Med Food 6(1):51–56. https://doi.org/10.1089/109662003765184741

Stewart JS, Lignell Å, Pettersson A, Elfving E, Soni MG (2008) Safety assessment of astaxanthin-rich microalgae biomass: Acute and subchronic toxicity studies in rats. Food Chem Toxicol 46(9):3030–3036. https://doi.org/10.1016/J.FCT.2008.05.038

Stivala L, Savio M, Quarta S, Scotti C, Cazzalini O, Rossi L et al (2000) The antiproliferative effect of b-carotene requires p21(Waf1/cip1) in normal human fibroblasts. Eur J Biochem FEBS 267: 2290–2296. https://doi.org/10.1046/j.1432-1327.2000.01236.x

Suganuma K, Nakajima H, Ohtsuki M, Imokawa G (2010) Astaxanthin attenuates the UVA-induced up-regulation of matrix-metalloproteinase-1 and skin fibroblast elastase in human dermal fibroblasts. J Dermatol Sci 58(2):136–142. https://doi.org/10.1016/J.JDERMSCI.2010.02.009

Sung H, Ferlay J, Siegel RL, Laversanne M, Soerjomataram I, Jemal A, Bray F (2021) Global cancer statistics 2020: GLOBOCAN estimates of incidence and mortality worldwide for 36 cancers in 185 countries. CA Cancer J Clin 71(3):209–249. https://doi.org/10.3322/CAAC.21660

Szentesi P, Csernoch L, Dux L, Keller-Pintér A (2019) Changes in redox signaling in the skeletal muscle with aging. Oxid Med Cell Longev 2019:1–12

Sztretye M, Dienes B, Gönczi M, Czirják T, Csernoch L, Dux L, Szentesi P, Keller-Pintér A (2019) Astaxanthin: a potential mitochondrial-targeted antioxidant treatment in diseases and with aging. Oxid Med Cell Longev 2019:3849692

Tachaprutinun A, Udomsup T, Luadthong C, Wanichwecharungruang S (2009) Preventing the thermal degradation of astaxanthin through nanoencapsulation. Int J Pharm 374(1–2):119–124. https://doi.org/10.1016/J.IJPHARM.2009.03.001

Talukdar J, Bhadra B, Dattaroy T, Nagle V, Dasgupta S (2020) Potential of natural astaxanthin in alleviating the risk of cytokine storm in COVID-19. Biomed Pharmacother 132:110886

Tanaka T, Makita H, Ohnishi M, Mori H, Satoh K, Hara A (1995) Chemoprevention of rat oral carcinogenesis by naturally occurring xanthophylls, astaxanthin and canthaxanthin. Cancer Res 55(18):4059–4064

Tian L, Wen Y, Li S, Zhang P, Wang Y, Wang J, Cao K, Du L, Wang N, Jie Y (2022) Benefits and safety of astaxanthin in the treatment of mild-to-moderate dry eye disease. Front Nutr 8:796951. https://doi.org/10.3389/fnut.2021.796951

Torrissen OJ (2000) Dietary delivery of carotenoids. In: Decker E, Faustman C, Lopez-Bote CJ (eds) Antioxidants in muscle foods: nutritional strategies to improve quality. Wiley, Hoboken, NJ, pp 289–313

Tseng CC, Lin YJ, Liu W, Lin HY, Chou HY, Thia C, Wu JH, Chang JS, Wen ZH, Chang JJ, David Wang HM (2020) Metabolic engineering probiotic yeast produces 3S, 3′S-astaxanthin to inhibit B16F10 metastasis. Food Chem Toxicol 135:110993. https://doi.org/10.1016/J.FCT.2019.110993

Turck D, Castenmiller J, de Henauw S, Hirsch-Ernst KI, Kearney J, Maciuk A, Mangelsdorf I, McArdle HJ, Naska A, Pelaez C, Pentieva K, Siani A, Thies F, Tsabouri S, Vinceti M, Cubadda F, Engel KH, Frenzel T, Heinonen M et al (2020) Safety of astaxanthin for its use as a novel food in food supplements. EFSA J 18(2):e05993. https://doi.org/10.2903/J.EFSA.2020.5993

Uchiyama K, Naito Y, Hasegawa G, Nakamura N, Takahashi J, Yoshikawa T (2002) Astaxanthin protects beta-cells against glucose toxicity in diabetic db/db mice. Redox Rep 7(5):290–293. https://doi.org/10.1179/135100002125000811

Valko M, Leibfritz D, Moncol J, Cronin MTD, Mazur M, Telser J (2007) Free radicals and antioxidants in normal physiological functions and human disease. Int J Biochem Cell Biol 39(1):44–84. https://doi.org/10.1016/J.BIOCEL.2006.07.001

Vasudevan S, Venkatraman A, Yahoob SAM, Jojula M, Sundaram R, Boomi P (2021) Biochemical evaluation and molecular docking studies on encapsulated astaxanthin for the growth inhibition of Mycobacterium tuberculosis. J Appl Biol Biotechnol 9:31–39

Veeruraj A, Liu L, Zheng J, Wu J, Arumugam M (2019) Evaluation of astaxanthin incorporated collagen film developed from the outer skin waste of squid Doryteuthis singhalensis for wound healing and tissue regenerative applications. Mater Sci Eng C Mater Biol Appl 95:29–42. https://doi.org/10.1016/j.msec.2018.10.055

Venza M, Visalli M, Beninati C, De Gaetano GV, Teti D, Venza I (2015) Cellular mechanisms of oxidative stress and action in melanoma. Oxid Med Cell Longev 2015:481782. https://doi.org/10.1155/2015/481782

Verschooten L, Declercq L, Garmyn M (2006) Adaptive response of the skin to UVB damage: role of the p53 protein. Int J Cosmet Sci 28(1):1–7. https://doi.org/10.1111/j.1467-2494.2006.00299.x

Vieira MV, Pastrana LM, Fuciños P (2020) Microalgae encapsulation systems for food, pharmaceutical and cosmetics applications. Mar Drugs 18(12):644. https://doi.org/10.3390/md18120644

Vieira MV, Derner RB, Lemos-Senna E (2021) Preparation and characterization of *Haematococcus pluvialis* carotenoid-loaded PLGA nanocapsules in a gel system with antioxidant properties for topical application. J Drug Deliv Sci Technol 61:102099. https://doi.org/10.1016/j.jddst.2020.102099

Votruba M, Gregor Z (2001) Neovascular age-related macular degeneration: present and future treatment options. Eye 15(3):424–429. https://doi.org/10.1038/eye.2001.147

Wakabayashi K, Hamada C, Kanda R, Nakano T, Io H, Horikoshi S, Tomino Y (2015) Oral astaxanthin supplementation prevents peritoneal fibrosis in rats. Perit Dial Int 35:506–516. https://doi.org/10.3747/pdi.2013.00317

Waldenstedt L, Inborr J, Hansson I, Elwinger K (2003) Effects of astaxanthin-rich algal meal (*Haematococcus pluvalis*) on growth performance, caecal campylobacter and clostridial counts and tissue astaxanthin concentration of broiler chickens. Anim Feed Sci Technol 108(1–4): 119–132. https://doi.org/10.1016/S0377-8401(03)00164-0

Wang HMD, Chen CC, Huynh P, Chang JS (2015) Exploring the potential of using algae in cosmetics. Bioresour Technol 184:355–362. https://doi.org/10.1016/J.BIORTECH.2014.12.001

Wang HMD, Li XC, Lee DJ, Chang JS (2017a) Potential biomedical applications of marine algae. Bioresour Technol 244(Pt 2):1407–1415. https://doi.org/10.1016/J.BIORTECH.2017.05.198

Wang Q, Zhao Y, Guan L, Zhang Y, Dang Q, Dong P, Li J, Liang X (2017b) Preparation of astaxanthin-loaded DNA/chitosan nanoparticles for improved cellular uptake and antioxidation capability. Food Chem 227:9–15. https://doi.org/10.1016/J.FOODCHEM.2017.01.081

WHO (2022) WHO fact sheets (World Health Organization, 2021). https://www.who.int/news-room/fact-sheets/detail/cardiovascular-diseases-(cvds). Accessed 24 Apr 2022

Wu C, Cederbaum AI (2003) Alcohol, oxidative stress, and free radical damage. Alcohol Res Health 27(4):277–284

Xu L, Yu H, Sun H, Yu X, Tao Y (2019) Optimized nonionic emulsifier for the efficient delivery of astaxanthin nanodispersions to retina: *in vivo* and *ex vivo* evaluations. Drug Deliv 26(1): 1222–1234. https://doi.org/10.1080/10717544.2019.1682718

Yan N, Fan C, Chen Y, Hu Z (2016) The potential for microalgae as bioreactors to produce pharmaceuticals. Int J Mol Sci 17(6):962. https://doi.org/10.3390/IJMS17060962

Yan T, Li H, Wu J et al (2017) Astaxanthin inhibits gemcitabine-resistant human pancreatic cancer progression through EMT inhibition and gemcitabine resensitization. Oncol Lett 14(5): 5400–5408

Yang YX, Kim YJ, Jin Z, Lohakare JD, Kim CH, Ohh SH, Lee SH, Choi JY, Chae BJ (2006) Effects of dietary supplementation of astaxanthin on production performance, egg quality in layers and meat quality in finishing pigs. Asian Australas J Anim Sci 19(7):1019–1025. https://doi.org/10.5713/AJAS.2006.1019

Yang Y, Seo JM, Nguyen A, Pham TX, Park HJ, Park Y, Kim B, Bruno RS, Lee J (2011) Astaxanthin-rich extract from the green alga *Haematococcus pluvialis* lowers plasma lipid concentrations and enhances antioxidant defense in apolipoprotein E knockout mice. J Nutr 141(9):1611–1617. https://doi.org/10.3945/jn.111.142109

Yang M, Chen Y, Zhao T, Wang Z (2020) Effect of astaxanthin on metabolic cataract in rats with type 1 diabetes mellitus. Exp Mol Pathol 113:104372. https://doi.org/10.1016/j.yexmp.2020.104372

Yoshida H, Yanai H, Ito K, Tomono Y, Koikeda T, Tsukahara H, Tada N (2010) Administration of natural astaxanthin increases serum HDL cholesterol and adiponectin in subjects with mild hyperlipidemia. Atherosclerosis 209:520–523

Zanoni F, Vakarelova M, Zoccatelli G (2019) Development and characterization of astaxanthin-containing whey protein-based nanoparticles. Mar Drugs 17(11):627. https://doi.org/10.3390/MD17110627

Zhang J, Sun Z, Sun P, Chen T, Chen F (2014) Microalgal carotenoids: beneficial effects and potential in human health. Food Funct 5(3):413–425. https://doi.org/10.1039/C3FO60607D

Zhao T, Yan X, Sun L, Yang T, Hu X, He Z, Liu F, Liu X (2019) Research progress on extraction, biological activities and delivery systems of natural astaxanthin. Trends Food Sci Technol 91: 354–361. https://doi.org/10.1016/J.TIFS.2019.07.014

Zhuge F, Ni Y, Wan C, Liu F, Fu Z (2021) Anti-diabetic effects of astaxanthin on an STZ-induced diabetic model in rats. Endocr J 68(4):451–459. https://doi.org/10.1507/endocrj.EJ20-0699

Zuluaga M, Gueguen V, Pavon-Djavid G, Letourneur D (2017) Carotenoids from microalgae to block oxidative stress. Bioimpacts 7:1–3

Chapter 13
Clinical Applications of *Haematococcus*

Mohammed Hassi, Oumaima Ouaddi, Raja ElKheloui, Rachid Bouharroud, Fatima Hamadi, and Mohamed Alouani

13.1 Introduction

Microalgae represent an important source for several bioactive substances that give them a great potential to be used for human health and medicine (Alam et al. 2020). Among microalgae bioactive molecules, we find poly-unsaturated fatty acids (PUFAs), polysaccharides, carotenoids, vitamins, sterols, proteins, enzymes, and others. These metabolites have broad therapeutic applications as antioxidant, anti-inflammatory, anticancer, antidiabetic, antimicrobial, antiviral, and anti-allergy agents (Zhuang et al. 2022; Khavari et al. 2021). For example, many species of microalgae are regarded to be a promising source of PUFAs such as omega 6 fatty acids and omega 3 fatty acids (Zhou et al. 2022; Borowitzka 2013). They are involved in the inhibition and action of numerous diseases such as thrombosis, atherosclerosis, arthritis, and the risk of heart disease (Zhou et al. 2022; Rahman 2020).

M. Hassi (✉)
Faculty of Sciences, Laboratory of Microbial Biotechnology and Plants Protection, Ibn Zohr University, Agadir, Morocco

Taroudant's Polydisciplinary Faculty, Ibn Zohr University, Taroudant, Morocco
e-mail: m.hassi@uiz.ac.ma

O. Ouaddi · R. ElKheloui · F. Hamadi
Faculty of Sciences, Laboratory of Microbial Biotechnology and Plants Protection, Department of Biology, Ibn Zohr University, Agadir, Morocco

R. Bouharroud
Regional Center of Agricultural Research of Agadir, National Institute for Agricultural Research, Rabat, Morocco

M. Alouani
Laboratory of Biotechnologies and Valorization of Natural Resources, Faculty of Sciences - Agadir, Ibn Zohr University, Agadir, Morocco

Faculty of Applied Sciences - Ait Melloul, Ibn Zohr University, Agadir, Morocco

R. Raja et al. (eds.), *Haematococcus*, https://doi.org/10.1007/978-981-99-2901-6_13

Many species of microalgae are able to produce PUFAs, the most important ones are: *Arthrospira platensis, Isochrysis galbana, Odontella, Porphyridium cruentum, Cryptothecodinium cohni,* etc. (Zhou et al. 2022; Guedes et al. 2011; da Silva et al. 2006). Moreover, microalgae are also high producers of polysaccharides which are also largely found in all living organisms (plants, animals, macroalgae, and microorganisms). However, the polysaccharides of microalgae have attracted more attention in recent years due to their various applications in food, cosmetics, and especially health (Caetano et al. 2022). The most studied microalgae for polysaccharides production are *Chlamydomonas reinhardtii, Spirulina platensis, Scenedesmus* sp., *Isochrysis* sp., *Tetraselmis* sp., *Rhodella reticulata, Porphyridium* spp., *Porphyridium cruentum, Porphyridium aerugineum, Chlorella* sp., *Dunaliella salina, Anabaena* sp., *Nostoc* sp., *Botryococcus braunii, Haematococcus pluvialis,* and *Chlorella vulgaris* (Morais et al. 2022; Costa et al. 2021; Delattre et al. 2016).

Some prominent pharmacological properties attributed to microalgae polysaccharides are their immunomodulatory, anti-inflammatory, antitumor, prebiotic, anticoagulant, antibacterial, hypoglycemic, antioxidant, and enhanced skin tissue regeneration abilities showed (Lekshmi et al. 2019; de Jesus Raposo et al. 2014). β-carotene, synthesized by many microalgae such *Dunaliella salina, Chlamydomonas reinhardtii,* and *Isochrysis galbana,* has demonstrated significant anti-inflammatory and immunomodulatory effects and can be used in the prevention and treatment of colon and prostate carcinoma (Ávila-Román et al. 2021; Sathasivam and Ki 2018). Astaxanthin (AST) is another carotenoid present in a variety of microalgae such *Haematococcus lacustris, Chlorella vulgaris, Chromochloris zofingiensis, Chlorococcum sp., Dunaliella salina,* and *Tetraselmis suecica* (Ávila-Román et al. 2021; Sathasivam et al. 2019). It is one of the most abundant aquatic carotenoids (Ambati et al. 2014) which the annual profit is over $200 million (Zhang et al. 2016).

AST is known as an effective antioxidant, which has very important medical applications such as protecting the skin, improving immune function, and protecting against cancer and inflammatory diseases (Deniz et al. 2017; Hamed 2016; Chu 2012). In microalgae, *Haematococcus lacustris (H. lacustris),* also named *H. pluvialis* Flotow in the scientific literature (Göksan et al. 2010), is the richest source of natural AST and is now cultivated at industrial scale (Guerin et al. 2003). This species is the most important in the genus *Haematococcus* and it is the most studied and described. Recently, more consideration has been given to the medical applications of *Haematococcus* AST due to its unique structural and chemical properties and also for its safety which has been demonstrated by numerous clinical and preclinical studies (Miyakawa 2021; Capelli et al. 2019). This chapter presents a review of current scientific works regarding the medical applications of microalgae of the genus *Haematococcus* and mainly *H. lacustris* which is considered as the richest source of natural AST. The areas covered in this study are as follows:

- *H. lacustris* and AST production;
- *Antioxidant activity of Haematococcus* AST;
- Anticancer activity of *Haematococcus* AST;

- *Haematococcus* AST for skin health;
- *Haematococcus* AST and neurodegenerative diseases;
- The anti-inflammation mechanisms of AST;
- *Haematococcus* AST and eye health.

13.2 *Haematococcus lacustris* and Astaxanthin Production

The genus *Haematococcus* (Chlorophyceae, Volvocales) is a unicellular green microalga that is cosmopolitan and reported from all continents (Klochkova et al. 2013). Eight species have been assigned to this genus based on a recent genetic classification and among them, *H. lacustris* (previously named *H. pluvialis*) is the most studied and is considered as the type species of *Haematococcus* (Kim et al. 2021; Mazumdar et al. 2018; Nakada and Ota 2016; Klochkova et al. 2013). Moreover, This species is considered as the most hopeful microalgae for production of natural AST (Oslan et al. 2022). It is also able to synthesize others compounds with a wide range biological activities, such as carotenoids, proteins, and fatty acids (Darwesh et al. 2022; Nishshanka et al. 2022). *H. lacustris* has a unique life cycle which can be divided into two-stages: namely the "green stage" and the "red stage (Fábregas et al. 2001) (Fig. 13.1). During the first phase, favorable growth conditions for *H. lacustris* strains are exposed to facilitate biomass growth. *H. lacustris* accumulates for AST occurs in the red stage under stress conditions (e.g., nutrient depletion, light stress, high acidity, and temperature fluctuations) (Li et al. 2020a; Xi et al. 2016).

AST is a red pigment classified as xantophyll that are oxygenated derivatives of carotenoids. The molecular formula of AST is $C_{40}H_{52}O_4$. It has two hydroxyls and

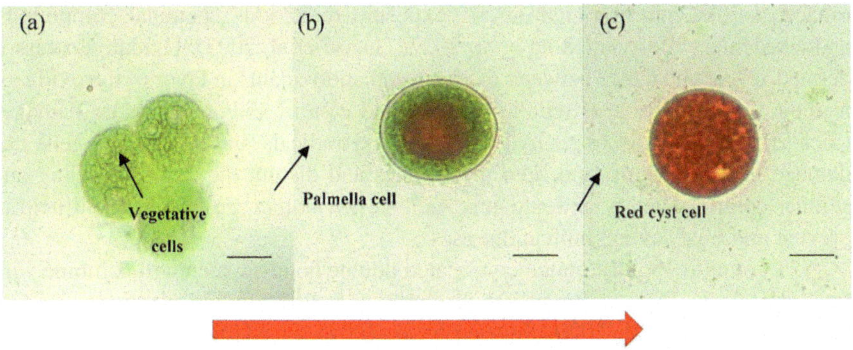

Fig. 13.1 Light microscopic images of *H. lacustris* cells and the astaxanthin accumulation (**a**) vegetative cells (at the beginning of the culture), (**b**) astaxanthin-accumulating (the intermediate stage), (**c**) red cyst cell (at the end of the culture). Scalebar: 10 μm (Oslan et al. 2021). (Figure from an OA article licensed under a Creative Commons Attribution-NonCommercial-No Derivatives 4.0 International License)

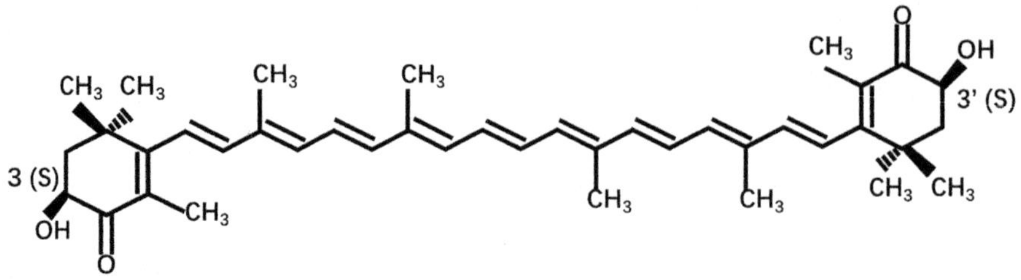

Fig. 13.2 Astaxanthin 3S, 3′S (3,3′-dihydroxy-4,4′-diketo-β-carotene)

two keto groups at 3,3′ and 4,4′ positions of the β-ionone rings, respectively, and its chemical formula is 3,3′-dihydroxy-β, β′-carotene-4,4′-dione (Fig. 13.2) (Patel et al. 2022). The presence of the hydroxyl and keto endings (Fig. 13.2) on each ionone ring, explains some unique features, such as the superior antioxidant activity. This antioxidative property is 1000 times higher than vitamin E, and ten times higher than the other carotenoids such as lutein, canthaxanthin, and β-carotene (Han et al. 2020; Zhao et al. 2019). However, the biosynthesis of AST in *H. lacustris* is a complex process that is highly upregulated under stress conditions (Shah et al. 2016).

13.3 Antioxidant Activity of *Haematococcus* Astaxanthin

In recent years, a high prevalence of non-communicable diseases such as aging, cancer, diabetes, neurodegenerative diseases, and cardiovascular events has affected the world population (Mukherjee et al. 2015; Rains and Jain 2011). These diseases are mainly caused by oxidative stress (Sharifi-Rad et al. 2020) which is linked to the modern lifestyle, including large consumption of processed food, air pollution, tobacco smoke, physiological stress, exposure to various chemical compounds, and a sedentary lifestyle (Mota et al. 2022; Guerin et al. 2003). Oxidative stress is defined as "an imbalance between oxidants and antioxidants in favor of the oxidants, leading to a disruption of redox signaling and control and/or molecular damage" (Sies 2020). Oxidants (e.g., hydroxyl, peroxyl radicals and singlet oxygen) can damage DNA, proteins, and lipid membranes and disrupt the redox signaling and cellular control. Dietary antioxidants, such as *Haematococcus* AST, could help to prevent and treat several human diseases.

AST contains 13 unsaturated conjugated double bonds, which will combine with free radicals to form non-reactive products and also provide electrons for free radicals (Pereira et al. 2020; Seifried et al. 2007). In addition, these conjugated bonds in the central non-polar region of AST allow the molecule to scavenge free radicals from within the cell by transporting them along its own carbon chain to the outside of the cell membrane where they are neutralized by other antioxidants, such as vitamin C (Fig. 13.3) (Pereira et al. 2020; Liu and Osawa 2007; Hussein et al. 2006). The stronger antioxidant activity of AST is also attributed to its superior

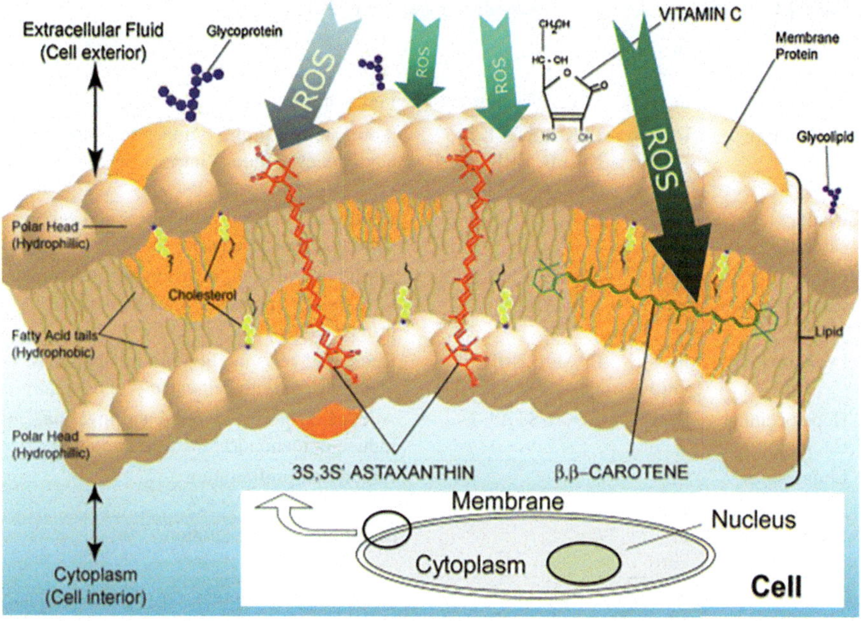

Fig. 13.3 Superior position of astaxanthin in cell membrane (Yamashita 2013)

position in cell membrane (Fig. 13.3). AST is the only one that can cross the bilayer cell membrane. Whereas, β-carotene and vitamin C only reside inside and outside the lipid bilayer membrane, respectively. Thus, AST can take a transmembrane alignment in biological membranes, helping to maintain membrane structure and decrease membrane fluidity, and acting as an antioxidant that provides better overall protection (Yamashita 2013; Park et al. 2010; Mcnulty et al. 2007; Goto et al. 2001). The antioxidant activities of AST are believed to have a key role in several other properties such as protection against UV-light photooxidation, inflammation, cancer, ulcers, aging, and age-related diseases, or the promotion of the immune response, liver function and heart, and eye health (Sheikhzadeh et al. 2011).

13.4 Anticancer Activity of *Haematococcus* Astaxanthin

Haematococcus AST have also been used in the treatment of different types of cancer including breast, colon, oral, lung, bladder, gastric, as well as melanoma, leukemia, and hepatocellular carcinoma (Demirel and Tuna 2021; Faraone et al. 2020; Zhang and Wang 2015). Various pathways are involved in the anticancer effects of AST, including apoptosis, inflammation, and cell junction (Table 13.1) (Zhuang et al. 2022; Faraone et al. 2020). The signal transducer and activator of

Table 13.1 Effects of AST on cancers (Zhang and Wang 2015)

Cancers	In vitro/ in vivo	Molecular targets	Functions
Oral cancer	In vivo	JAK-2/STAT-3, NF-κB, ERK, AKT (PKB)	Abrogate cell proliferation, invasion and angiogenesis, induce intrinsic apoptosis
Bladder carcinogenesis	In vivo	/	Reduce the incidence of cancer and suppression of cell proliferation
Colon carcinogenesis	Both	NF-κB, ERK, JNK, p38, AKT	Inhibit cell growth, invasion and inflammation, induce apoptosis, arrest cell cycle progression
Leukemia	In vitro	PPARγ, Nrf2	Decreased cell viability, induce apoptosis, and interfere with cell cycle progression
Hepatocellular carcinoma	Both	JAK-1/STAT-3	Attenuate cell proliferation and invasion, induce mitochondria-mediated apoptosis
Lung cancer	In vitro	/	Inhibit cell proliferation
Breast cancer	In vitro	/	Suppress cell proliferation

JAK janus kinase, *STAT* signal transducers and activators of transcription, *NF-κB* nuclear factor kappa-light-chain-enhancer of activated B cells, *ERK* extracellular signal-regulated kinase, *JNK* c-Jun *N*-terminal kinases, *PKB* protein Kinase B, *PPARγ* peroxisome proliferator-activated receptor gamma, *Nrf2* NF-E2-related factor 2

transcription 3 (STAT3), the nuclear factor kappa-light-chain-enhancer of activated B cells (NF-κB), and the peroxisome proliferator-activated receptor gamma (PPARγ) are the most important molecules and pathways implicated in the anticancer effects of AST (Zhang and Wang 2015).

Ramamoorthy (2020) reported the anticancer effects in A549 lung cancer cells by AST from *H. lacustris.* The anticancer potential of AST is found to be mediated via its ability to induce cell cycle arrest at the G2/M phase, lysosomal acidification, and apoptotic induction. Furthermore, AST mono- and diesters from the green alga *H. pluvialis* have shown effective inhibition of induced skin cancer in rat model (Rao et al. 2013). Combined actions were suggested to explain the mechanism of protection of this induced cancer such as inhibition of accumulation of reactive oxygen species (ROS) and inhibition of tyrosinase enzyme activity (Rao et al. 2013). Subsequently, the mentioned properties may prevent uncontrolled proliferation of melanocytes and prevention of accumulation of melanocytes and melanin pigments in addition to the inhibition of polyamine accumulation (Rao et al. 2013). Chen et al. (2017) reported that AST inhibit the expression of matrix metalloproteinase 1, 2, and 9 and affect the migration of melanoma cells then result to ultimately inhibit tumor growth. The growth-inhibitory effects of the AST-rich *H. pluvialis* were studied in HCT-116 colon cancer cells.

AST was added to the extract and was responsible for arresting the progression of the cell cycle and promoting the apoptosis (Palozza et al. 2009). AST showed higher antitumor activity than other carotenoids, including β-carotene and canthaxanthin (Chew and Park 2004). Thus, a study conducted by Chew et al. (1999) has demonstrated that a dietary AST is more effective against breast cancer than β-carotene and canthaxanthin. It leads to a growth reduction of induced mammary tumors by 50%. Furthermore, AST protected mice from bladder carcinogenesis by reducing the incidence of chemically induced bladder carcinoma and by eliminating the cell proliferation (Zhang and Wang 2015). Other studies using natural AST have also demonstrated its anticancer effects in mammals, however, its underlying mechanisms remain to be elucidated (Ramamoorthy 2020; Pinto et al. 2018; Shao et al. 2016; Song et al. 2011).

13.5 *Haematococcus* Astaxanthin for Skin Health

Excessive exposure of unprotected skin to UV irradiation leads to modifications in the skin structure resulting in pigmentation, loss of skin elasticity, and wrinkles, which accelerate skin aging and even carcinogenesis of skin cells (Singh et al. 2019a; Hart and Norval 2018). Recently, various studies have highlighted the positive effects (antioxidant, anti-inflammatory, immunomodulatory, and DNA repair properties) of *Haematococcus* AST on the human skin (Zhou et al. 2021; Davinelli et al. 2018). These properties have encouraged its use in maintaining skin health and fighting skin damage (Zhou et al. 2021; Yoshihisa et al. 2016). Two clinical studies conducted by Tominaga et al. (2012) showed that AST derived from *H. pluvialis* improve skin wrinkle, age spot size, elasticity, skin texture, moisture content of corneocyte layer and corneocyte condition by combining both oral supplementation and topical treatment for 8 weeks. Furthermore, dietary *H. pluvialis* AST supplementation could lower wrinkle formations, decreased transepidermal water loss, inhibited the generation of ROS, increased collagen fibers in the skin and maintain epidermal barrier functions significantly in the dorsal skin of HR-1 hairless mice under UVA exposure by its antioxidative effects (Zhou et al. 2021; Li et al. 2020c). Multiple human intervention studies on skin and AST have been conducted and have demonstrated that AST is a promising compound in the field of dermatology (Davinelli et al. 2018). Thus, various dermatological indications can be proposed such as aging skin, dry skin, prevention of UV-induced skin damage and as an adjunct in skin diseases related to oxidative stress (such as melasma and psoriasis) (Singh et al. 2019a). Moreover, in order to fully elucidate the different mode of actions of AST in skin protection, additional comprehensive experiments will be necessary.

13.6 *Haematococcus* Astaxanthin and Neurodegenerative Diseases

Neurodegenerative diseases are common diseases of the central nervous characterized by a progressive, selective loss of anatomically or physiologically related neuronal systems (Donoso et al. 2021; Cummings 2017). Generally, neurodegenerative diseases are incurable (Gitler et al. 2017) and are usually accompanied by cognitive and motor dysfunction (de Tommaso et al. 2016). There are many types of neurodegenerative diseases, including Alzheimer's disease, Parkinson's disease, amyotrophic lateral sclerosis, and Huntington's disease (Thompson et al. 2020; Delgado-Morales et al. 2017). The prevalence and incidence of neurodegenerative diseases increase dramatically with age, even though people of all ages are at risk of developing these diseases (Mayeux 2003). Thus, oxidative stress plays a key role because it often leads to cell alterations. These alterations are manifested by disturbances in the DNA repair system, mutations, and dysfunction of mitochondria (Kim et al. 2015). Considering the role of oxidative stress in neurodegenerative diseases, there is a growing interest in antioxidant therapies.

AST has been suggested as novel therapeutic molecule against these diseases associating its benefits to anti-oxidation, anti-inflammation, and anti-apoptotic mechanisms (Fakhri et al. 2018, 2019; Wu et al. 2015; Yamagishi and Aihara 2014). However, few studies have evaluated the effect of natural AST (Imai et al. 2018; Ito et al. 2018; Zanotta et al. 2014; Katagiri et al. 2012). Katagiri et al. (2012) studied, in a double-blind, placebo-controlled study, the effects of ingestion of AST-rich *H. pluvialis* extract on cognitive function of 96 elderly subjects. They found that healthy elderly people cognitive function was improved with AST, as the treated subjects made fewer errors in the maze tests and demonstrated faster reaction times in CogHealth (a computer-based test of memory and thinking capability tests). Zanotta et al. (2014) also reported that a dietary supplement made from the extract of *Bacopa monnieri, H. pluvialis* AST, phosphatidylserine, and vitamin E has allowed the counter acting of cognitive impairments in subjects with mild cognitive impairment. The effects of *H. pluvialis* AST-rich oil on brain function in healthy older adults with age-related forgetfulness have also been studied (Satoh et al. 2009). The findings suggest that administration of AST might improve higher brain function including cognition, attention, memory, information processing, and resultant behaviors in older persons (Satoh et al. 2009). Zhang et al. (2007) reported that *H. pluvialis* AST-enriched powder also significantly improved memory in BALB/c mice. The effects of supplementation with natural AST on mental and physical fatigue have also been studied, reporting a reduction in the daily sense of fatigue caused by both mental and physical loads after this treatment (Donoso et al. 2021; Imai et al. 2018; Hongo et al. 2017).

13.7 The Anti-inflammation Mechanisms of Astaxanthin

Inflammation is the body's reaction to hazardous stimuli, such as pathogens, damaged cells, toxic compounds, irradiation, or other types of trauma. In order to stop these traumas, the inflammatory response of the body is stimulated via different defense mechanisms which allows to restore the homeostasis of the attacked tissues (Ávila-Román et al. 2021; Singh et al. 2019b). However, nonresolving acute inflammation may lead to chronic inflammation, which is involved in the development of many chronic diseases (Ávila-Román et al. 2021; Nathan and Ding 2010). In recent decades, massive evidences in vitro and in many clinical studies have reported the anti-inflammatory effects and mechanisms of AST (Kohandel et al. 2022; Chang and Xiong 2020). Several studies have demonstrated the effectiveness of AST in alleviating acute and chronic inflammation in several diseases, particularly neurodegenerative diseases (Donoso et al. 2021), diabetes (Landon et al. 2020), cancer (Ávila-Román et al. 2021), gastrointestinal disease (Sakai et al. 2019; Yasui et al. 2011), renal inflammation (Liu et al. 2015), as well as skin and eye diseases (Li et al. 2020a, b, c; Yoon et al. 2014).

AST has been shown to provide protection against Alzheimer's disease. Supplementation with enriched-AST in a double-transgenic mouse model, APP/PSEN1 of Alzheimer's disease modulated oxidative stress, and attenuated neuroinflammation which alleviated cognitive dysfunction (Che et al. 2018). In addition, AST has shown protection against Parkinson's disease via several mechanisms including attenuated dopaminergic neurodegeneration (Shen et al. 2021). As a powerful antioxidant without side effects, AST has been suggested to contribute to the prevention and treatment of diabetes mellitus-associated pathologies (Landon et al. 2020). AST reduces inflammation, oxidative stress (OS), and apoptosis by regulating different OS pathways (Landon et al. 2020). The use of AST in the treatment of diabetic retinopathy and neuropathy has also shown great efficacy. Its action involves the activation of the NF-κB pathway, suppression of microvascular damage through the generation of vascular endothelial growth factors and by exerting anti-apoptotic activity via modulation of the MAPK and PI3K/Akt pathways. Furthermore, AST inhibited NF-κB translocation, transforming growth factor-beta (TGF-β) generation, inflammation, and fibrosis (Landon et al. 2020). *Helicobacter pylori* (*H. pylori*) is a pathogenic bacteria involved in several gastrointestinal diseases, which stimulates the production of ROS and leads to the stimulation of inflammatory reactions in tissues (Polk and Peek 2010). More recently, the use of AST in human gastric epithelial cells has activated PPAR which inhibited *H. pylori*-induced mitochondrial dysfunction and ROS-mediated IL-8 expression (Kim et al. 2018). Dietary AST has also been found to treat *H. pylori* ulcer disease. AST reduced symptoms of gastric inflammation and was also associated with changes in the inflammation response (Bennedsen et al. 1999).

13.8 *Haematococcus* Astaxanthin and Eye Health

Oxidative damage has been reported to play an essential role in several ocular diseases by promoting membrane degradation (Donoso et al. 2021). The most common of these diseases are age-related macular degeneration (Imamura et al. 2006), age-related cataract (Spector 1995), corneal inflammation (Alio et al. 1995a), and keratitis (Alio et al. 1995b). Chitchumroonchokchai et al. (2004) have demonstrated that AST provides significant protection of ocular epithelial cells from UVB-induced stress. In addition, AST supplementation significantly improves tear production and tears film stability in ocular surface studies (Hou et al. 2016). In an in vitro study, AST has demonstrated a potential therapeutic effect on mild-to-moderate dry eye disease (DED). It was able to reduce age-related oxidative stress markers in corneal epithelial cells (Shimokawa et al. 2019) and the ocular surface inflammation by decreasing the expression of two important inflammatory cytokines (TNF-α and IL-1β) and the high mobility group box 1 protein (Li et al. 2020a, b, c). The same results have been approved by a study conducted by Tian et al. (2022) in humans. They assessed the benefits and safety of *H. pluvialis* AST in the treatment of DED in middle-aged and elderly patients. They found that AST oral administration ensure an effective treatment of DED by different ways including the improvement of the stability of the tear film, the repair of corneal and conjunctival epithelial cells, and the secretion function of the meibomian gland, which helps to reduce dry eyes. Furthermore, *H. pluvialis* AST present a neuroprotective potential against optic nerve ischemia. It was demonstrated in a study conducted by Lin et al. (2020). Their findings indicate that AST can preserve visual function and reduce retinal ganglion cells apoptosis after ischemic insults. *H. pluvialis* AST can also be used for the treatment of eye fatigue. It was approved in ten clinical studies (Table 13.2). All of the studies have shown that AST reduces eye strain, as well as irritability, watery

Table 13.2 Clinical studies on eye fatigue (Yamashita 2021)

References	Dose: mg/day; (): number of subjects; duration	Study design
Nagaki et al. (2002)	0(13), 5(13); 4 week	DB-PC[a]
Nakamura et al. (2004)	0(10), 2(12), 4(14), 12(10); 4 week	DB-PC
Nitta et al. (2005)	0(10), 6(10), 12(10); 4 week	DB-PC
Shiratori et al. (2005)	0(19), 6(20); 4 week	DB-PC
Takahashi and Kajita (2005)	6(9); 2 week	Open label
Iwasaki and Tawara (2006)	0(10), 6(10); 2 week	DB-PC, crossover
Nagaki et al. (2006)	0(23), 6(25); 4 week	DB-PC
Kajita et al. (2009)	6(22); 4 week	Open label
Seya et al. (2009)	6(10); 4 week	Open label
Nagaki et al. (2010)	0(40), 9(42); 4 week	DB-PC

[a]*DB-PC* randomized double-blind placebo-controlled

eyes, red eyes, stiff shoulders and back, heavy head, and headaches (Yamashita 2021).

13.9 Conclusion

Based on the recent published studies, we conclude that *Haematococcus* AST might be extremely beneficial in working against several diseases such as cancer, skin diseases, diabetes, neurodegenerative diseases, and eye diseases. These beneficial effects are mainly due to the strong antioxidant activity of AST which is used to protect cells, tissues, and organs from oxidative damage. Moreover, the use of *Haematococcus* AST in human health management is due to its unique structural and chemical properties and also for its safety which has long been established through numerous clinical and preclinical safety evaluations. Thus, the administration of *Haematococcus* AST will be of great value to prevent health issues.

References

Alam MA, Xu J-L, Wang Z (eds) (2020) Microalgae biotechnology for food, health and high value products. https://doi.org/10.1007/978-981-15-0169-2

Alio JL, Ayala MJ, Mulet ME, Artola A, Ruiz JM, Bellot J (1995a) Antioxidant therapy in the treatment of experimental acute corneal inflammation. Ophthalmic Res 27(3):136–143. https://doi.org/10.1159/000267648

Alio JL, Artola A, Serra A, Ayala MJ, Mulet ME (1995b) Effect of topical antioxidant therapy on experimental infectious keratitis. Cornea 14(2):175–179. https://doi.org/10.1097/00003226-199503000-00011

Ambati R, Phang S-M, Ravi S, Aswathanarayana R (2014) Astaxanthin: sources, extraction, stability, biological activities and its commercial applications—a review. Mar Drugs 12(1): 128–152. https://doi.org/10.3390/md12010128

Ávila-Román J, García-Gil S, Rodríguez-Luna A, Motilva V, Talero E (2021) Anti-inflammatory and anticancer effects of microalgal carotenoids. Mar Drugs 19(10):531. https://doi.org/10.3390/md19100531

Bennedsen M, Wang X, Willén R, Wadström T, Andersen LP (1999) Treatment of H. pylori infected mice with antioxidant astaxanthin reduces gastric inflammation, bacterial load and modulates cytokine release by splenocytes. Immunol Lett 70(3):185–189. https://doi.org/10.1016/s0165-2478(99)00145-5

Borowitzka MA (2013) High-value products from microalgae—their development and commercialisation. J Appl Phycol 25(3):743–756. https://doi.org/10.1007/s10811-013-9983-9

Caetano PA, do Nascimento TC, Fernandes AS, Nass PP, Vieira KR, Maróstica Junior MR, Jacob-Lopes E, Zepka LQ (2022) Microalgae-based polysaccharides: insights on production, applications, analysis, and future challenges. Biocatal Agric Biotechnol 45:102491. https://doi.org/10.1016/j.bcab.2022.102491

Capelli B, Talbott S, Ding L (2019) Astaxanthin sources : suitability for human health and nutrition. Funct Foods Health Dis 9(6):430. https://doi.org/10.31989/ffhd.v9i6.584

Chang MX, Xiong F (2020) Astaxanthin and its effects in inflammatory responses and inflammation-associated diseases: recent advances and future directions. Molecules 25(22): 5342. https://doi.org/10.3390/molecules25225342
Che H, Li Q, Zhang T, Wang D, Yang L, Xu J, Yanagita T, Xue C, Chang Y, Wang Y (2018) Effects of Astaxanthin and docosahexaenoic-acid-acylated astaxanthin on Alzheimer's disease in APP/PS1 double-transgenic mice. J Agric Food Chem 66(19):4948–4957. https://doi.org/10.1021/acs.jafc.8b00988
Chen Y-T, Kao C-J, Huang H-Y, Huang S-Y, Chen C-Y, Lin Y-S, Wen Z-H, Wang H-MD (2017) Astaxanthin reduces MMP expressions, suppresses cancer cell migrations, and triggers apoptotic caspases of in vitro and in vivo models in melanoma. J Funct Foods 31:20–31. https://doi.org/10.1016/j.jff.2017.01.005
Chew BP, Park JS (2004) Carotenoid action on the immune response. J Nutr 134(1):257S–261S. https://doi.org/10.1093/jn/134.1.257s
Chew BP, Park JS, Wong MW, Wong TS (1999) A comparison of the anticancer activities of dietary Beta-carotene, canthaxanthin and astaxanthin in mice in vivo. Anticancer Res 19(3A):1849–1853. PMID: 10470126
Chitchumroonchokchai C, Bomser JA, Glamm JE, Failla ML (2004) Xanthophylls and α-tocopherol decrease UVB-induced lipid peroxidation and stress signaling in human lens epithelial cells. J Nutr 134(12):3225–3232. https://doi.org/10.1093/jn/134.12.3225
Chu W-L (2012) Biotechnological applications of microalgae. Int E-J Sci Med Educ 6(Suppl1): S24–S37. https://doi.org/10.56026/imu.6.suppl1.s24
Costa JAV, Lucas BF, Alvarenga AGP, Moreira JB, de Morais MG (2021) Microalgae polysaccharides: an overview of production, characterization, and potential applications. Polysaccharides 2(4):759–772. https://doi.org/10.3390/polysaccharides2040046
Cummings J (2017) Disease modification and neuroprotection in neurodegenerative disorders. Transl Neurodegeneration 6(1):25. https://doi.org/10.1186/s40035-017-0096-2
Darwesh OM, Mahmoud RH, Abdo SM, Marrez DA (2022) Isolation of Haematococcus lacustris as source of novel anti-multi-antibiotic resistant microbes agents; fractionation and identification of bioactive compounds. Biotechnol Rep 35:e00753. https://doi.org/10.1016/j.btre.2022.e00753
Davinelli S, Nielsen M, Scapagnini G (2018) Astaxanthin in skin health, repair, and disease: a comprehensive review. Nutrients 10(4):522. https://doi.org/10.3390/nu10040522
Delattre C, Pierre G, Laroche C, Michaud P (2016) Production, extraction and characterization of microalgal and cyanobacterial exopolysaccharides. Biotechnol Adv 34(7):1159–1179. https://doi.org/10.1016/j.biotechadv.2016.08.001
Delgado-Morales R, Agís-Balboa RC, Esteller M, Berdasco M (2017) Epigenetic mechanisms during ageing and neurogenesis as novel therapeutic avenues in human brain disorders. Clin Epigenetics 9(1):67. https://doi.org/10.1186/s13148-017-0365-z
Demirel PB, Tuna BG (2021) Anticancer properties of astaxanthin: a molecule of great promise. In: Global perspectives on astaxanthin. Academic, Cambridge, pp 427–445. https://doi.org/10.1016/b978-0-12-823304-7.00003-9
Deniz I, García-Vaquero M, Imamoglu E (2017) Trends in red biotechnology. In: Microalgae-based biofuels and bioproducts, pp 429–460. https://doi.org/10.1016/b978-0-08-101023-5.00018-2
Donoso A, González-Durán J, Muñoz AA, González PA, Agurto-Muñoz C (2021) Therapeutic uses of natural astaxanthin: an evidence-based review focused on human clinical trials. Pharmacol Res 166:105479. https://doi.org/10.1016/j.phrs.2021.105479
Fábregas J, Otero A, Maseda A, Domínguez A (2001) Two-stage cultures for the production of astaxanthin from Haematococcus pluvialis. J Biotechnol 89(1):65–71. https://doi.org/10.1016/s0168-1656(01)00289-9
Fakhri S, Dargahi L, Abbaszadeh F, Jorjani M (2018) Astaxanthin attenuates neuroinflammation contributed to the neuropathic pain and motor dysfunction following compression spinal cord injury. Brain Res Bull 143:217–224. https://doi.org/10.1016/j.brainresbull.2018.09.011

Fakhri S, Dargahi L, Abbaszadeh F, Jorjani M (2019) Effects of astaxanthin on sensory-motor function in a compression model of spinal cord injury: involvement of ERK and AKT signalling pathway. Eur J Pain 23(4):750–764. Portico. https://doi.org/10.1002/ejp.1342

Faraone I, Sinisgalli C, Ostuni A, Armentano MF, Carmosino M, Milella L, Russo D, Labanca F, Khan H (2020) Astaxanthin anticancer effects are mediated through multiple molecular mechanisms: a systematic review. Pharmacol Res 155:104689. https://doi.org/10.1016/j.phrs.2020.104689

Gitler AD, Dhillon P, Shorter J (2017) Neurodegenerative disease: models, mechanisms, and a new hope. Dis Models Mech 10(5):499–502. https://doi.org/10.1242/dmm.030205

Göksan T, Ak I, Gokpinar S (2010) An alternative approach to the traditional mixotrophic cultures of Haematococcus pluvialis flotow (Chlorophyceae). J Microbiol Biotechnol 20(9):1276–1282. https://doi.org/10.4014/jmb.0909.09005

Goto S, Kogure K, Abe K, Kimata Y, Kitahama K, Yamashita E, Terada H (2001) Efficient radical trapping at the surface and inside the phospholipid membrane is responsible for highly potent antiperoxidative activity of the aarotenoid astaxanthin. Biochim Biophys Acta Biomembr 1512(2):251–258. https://doi.org/10.1016/s0005-2736(01)00326-1

Guedes AC, Amaro HM, Malcata FX (2011) Microalgae as sources of high added-value compounds-a brief review of recent work. Biotechnol Prog 27(3):597–613. https://doi.org/10.1002/btpr.575

Guerin M, Huntley ME, Olaizola M (2003) Haematococcus astaxanthin: applications for human health and nutrition. Trends Biotechnol 21(5):210–216. https://doi.org/10.1016/s0167-7799(03)00078-7

Hamed I (2016) The evolution and versatility of microalgal biotechnology: a review. Compr Rev Food Sci Food Saf 15(6):1104–1123. Portico. https://doi.org/10.1111/1541-4337.12227

Han S-I, Chang SH, Lee C, Jeon MS, Heo YM, Kim S, Choi Y-E (2020) Astaxanthin biosynthesis promotion with pH shock in the green microalga, Haematococcus lacustris. Bioresour Technol 314:123725. https://doi.org/10.1016/j.biortech.2020.123725

Hart PH, Norval M (2018) Ultraviolet radiation-induced immunosuppression and its relevance for skin carcinogenesis. Photochem Photobiol Sci 17(12):1872–1884. https://doi.org/10.1039/c7pp00312a

Hongo N, Fujishita M et al (2017) Daily fatigue-reducing effect of astaxanthin—a randomized, placebo-controlled, double-blind, parallel-group study. Jpn Pharmacol Ther 45(1):67–72

Hou Y-C, Huang J-Y, Yeh P-T (2016) A randomized, double-blind, placebo-controlled study of oral antioxidant supplement therapy in patients with dry eye syndrome. Clin Ophthalmol 10:813. https://doi.org/10.2147/opth.s106455

Hussein G, Sankawa U, Goto H, Matsumoto K, Watanabe H (2006) Astaxanthin, a carotenoid with potential in human health and nutrition. J Nat Prod 69(3):443–449. https://doi.org/10.1021/np050354

Imai A, Oda Y, Ito N, Seki S, Nakagawa K, Miyazawa T, Ueda F (2018) Effects of dietary supplementation of astaxanthin and sesamin on daily fatigue: a randomized, double-blind, placebo-controlled, two-way crossover study. Nutrients 10(3):281. https://doi.org/10.3390/nu10030281

Imamura Y, Noda S, Hashizume K, Shinoda K, Yamaguchi M, Uchiyama S, Shimizu T, Mizushima Y, Shirasawa T, Tsubota K (2006) Drusen, choroidal neovascularization, and retinal pigment epithelium dysfunction in SOD1-deficient mice: a model of age-related macular degeneration. Proc Natl Acad Sci 103(30):11282–11287. https://doi.org/10.1073/pnas.0602131103

Ito N, Saito H, Seki S, Ueda F, Asada T (2018) Effects of composite supplement containing astaxanthin and sesamin on cognitive functions in people with mild cognitive impairment: a randomized, double-blind, placebo-controlled trial. J Alzheimers Dis 62(4):1767–1775. https://doi.org/10.3233/jad-170969

Iwasaki T, Tawara A (2006) Effect of astaxanthin on eye strain induced by accommodative dysfunction. AtarashiiGanka (J Eye) 23(6):829–834 de Jesus Raposo MF, de Morais AMMB, de Morais RMSC (2014) Influence of sulphate on the composition and antibacterial and antiviral properties of the exopolysaccharide from Porphyridium cruentum. Life Sci 101(1–2):56–63. https://doi.org/10.1016/j.lfs.2014.02.013

Kajita M, Tsukahara H, Kato M (2009) The effect of astaxanthin of a dietary supplement containing astaxanthin on the accommodation function of the eye in middle-aged and older people. Med Consult New Rem 46(3):89–93

Katagiri M, Satoh A, Tsuji S, Shirasawa T (2012) Effects of astaxanthin-rich Haematococcus pluvialis extract on cognitive function: a randomised, double-blind, placebo-controlled study. J Clin Biochem Nutr 51(2):102–107. https://doi.org/10.3164/jcbn.d-11-00017

Khavari F, Saidijam M, Taheri M, Nouri F (2021) Microalgae: therapeutic potentials and applications. Mol Biol Rep 48:4757. https://doi.org/10.1007/s11033-021-06422-w

Kim GH, Kim JE, Rhie SJ, Yoon S (2015) The role of oxidative stress in neurodegenerative diseases. Exp Neurobiol 24(4):325–340. https://doi.org/10.5607/en.2015.24.4.325

Kim S, Lim J, Kim H (2018) Astaxanthin inhibits mitochondrial dysfunction and interleukin-8 expression in helicobacter pylori-infected gastric epithelial cells. Nutrients 10(9):1320. https://doi.org/10.3390/nu10091320

Kim TY, Lee S-H, Lee S-Y (2021) Two newly identified Haematococcus strains efficiently accumulated radioactive cesium over higher astaxanthin production. Environ Res 199: 111301. https://doi.org/10.1016/j.envres.2021.111301

Klochkova TA, Kwak MS, Han JW, Motomura T, Nagasato C, Kim GH (2013) Cold-tolerant strain of Haematococcus pluvialis (Haematococcaceae, Chlorophyta) from Blomstrandhalvøya (Svalbard). Algae 28(2):185–192. https://doi.org/10.4490/algae.2013.28.2.185

Kohandel Z, Farkhondeh T, Aschner M, Pourbagher-Shahri AM, Samarghandian S (2022) Anti-inflammatory action of astaxanthin and its use in the treatment of various diseases. Biomed Pharmacother 145:112179. https://doi.org/10.1016/j.biopha.2021.112179

Landon R, Gueguen V, Petite H, Letourneur D, Pavon-Djavid G, Anagnostou F (2020) Impact of astaxanthin on diabetes pathogenesis and chronic complications. Mar Drugs 18(7):357. https://doi.org/10.3390/md18070357

Lekshmi VS, Rauf AA, Muraleedhara Kurup G (2019) Toxicological evaluation of bioactive sulfated polysaccharides from the edible marine algae Padina tetrastromatica and Ulva fasciatain Sprague-Dawley rats. J Biol Active Prod 9(2):79–96. https://doi.org/10.1080/22311866.2019.1599730

Li F, Cai M, Lin M, Huang X, Wang J, Ke H, Wang C, Zheng X, Chen D, Yang S (2020a) Enhanced biomass and astaxanthin production of Haematococcus pluvialis by a cell transformation strategy with optimized initial biomass density. Mar Drugs 18(7):341. https://doi.org/10.3390/md18070341

Li H, Li J, Hou C, Li J, Peng H, Wang Q (2020b) The effect of astaxanthin on inflammation in hyperosmolarity of experimental dry eye model in vitro and in vivo. Exp Eye Res 197:108113. https://doi.org/10.1016/j.exer.2020.108113

Li X, Matsumoto T, Takuwa M, Saeed Ebrahim Shaiku Ali M, Hirabashi T, Kondo H, Fujino H (2020c) Protective effects of astaxanthin supplementation against ultraviolet-induced photoaging in hairless mice. Biomedicine 8(2):18. https://doi.org/10.3390/biomedicines8020018

Lin W-N, Kapupara K, Wen Y-T, Chen Y-H, Pan I-H, Tsai R-K (2020) Haematococcus pluvialis-derived astaxanthin is a potential neuroprotective agent against optic nerve ischemia. Mar Drugs 18(2):85. https://doi.org/10.3390/md18020085

Liu X, Osawa T (2007) Cis astaxanthin and especially 9-cis astaxanthin exhibits a higher antioxidant activity in vitro compared to the all-trans isomer. Biochem Biophys Res Commun 357(1): 187–193. https://doi.org/10.1016/j.bbrc.2007.03.120

Liu G, Shi Y, Peng X, Liu H, Peng Y, He L (2015) Astaxanthin attenuates adriamycin-induced focal segmental glomerulosclerosis. Pharmacology 95(3–4):193–200. Portico. https://doi.org/10.1159/000381314

Mayeux R (2003) Epidemiology of neurodegeneration. Annu Rev Neurosci 26(1):81–104. https://doi.org/10.1146/annurev.neuro.26.043002.094919

Mazumdar N, Gopalakrishnan KK, Visnovsky G, Novis PM (2018) A novel alpine species of Haematococcus (Chlamydomonadales: Chlorophyta) from New Zealand. N Z J Bot 56(2): 216–226. https://doi.org/10.1080/0028825x.2018.1458737

Mcnulty HP, Byun J, Lockwood SF, Jacob RF, Mason RP (2007) Differential effects of carotenoids on lipid peroxidation due to membrane interactions: X-ray diffraction analysis. Biochim Biophys Acta Biomembr 1768(1):167–174. https://doi.org/10.1016/J.Bbamem.2006.09.010

Miyakawa K (2021) Commercial production of astaxanthin from the green alga Haematococcus pluvialis. Carotenoids: biosynthetic and biofunctional approaches. Adv Exp Med Biol 1261:3–10. https://doi.org/10.1007/978-981-15-7360-6_1

Morais MG, Rosa GM, Moraes L, Alvarenga AGP, da Silva JLV, Costa JAV (2022) Microalgae polysaccharides with potential biomedical application. In: Oliveira JM, Radhouani H, Reis RL (eds) Polysaccharides of microbial origin. Springer, Cham. https://doi.org/10.1007/978-3-030-42215-8_20

Mota JC, Almeida PP, Freitas MQ, Stockler-Pinto MB, Guimarães JT (2022) Far from being a simple question: the complexity between in vitro and in vivo responses from nutrients and bioactive compounds with antioxidant potential. Food Chem 402:134351. https://doi.org/10.1016/j.foodchem.2022.134351

Mukherjee K, Chio TI, Sackett DL, Bane SL (2015) Detection of oxidative stress-induced carbonylation in live mammalian cells. Free Radic Biol Med 84:11–21. https://doi.org/10.1016/j.freeradbiomed.2015.03.011

Nagaki Y, Hayasaka S, Yamada T, Hayasaka Y, Sanada M, Uonomi T (2002) Effects of astaxanthin on accommodation, critical flicker fusion, and pattern visual evoked potential in visual display terminal workers. J Tradit Med 19(5):170–173

Nagaki Y, Mihara M, Tsukahara H, Ohno S (2006) The supplementation effect of astaxanthin on accomodation and asthenopia. J Cli Ther Med 22(1):41–54

Nagaki Y, Tsukahara H, Yoshimoto T, Masuda K (2010) Effect of astaxanthin on accommodation and asthenopia. Folia Ophthalmol Jpn 3(5):461–468

Nakada T, Ota S (2016) What is the correct name for the type of Haematococcus Flot (Volvocales, Chlorophyceae)? TAXON 65(2):343–348. Portico. https://doi.org/10.12705/652.11

Nakamura A, Isobe R, Otaka Y, Abematsu Y, Nakata D, Honda C, Sakurai S, Shimada Y, Horiguchi M (2004) Changes in visual function following peroral astaxanthin. Jpn J Clin Ophthalmol 58(6):1051–1054

Nathan C, Ding A (2010) Nonresolving inflammation. Cell 140(6):871–882. https://doi.org/10.1016/j.cell.2010.02.029

Nishshanka GKSH, Liyanaarachchi VC, Nimarshana PHV, Ariyadasa TU, Chang J-S (2022) Haematococcus pluvialis: a potential feedstock for multiple-product biorefining. J Clean Prod 344:131103. https://doi.org/10.1016/j.jclepro.2022.131103

Nitta T, Ohgami K, Shiratori K, Shinmei Y, Chin S, Yoshida K, Tsukahara H, Ohno S (2005) Effects of astaxanthin on accommodation and asthenopia-dose finding study in healthy volunteers. J Clin Ther Med 21(5):79–92

Oslan SNH, Shoparwe NF, Yusoff AH, Rahim AA, Chang CS, Tan JS, Oslan SN, Arumugam K, Ariff AB, Sulaiman AZ, Mohamed MS (2021) A review on *Haematococcus pluvialis* bioprocess optimization of green and red stage culture conditions for the production of natural astaxanthin. Biomol Ther 11(2):256. https://doi.org/10.3390/biom11020256

Oslan SNH, Oslan SN, Mohamad R, Tan JS, Yusoff AH, Matanjun P, Mokhtar RAM, Shapawi R, Huda N (2022) Bioprocess strategy of Haematococcus lacustris for biomass and astaxanthin production keys to commercialization: perspective and future direction. Fermentation 8(4):179. https://doi.org/10.3390/fermentation8040179

Palozza P, Torelli C, Boninsegna A, Simone R, Catalano A, Mele MC, Picci N (2009) Growth-inhibitory effects of the astaxanthin-rich alga *Haematococcus pluvialis* in human colon cancer cells. Cancer Lett 283(1):108–117. https://doi.org/10.1016/j.canlet.2009.03.031

Park JS, Chyun JH, Kim YK, Line LL, Chew BP (2010) Astaxanthin decreased oxidative stress and inflammation and enhanced immune response in humans. Nutr Metab 7(1):18. https://doi.org/10.1186/1743-7075-7-18

Patel AK, Tambat VS, Chen C-W, Chauhan AS, Kumar P, Vadrale AP, Huang C-Y, Dong C-D, Singhania RR (2022) Recent advancements in astaxanthin production from microalgae: a review. Bioresour Technol 364:128030. https://doi.org/10.1016/j.biortech.2022.128030

Pereira C, Souza A, Vasconcelos A, Prado P, Name J (2020) Antioxidant and anti-inflammatory mechanisms of action of astaxanthin in cardiovascular diseases (review). Int J Mol Med 47(1): 37–48. https://doi.org/10.3892/ijmm.2020.4783

Pinto A, Bonucci A, Maggi E, Corsi M, Businaro R (2018) Anti-oxidant and anti-inflammatory activity of ketogenic diet: new perspectives for neuroprotection in Alzheimer's disease. Antioxidants 7(5):63. https://doi.org/10.3390/antiox7050063

Polk DB, Peek RM (2010) Helicobacter pylori: gastric cancer and beyond. Nat Rev Cancer 10(6): 403–414. https://doi.org/10.1038/nrc2857

Rahman KM (2020) Food and high value products from microalgae: market opportunities and challenges. In: Microalgae biotechnology for food, health and high value products. Springer, Singapore, pp 3–27. https://doi.org/10.1007/978-981-15-0169-2_1

Rains JL, Jain SK (2011) Oxidative stress, insulin signaling, and diabetes. Free Radic Biol Med 50(5):567–575. https://doi.org/10.1016/j.freeradbiomed.2010.12.006

Ramamoorthy K (2020) Anticancer effects and lysosomal acidification in A549 cells by astaxanthin from Haematococcus lacustris. Bioinformation 16(11):965–973. https://doi.org/10.6026/97320630016965

Rao AR, Sindhuja HN, Dharmesh SM, Sankar KU, Sarada R, Ravishankar GA (2013) Effective inhibition of skin cancer, tyrosinase, and antioxidative properties by astaxanthin and astaxanthin esters from the green alga Haematococcus pluvialis. J Agric Food Chem 61(16):3842–3851. https://doi.org/10.1021/jf304609j

Sakai S, Nishida A, Ohno M, Inatomi O, Bamba S, Sugimoto M, Kawahara M, Andoh A (2019) Astaxanthin, A xanthophyll carotenoid, prevents development of dextran Sulphate sodium-induced murine colitis. J Clin Biochem Nutr 64(1):66–72. https://doi.org/10.3164/jcbn.18-47

Sathasivam R, Ki J-S (2018) A review of the biological activities of microalgal carotenoids and their potential use in healthcare and cosmetic industries. Mar Drugs 16(1):26. https://doi.org/10.3390/md16010026

Sathasivam R, Radhakrishnan R, Hashem A, Abd Allah EF (2019) Microalgae metabolites: a rich source for food and medicine. Saudi J Biol Sci 26(4):709–722. https://doi.org/10.1016/j.sjbs.2017.11.003

Satoh A, Tsuji S, Okada Y, Murakami N, Urami M, Nakagawa K, Ishikura M, Katagiri M, Koga Y, Shirasawa T (2009) Preliminary clinical evaluation of toxicity and efficacy of a new Astaxanthin-rich Haematococcus pluvialis extract. J Clin Biochem Nutr 44(3):280–284. https://doi.org/10.3164/jcbn.08-238

Seifried HE, Anderson DE, Fisher EI, Milner JA (2007) A review of the interaction among dietary antioxidants and reactive oxygen species. J Nutr Biochem 18(9):567–579. https://doi.org/10.1016/j.jnutbio.2006.10.007

Seya Y, Takahashi J, Imanaka K (2009) Relationship between visual and reaction times –effects of a repetition of A visual task and long-term intake of a supplement food including Astaxanthin on reaction time-Japanese. Jpn J Physiol Anthropol 14(2):17–24

Shah MMR, Liang Y, Cheng JJ, Daroch M (2016) Astaxanthin-producing green microalga Haematococcus pluvialis: from single cell to high value commercial products. Front Plant Sci 7:531. https://doi.org/10.3389/fpls.2016.00531

Shao Y, Ni Y, Yang J, Lin X, Li J, Zhang L (2016) Astaxanthin inhibits proliferation and induces apoptosis and cell cycle arrest of mice H22 hepatoma cells. Med Sci Monit 22:2152–2160. https://doi.org/10.12659/msm.899419

Sharifi-Rad M, Anil Kumar NV, Zucca P, Varoni EM, Dini L, Panzarini E, Rajkovic J, Tsouh Fokou PV, Azzini E, Peluso I, Prakash Mishra A, Nigam M, El Rayess Y, Beyrouthy ME,

Polito L, Iriti M, Martins N, Martorell M, Docea AO et al (2020) Lifestyle, oxidative stress, and antioxidants: back and forth in the pathophysiology of chronic diseases. Front Physiol 11:694. https://doi.org/10.3389/fphys.2020.00694

Sheikhzadeh N, Tayefi-Nasrabadi H, Khani Oushani A, Najafi Enferadi MH (2011) Effects of Haematococcus pluvialis supplementation on antioxidant system and metabolism in rainbow trout (Oncorhynchus Mykiss). Fish Physiol Biochem 38(2):413–419. https://doi.org/10.1007/s10695-011-9519-7

Shen DF, Qi HP, Ma C, Chang MX, Zhang WN, Song RR (2021) Astaxanthin suppresses endoplasmic reticulum stress and protects against neuron damage in Parkinson's disease by regulating Mir-7/SNCA Axis. Neurosci Res 165:51–60. https://doi.org/10.1016/j.neures.2020.04.003

Shimokawa T, Yoshida M, Fukuta T, Tanaka T, Inagi T, Kogure K (2019) Efficacy of high-affinity liposomal astaxanthin on up-regulation of age-related markers induced by oxidative stress in human corneal epithelial cells. J Clin Biochem Nutr 64(1):27–35. https://doi.org/10.3164/jcbn.18-27

Shiratori K, Ogami K, Nitta T, Shinmei Y, Chin S, Yoshida K, Tsukahara H, Takehara I, Ohno S (2005) Effect of astaxanthin on accommodation and asthenopia-efficacy identification study in healthy volunteers. J Clin Ther Med 21(6):65–78

Sies H (2020) Oxidative stress: concept and some practical aspects. Antioxidants 9(9):852. https://doi.org/10.3390/antiox9090852

da Silva TL, Mendes A, Mendes RL, Calado V, Alves SS, Vasconcelos JMT, Reis A (2006) Effect of n-Dodecane on Crypthecodinium cohnii fermentations and DHA production. J Ind Microbiol Biotechnol 33(6):408–416. https://doi.org/10.1007/s10295-006-0081-8

Singh KN, Patil S, Barkate H (2019a) Protective effects of astaxanthin on skin: recent scientific evidence, possible mechanisms, and potential indications. J Cosmet Dermatol 19(1):22–27. Portico. https://doi.org/10.1111/jocd.13019

Singh N, Baby D, Rajguru J, Patil P, Thakkannavar S, Pujari V (2019b) Inflammation and cancer. Ann Afr Med 18(3):121. https://doi.org/10.4103/aam.aam_56_18

Song X, Zhang J, Wang M, Liu W, Gu X, Lv C (2011) Astaxanthin induces mitochondria-mediated apoptosis in rat hepatocellular carcinoma CBRH-7919 cells. Biol Pharm Bull 34(6):839–844. https://doi.org/10.1248/bpb.34.839

Spector A (1995) Oxidative stress-induced cataract: mechanism of action. FASEB J 9(12): 1173–1182. Portico. https://doi.org/10.1096/fasebj.9.12.7672510

Takahashi N, Kajita M (2005) Effects of astaxanthin on accommodative recovery. J Clin Ther Med 21(4):43–48

Thompson C, Otero P, Srinageshwar B, Petersen RB, Dunbar GL, Rossignol J (2020) Possible roles of epigenetics in stem cell therapy for Parkinson's disease. Epigenomics 12(7):647–656. https://doi.org/10.2217/epi-2019-0347

Tian L, Wen Y, Li S, Zhang P, Wang Y, Wang J, Cao K, Du L, Wang N, Jie Y (2022) Benefits and safety of astaxanthin in the treatment of mild-to-moderate dry eye disease. Front Nutr 8:796951. https://doi.org/10.3389/fnut.2021.796951

Tominaga K, Hongo N, Karato M, Yamashita E (2012) Cosmetic benefits of astaxanthin on humans subjects. Acta Biochim Pol 59(1):43. https://doi.org/10.18388/abp.2012_2168

de Tommaso M, Arendt-Nielsen L, Defrin R, Kunz M, Pickering G, Valeriani M (2016) Pain in neurodegenerative disease: current knowledge and future perspectives. Behav Neurol 2016:1–14. https://doi.org/10.1155/2016/7576292

Wu H, Niu H, Shao A, Wu C, Dixon B, Zhang J, Yang S, Wang Y (2015) Astaxanthin as a potential neuroprotective agent for neurological diseases. Mar Drugs 13(9):5750–5766. https://doi.org/10.3390/md13095750

Xi T, Kim DG, Roh SW, Choi J-S, Choi Y-E (2016) Enhancement of astaxanthin production using Haematococcus pluvialis with novel LED wavelength shift strategy. Appl Microbiol Biotechnol 100(14):6231–6238. https://doi.org/10.1007/s00253-016-7301-6

Yamagishi R, Aihara M (2014) Neuroprotective effect of astaxanthin against rat retinal ganglion cell death under various stresses that induce apoptosis and necrosis. Mol Vis 20:1796–1805. PMID: 25593507; PMCID: PMC4287717

Yamashita E (2013) Astaxanthin as a medical food. Funct Foods Health Dis 3(7):254. https://doi.org/10.31989/ffhd.v3i7.49

Yamashita E (2021) Extensive bioactivity of astaxanthin from *Haematococcus pluvialis* in human. Adv Exp Med Biol 1261:249–259. https://doi.org/10.1007/978-981-15-7360-6_23

Yasui Y, Hosokawa M, Mikami N, Miyashita K, Tanaka T (2011) Dietary astaxanthin inhibits colitis and colitis-associated colon carcinogenesis in mice via modulation of the inflammatory cytokines. Chem Biol Interact 193(1):79–87. https://doi.org/10.1016/j.cbi.2011.05.006

Yoon H-S, Cho HH, Cho S, Lee S-R, Shin M-H, Chung JH (2014) Supplementing with dietary Astaxanthin combined with collagen hydrolysate improves facial elasticity and decreases matrix Metalloproteinase-1 and -12 expression: a comparative study with placebo. J Med Food 17(7): 810–816. https://doi.org/10.1089/jmf.2013.3060

Yoshihisa Y, Andoh T, Matsunaga K, Rehman MU, Maoka T, Shimizu T (2016) Efficacy of astaxanthin for the treatment of atopic dermatitis in a murine model. PLoS One 11(3):e0152288. https://doi.org/10.1371/journal.pone.0152288

Zanotta D, Puricelli S, Bonoldi G (2014) Cognitive effects of a dietary supplement made from extract of Bacopa Monnieri, astaxanthin, phosphatidylserine, and vitamin E in subjects with mild cognitive impairment: a noncomparative, exploratory clinical study. Neuropsychiatr Dis Treat 10:225. https://doi.org/10.2147/ndt.s51092

Zhang L, Wang H (2015) Multiple mechanisms of anti-cancer effects exerted by astaxanthin. Mar Drugs 13(7):4310–4330. https://doi.org/10.3390/md13074310

Zhang X, Pan L, Wei X, Gao H, Liu J (2007) Impact of astaxanthin-enriched algal powder of Haematococcus pluvialis on memory improvement in BALB/C mice. Environ Geochem Health 29(6):483–489. https://doi.org/10.1007/s10653-007-9117-x

Zhang C, Zhang L, Liu J (2016) The role of photorespiration during Astaxanthin accumulation in Haematococcus pluvialis (Chlorophyceae). Plant Physiol Biochem 107:75–81. https://doi.org/10.1016/j.plaphy.2016.05.029

Zhao T, Yan X, Sun L, Yang T, Hu X, He Z, Liu F, Liu X (2019) Research progress on extraction, biological activities and delivery systems of natural astaxanthin. Trends Food Sci Technol 91: 354–361. https://doi.org/10.1016/j.tifs.2019.07.014

Zhou X, Cao Q, Orfila C, Zhao J, Zhang L (2021) Systematic review and meta-analysis on the effects of astaxanthin on human skin ageing. Nutrients 13(9):2917. https://doi.org/10.3390/nu13092917

Zhou L, Li K, Duan X, Hill D, Barrow C, Dunshea F, Martin G, Suleria H (2022) Bioactive compounds in microalgae and their potential health benefits. Food Biosci 49:101932. https://doi.org/10.1016/j.fbio.2022.101932

Zhuang D, He N, Khoo KS, Ng E-P, Chew KW, Ling TC (2022) Application progress of bioactive compounds in microalgae on pharmaceutical and cosmetics. Chemosphere 291:132932. https://doi.org/10.1016/j.chemosphere.2021.132932

Chapter 14
In Silico Exploration of Therapeutics in *Haematococcus pluvialis*

Dhamodharan Prabhu and Sundarraj Rajamanikandan

14.1 Introduction

Unicellular freshwater microalga *Haematococcus pluvialis* is widely known for being the bountiful source among the natural products with large number of valuable products. The most potent natural antioxidant found in *H. pluvialis* is 3, 3′--dihydroxy-β-carotene-4, 4′-dione (commonly referred as astaxanthin), which is an unlimited requirement of the aquaculture and nutraceutical industries (Guerin et al. 2003). However, *Haematococcus* produces astaxanthin to an amount of 4% from its cellular dry weight (DW) is substantially greater and increased production compared to aquaculture wastes, higher plants, and yeasts. Also, astaxanthin from *H. pluvialis* has been the one of the vital products with enormous significance in commercial market and recently received much interest among the scientific community (Gwak et al. 2014).

Transcriptome sequencing is a crucial and effective method for learning about the functional genomics of microalgae. In fact, recent years have seen advancements in the study of microalgal transcriptomes. For instance, *Dunaliella* microalga was sequenced and similarly de novo transcriptomes were done by several groups, which reported that *Dunaliella* microalga were found be medicinally important and commercially notable, especially the genes/transcriptomes were observed to involved in essential pathways and also crucial in manufacturing biofuels (Rismani-Yazdi et al. 2011). An Arctic cold *Chlamydomonas* sp. strain was used for the transcriptome study, demonstrating the genes in charge of fatty acid

D. Prabhu
Centre for Drug Discovery, Department of Biotechnology, Karpagam Academy of Higher Education, Coimbatore, India

S. Rajamanikandan (✉)
Centre for Drug Discovery, Department of Biochemistry, Karpagam Academy of Higher Education, Coimbatore, India

R. Raja et al. (eds.), *Haematococcus*, https://doi.org/10.1007/978-981-99-2901-6_14

(FA) biosynthesis, photosynthetic activity, and cold tolerance, as well as locating the probable antifreeze protein homologs in microalgae (Gao et al. 2015; Kim et al. 2013). Similarly, for an enhanced understanding of the functional genes and associated pathways from the transcriptome profiles of *Nannochloropsis* sp (Zheng et al. 2013). The sequences of transcriptomes possess the coding DNA regions only and are combined with an improved next-generation sequencing technique. Transcriptome sequencing can greatly aid the assembly of de novo contig transcriptomes and have huge transcriptome coverage depth with fewer sequencing requirements (Rismani-Yazdi et al. 2011).

A common deep sequencing technique for studying whole transcriptomes is RNA sequencing (RNA-seq). The high-throughput DNA sequencing technologies generates a library composed of cDNA reads, and these reads were aligned either to a reference transcriptome/genome or organism genome assembly from the scratch devoid of the knowledge (Gao et al. 2015). RNA-seq is therefore a very appealing process for the investigation of *H. pluvialis*. RNA-seq can provide the degree of transcriptional expression of each gene as well as a genome-scale transcription map of the transcriptome structure (Gwak et al. 2014; Wang et al. 2010). These methods can greatly advance our knowledge of the genetic regulation of astaxanthin production in *H. pluvialis*. The singletons and assembled sequences were searched for similarity in NCBI-BLAST tool and annotated using IDs from the Kyoto Orthology Genomes (KOG), Encyclopaedia of Genes and Genomes (KEGG), and Gene Ontology (GO). The GoPipe and BLAST results were utilized to functionally annotate the unigenes assembled using the NCBI database (Wang et al. 2010).

14.2 Protein Functional Annotation

The whole collection of proteins found or expressed in a cell, tissue, organism, or species is known as the proteome. This proteome information was normally translated from the genome that had been completely sequenced and interpreted. The complete proteome, which includes the proteome common to all strains, the dispensable proteome, which is only present in one or two strains, and the unique proteins found only in a particular strain that given its identity, is the total set of proteins translated by evolutionarily linked strains of the identical microbe. As a result, the availability of annotated genomes improves comparative genomics and pan-genomics studies to better comprehend the mode of disease pathogenesis. Understanding protein interactions and structures is essential to comprehending the pathological and physiological functions of a cell (Prabhu et al. 2020a, b). Hub proteins are proteins that interact with various protein partners and can be predicted by PPI network. The validation of anticipated protein function, the function of a particular protein in metabolic pathways, and the identification of functional protein associations are all made easier by the interaction information (Jeyakanthan et al. 2005, 2023). For the investigation of the core proteome, any strains of *H. pluvialis* that have a complete genome and the most recent assembly can be taken into consideration. The FTP NCBI Refseq site is frequently used to retrieve the proteome of various strains.

Core proteomics analysis involves the following steps: (1) using the CD-HIT programme to group related proteins, (2) Choosing a protein cluster (COP) that is shared by all *H. pluvialis* strains, (3) retrieving the proteins from the chosen *H. pluvialis* strain's COP, and (4) using the Kyoto Encyclopaedia of Genes and Genomes (KEGG) database to analyse the metabolic pathway. To retrieve nonhomologous proteins, functionally characterize fictitious proteins, and find viable drug targets coupled with physicochemical properties, subtractive genomics uses BLASTp against the human proteome, Database of Essential Genes (DEG), and Drug Bank Database, respectively. The specified protein's druggable score can be estimated using fpocket. Protein–protein interactions can be located and viewed using Cytoscape and the STRING database, respectively (Solanki and Tiwari 2018). The node degree allows for the identification of the hub proteins. Subtractive genomics has gained popularity in recent years as a flexible technique for locating therapeutic targets for the development of novel antibiotics against drug-resistant bacteria. Presently, the substantial use of computational methods for the discovery of prospective medication and vaccine candidates is made possible by the wealth of data available on the proteome and genomic characteristics of *H. pluvialis* strains.

A method known as subtractive genomics eliminates pathogenic genes that are required for the organism's survival but are absent in the host. The proteins that have been shortlisted can then be cloned, over-expressed, and purified using affinity chromatography, and therapeutic candidates can then be verified in in vivo animal models. In subtractive genomics, there are eight steps: (1) retrieval of proteome or genome data and phylogenetic analysis; (2) CD-HIT analysis; (3) screening of essential proteins; (4) analysis of virulence factors; (5) analysis of resistance proteins; (6) selection of nonhomologous proteins to host; (7) identification of pathogen-specific pathway; and (8) analysis of druggability. Redundant sequences in the bacterial proteome must be removed in order to make efficient use of the dataset and prevent producing redundant search results. The non-redundancy and redundancy values of each searched strain of *H. pluvialis* must therefore be used to differentiate them (Jeyakanthan et al. 2023; Kaur et al. 2013; Solanki and Tiwari 2018).

The International Nucleotide Sequence Database Collaboration numbers can be used for whole genomes, whereas whole genome sequencing numbers can be used for draught sequences. Genome information for a few strains of *H. pluvialis* can be accessed from NCBI. RNAmmer to look for rRNA sequences in the entire genome's DNA sequence. One 16s rRNA gene can be randomly sampled for each strain of *H. pluvialis* because there is very little sequence variation among 16s rRNA genes within the same species' genome. The 16srRNA sequences from all *H. pluvialis* strains are used in the creation of a phylogenetic tree using the MEGA6 tool, and ClustalW is frequently used for multiple sequence alignments of the sequence. The global sequence identity algorithm and CD-HIT clustering approaches can be used to identify duplicate proteins and align their amino acid sequences, respectively. Potential therapeutic targets include the target proteins that have homologous matches in the essential gene database. The key proteins in the proteome of *H. pluvialis* are found using BLASTp searches against DEG bacterial proteins using cut-off criteria like *E*-value and bit score. The adhesion, colonization, and invasion processes, which are mediated by the virulence factors, allow bacteria to change or impair the host's defence mechanism.

Proteins may be categorized into four groups using the virulence factor database: aggressive, defensive, non-specific, and virulence-associated proteins (Kaur et al. 2013). It is possible to identify potential and current new antibiotic resistance genes or proteins in *H. pluvialis* using the antibiotic resistance gene-ANNOTation database. The proteins causing antibiotic resistance were predicted using a BLAST search of the *H. pluvialis* proteome against the antibiotic resistance gene-ANNOTation database. The first steps in finding therapeutic targets to develop pathogen-specific medicines involve manual comparative study of virulence, essential, and resistance proteins to discover nonhomologous to host proteins. Nonhomologous proteins' metabolic pathways can be found using the KEGG database. The pathways involved in the metabolic process were identified exclusively in the organism and not in humans, the metabolic pathway of the host may also be listed with a K number and compared to the metabolic pathways of *H. pluvialis*. The proteins can also be divided into groups according to their functions, such as common pathways shared by both pathogens and hosts as well as particular, distinct pathways found only in pathogens. The Drug Bank database may then be searched against all potential therapeutic targets to determine the druggability of the target proteins in the unique pathway (Jeyakanthan et al. 2023).

The chosen proteins are classed as prospective therapeutic targets based on the consensus location found by employing online servers such as PSORTb and CELLOv2.5 that have bit scores larger than 100 and *E*-values less than 0.005. The accessibility value of therapeutic protein targets can be improved by the presence of low-molecular-weight proteins, typically about 100–110 kDa (Prabhu et al. 2020a, b). The proteins that are essential to a particular biosynthetic pathway can then be chosen for further in silico investigation. To give a rigorous evaluation and integration of protein–protein association, including direct or indirect interactions, the metabolic functional interaction between the proteins must be done for all of the specified target proteins using the protein interaction database STRING. The pharmacological targets in A. baumannii may be built into 3D structural models using the homology modelling approach. To choose the best template for modelling that has more than 30% similarity, the query protein sequences can be subjected to BLASTp examination in PDB (Surekha et al. 2017). Through the use of ligand modelling, the coordinates of the bound ligand or lead molecules may be transferred from the models created by software such as MODELLER (Prabhu et al. 2021; Rajamanikandan and Srinivasan 2017).

The created model that has the highest Discrete Optimized Protein Energy and Molpdf scores are eligible for additional validation. Ramachandran plot may be used to assess the stereochemical quality of a rudimentary homology model to gain access to the overall structural geometry of the drug targets. The steric hindrances in the amino acid side chains that can be verified by PROCHEK and ERRAT plot may be checked using the ERRAT plot (Prabhu et al. 2021; Rajamanikandan and Srinivasan 2017; Surekha et al. 2017). CastP can be used to locate the amino acid residues that make up the active site. Using ClustalX, the target proteins' amino acid sequence may be aligned with the amino acid sequences of other bacterial proteins to determine which amino acid residues are conserved (Prabhu et al. 2020b). To find the ligands with the highest minimum inhibitory activity reported against the same targets of various bacteria, a literature search may be done. By using molecular modelling, the 3D structure of certain lead molecules may be created. Gastieger-Huckel partial charges and the Tripos force field, respectively, may be used to evaluate the geometry optimization and minimization of 3D structures. Furthermore, molecular docking mediated identification of lead molecules may be docked in the protein's active site or binding site, and the ligands can be rated according to how well they interact with the protein there. The preparation of ligands and proteins, grid creation, and ligand docking are the three crucial phases in molecular docking. The ADME characteristics of the best docked lead compound may be researched. The QikProp module of Schrodinger can be used to study the ADME characteristics and drug likelihood. It is possible to forecast the stability of the improved homology model of the target proteins using molecular dynamics (MD) simulation. Using molecular mechanics' generalized Boltzmann surface area approach, the specific trajectories produced from the MD simulation of the protein–ligand complex were employed for investigation of hydrogen bonding and binding free energy calculation (Choubey and Jeyaraman 2016; Prabhu et al. 2021; Rajamanikandan and Srinivasan 2017).

14.3 Gene Annotation

All of the hypothetical or putative genes were searched against several databases for functional annotation, including PFAM (Protein family) (Mistry et al. 2021), KO (KEGG ortholog) (Kanehisa et al. 2008), non-redundant NCBI protein sequence database, non-redundant NCBI nucleotide sequence database (Benson et al. 2018), KOG, a manually curated and evaluated protein sequence in SwissProt database (Bairoch and Apweiler 2000), and gene ontology (Young et al. 2010). There were about 77,000 genes (49%) were interpreted in all databases, and about 5900 genes (4%) were interpreted at least in one of seven databases. In particular, non-redundant, KO, protein family, gene ontology, and KOG annotated 55,835 genes, 23,600 genes, 57,998 genes, 60,125 genes, and 21,500 genes, respectively (Table 14.1). Cellular component (CC), biological process (BP), and molecular function are the three gene ontology categories used to classify the annotated genes. Fifty-six sub-categories were created for the 60,125 annotated genes in the GO database. The metabolic process, single-organism process, and cellular process were the three top sub-categories for the biological process, with 35,416 genes (58% of around 60,900 genes annotated in GO), 32,300 genes (~53%), and 26,600 (~43%) genes, respectively. Organelle, cell component and cell were the three top

Table 14.1 An overview of the function annotation of *H. pluvialis* hypothetical or putative genes

Databases used for annotation	Total genes annotated (~)	Approximate annotation in Percentage (~)
Non-redundant NCBI database of protein sequences (NR)	55,835	35.56
Non-redundant NCBI database of nucleotide sequences (NT)	13,500	8.50
KEGG Ortholog database (KO)	23,600	15
SwissProt	36,422	24.21
Protein family database	57,998	36.95
Gene ontology database	60,125	37.98
Clusters of orthologous groups of proteins database	21,500	13.29
Hits found in all above databases	5920	3.89
Hits found in minimum one database	75,852	47.99
Total gene (hypothetical/ putative)	156,992	100

sub-categories under cellular component, with 20,557 (33.79%), 20,577 (33.79%), and 7693 (12.64%) genes, respectively (He et al. 2018).

Twenty-five groups were created from the 21,500 annotated genes in the KOG database. Post-translational modification, protein turnover, and chaperones ranked first in the KOG categories (2931, 13.58%), followed by "General function prediction only" (2557, 11.85%). "Cell motility" (21, 0.09%) and "Extracellular structures" (21, 0.09%) were the two smallest groups (He et al. 2018).

14.4 Molecular Docking of Astaxanthin to β-Glucuronidase

Several valuable compounds can be found in microalgae. Antioxidants, enzymes, carotenoids, fatty acids, toxins, sterols, peptides, and polymers, are just a few of the unusual products that the majority of microalgae species create. Because of their strong anti-inflammatory and antioxidant properties, carotenoids, especially astaxanthin, are widely known for their therapeutic benefits in the treatment of numerous diseases and the ageing process (Safarov et al. 2018). There are structural variations between bacterial and human glucoronidases. One strategy for treating neoplastic alterations or getting rid of the negative effects of various chemotherapeutic methods is to suppress bacterial glucoronidase. The whole of the experimental data could serve as the foundation for the creation of fresh, efficient in silico inhibitors (Niwa et al. 1972).

Natural astaxanthin is a carotenoid that is mostly presents in the nautical creatures, mainly in crayfish, microalgae, trout, crabs, krill, salmon, and shrimp. Astaxanthin, found to been in red colour by secreted pigment in reddish have given salmon, trout, lobster, shrimp, and other shellfish its pink or red color (Pashkow et al. 2008). According to experts, *Haematococcus pluvialis*, a green microalga, is the best source of astaxanthin. In comparison to other naturally occurring inhibitors and standard inhibitor, astaxanthin may be possibly more effective (Niwa et al. 1972). With wogonoside and baicalin excluded, the molecule astaxanthin has exhibited the highest potential in terms of the inhibitory efficacy against eGUS. According to the findings, astaxanthin may have potential impacts on the suppression of eGUS in addition to its exceptional nutritional and antioxidant qualities. As a result, it may be utilized to treat the side effects of NSAIDS and chemotherapy. The studies demonstrated the potential of astaxanthin as an eGUS inhibitor using molecular docking. Thus, astaxanthin can be seen as a possible treatment for avoiding enteropathies brought on by some hazardous substances (Niwa et al. 1972; Pashkow et al. 2008; Safarov et al. 2018; Wallace 2000; Yang et al. 2009).

14.5 Computational Binding Analysis of Astaxanthin with Hepatocellular Carcinoma Proteins

Because it contains oxygen atoms in addition to carbon and hydrogen, astaxanthin is a xanthophyll. The terminal two rings connected by a chain of polyene make up astaxanthin. The astaxanthin molecule has a hydroxyl group (eOH) at each end and asymmetric two carbons at the locations of the -ionone ring (3, 3′). A monoester is created when a –OH group combines with one of the fatty acids, whereas the di-esterase created while both fatty acids and –OH group reacts. There are free, geometric isomers, esterified forms, and stereoisomers of astaxanthin (Higuera-Ciapara et al. 2006). In the treatment and prevention of numerous illnesses, such as gastrointestinal, cancer, cardiovascular, diabetes, neurodegenerative, liver, eye, skin, renal failures, and male infertility diseases, the molecule astaxanthin may have health-promoting effects (Dhankhar et al. 2012). According to several reports, astaxanthin's anti-cancer properties are explained by how it affects the abnormal development of malignant cells via several pathways, which includes the primary inflammation, cell junction, and apoptosis. By activating the transcription of genes Bcl2 and Bax, astaxanthin limits the growth of hepatocellular carcinoma cells and causes them to die.

Our research demonstrates that astaxanthin has better docking potential with five different apoptotic proteins than the effective medication sorafenib, including, Caspase-9, EGFR, VEGFR, Caspase-3, and Bcl-2. When medications interact with target polypeptide chains, the astaxanthin molecule binding mode of various element entities of the marked proteins demonstrated huge differences in the folding pattern. Astaxanthin was docked into the active site of brain derived neurotrophic factor protein and Ras homolog RHOD protein (Suganya and Anuradha 2019). Docking energies for astaxanthin and RHOD are as follows: $E_{max} = 88.39$, $E_{min} = 247.72$. Range of the docking energies for rhod and beta-carotene: $E_{max} = 86.55$, $E_{min} = 232.07$. The lowest minimal energy in docking is for highest affinity. It is determined that astaxanthin reported to have the strong binding with proteins (Guttula et al. 2011). The molecule astaxanthin reported to bind in the active site of the target protein monoamine oxidase A (MAO-A), according to molecular docking investigations, with a projected K_d of $4.4 \cdot 10^{-6}$ M. The astaxanthin may have a reversible and selective inhibitory effect on the target MAO-A. While non-irreversible and non-selective MAO-A inhibitors were ceased due to their harmful side effects brought on by their connections with additional medications and certain dietary elements, reversible, and selective inhibitors of MAO-A were still commonly in use as antidepressants and anxiolytics (Safarova et al. 2016).

The phytoconstituent of astaxanthin has been shown through in silico molecular docking experiments to have the capacity to suppress the phosphorylation hyperactivity on the protein Tau. An important factor in predicting the amino acid residues that will be present between a chosen protein and the inhibitors are determined by the interactive hydrogen bonds. The hydrogen bond interaction regions were revealed to have a collection of polar residues on the active site of the GSK-3 protein, including ASN-686, ASN-689, ASP-700, ARG-180, ARG-96, GLN-795, GLN-89, GLU-97, ILE-562, and LYS-585 (1J1B). Overall findings from numerous in silico predictions indicated that the substance astaxanthin is significantly preventing the hyper phosphorylation of Tau in relation to Alzheimer's disease (Praveen and Yellamma 2017). The effectiveness of astaxanthin and its derivatives against many apoptotic and receptor proteins has been demonstrated.

14.6 Computational Screening of Metabolites

The extracellular metabolites are known as a biosynthetic gene cluster (smBGC) that controls a specific network of metabolic action to synthesize metabolites. These genes carry out a wide range of functions, such as regulating product resistance, assembling and altering translation and post-translational products, assisting in the production of precursor molecules, and controlling expression. The control of the smBGCs network is impacted by biotic stress and abiotic factors present in the surroundings where the bacteria are situated (Skinnider et al. 2017). The generation of secondary metabolites in the laboratory is hindered by the difference in ecological conditions amid the laboratory and the natural ecology. Few unexpressed bacterial genes are "hidden," or "cryptic" in laboratory settings. The chemical, biological, and physical prerequisites of the gene translations were not satisfied. Utilizing merely laboratory detection techniques, it is challenging to confirm the existence of this gene. Therefore, a strategy that takes into account the target microorganism's whole metabolic capacity is necessary (Steven et al. 2022).

One of the most popular methods for overcoming this problem is genome mining. It is a technique for finding biosynthetic gene clusters that have not been discovered before. In silico and in vitro stages can be used to categorize genome mining. One of the in silico processes is the discovery and characterization of BGCs. Various bioinformatics analysis must be run on the target microorganisms' genome

Table 14.2 Databases those are presently available for screening secondary metabolites

S. No.	Database name	Website link
1.	MIBiG	https://mibig.secondarymetabolites.org/
2.	IMG-ABC	https://img.jgi.doe.gov/abc
3.	DoBISCUIT	http://www.bio.nite.go.jp/pks/
4.	ClustScan Database	http://csdb.bioserv.pbf.hr/csdb/ClustScanWeb.html
5.	ClusterMine360	http://www.clustermine360.ca/
6.	Bactibase	http://bactibase.pfba-lab-tun.org
7.	antiSMASH database	http://antismash-db.secondarymetabolites.org/

sequences during this step. Numerous aids promote the computational high-throughput screening of certain chemicals are offered through widely accessible databases and software. The overall process of in silico natural product screening, which makes use of a variety of freely accessible tools and databases, while the experimental step entails the characterization and identification of the biosynthetic gene cluster product utilizing in vitro methods. The sequences of the genes were expressed and depict in specific vectors when they are predicted to encode particular genes (Steven et al. 2022). The rule-based approach makes use of an alignment-based on the sequence profile of smBGCs in the Markov Model, which is included in the database and is associated with a particular biosynthetic route (Skinnider et al. 2017). Evidently, readily accessible resources and databases frequently include a range of innate yields or metabolites (Table 14.2). Currently, no carotenoid-specific tools or datasets accessible. As a result, selection of a particular group of innate yield may be done by altering the search criteria in pertinent databases or applications.

In addition, other gears are required for developing hidden Markov models and using additional computational techniques. With genome data as the query, quick, and simple secondary metabolite screening can be carried out utilizing publically accessible technologies (Table 14.2). Two gears that are frequently used to search for smBGCs using a rule-based approach are PRISM and antiSMASH (Blin et al. 2021; Cimermancic et al. 2014). These rule-based techniques make it simple to quickly search databases for hidden, related smBGCs. It is challenging to forecast fresh smBGCs are not equal to those in database because this technique only identifies smBGCs that are similar to those in databases.

The biosynthetic gene cluster sequences that may be found in open-access sources (Table 14.3) can be utilized as the training data set for tradition ML replica and algorithms that can be used in more sophisticated selection pipelines. A strategy based on ML-based techniques, such as DeepBGC and ClusterFinder, which is necessary for the discovery of novel smBGCs (Hannigan et al. 2019). The high false-positive rate of this machine learning-based approach, however, is a disadvantage. However, it is anticipated that the combination of these two methods will yield a precise and comprehensive list of smBGCs (Lee et al. 2020). Previously, two missing genes in the pathway for carotenoid production were found using genome-based in silico screening. The local protein database created with the help of the blastdb tool was compared using the BLAST+. In silico genome-based screening has

Table 14.3 Tools for the general biosynthetic gene cluster identification

S. No.	Webserver/database	Website link
1.	HMMER	http://www.ebi.ac.uk/Tools/hmmer/search/jackhmmer
2.	SMURF	https://www.jcvi.org/smurf
3.	NaPDoS	http://napdos.ucsd.edu/
4.	NP.Searcher	http://dna.sherman.lsi.umich.edu/
5.	MultiGeneBlast	http://multigeneblast.sourceforge.net/
6.	IMG-ABC	https://img.jgi.doe.gov/cgi-bin/abc/main.cgi
7.	FramePlot 4.0beta	http://nocardia.nih.go.jp/fp4
8.	eSNaPD	http://esnapd2.rockefeller.edu/
9.	ClusterMine 360	http://clustermine360.ca/
10.	ClusterFinder	http://github.com/petercim/ClusterFinder
11.	Artemis	http://www.sanger.ac.uk/science/tools/artemis
12.	antiSMASH 6.0	http://antismash.secondarymetabolites.org/

also been used in studies to identify aquatic bacterial carotenoid gene clusters using the techniques and databases listed in Table 14.3. The antiSMASH pipeline is the most often utilized instrument in these research projects. The "genome mining" methods are widely in use towards the gene clusters identification, forecast their resemblance to gene clusters that are already present in databases, and create a potential pathway for carotenoid production (Iftime et al. 2016; Ma et al. 2016; Myronovskyi et al. 2014; Niero et al. 2021; Rodriguez-Amaya 2016; Takemura et al. 2021; Xu et al. 2018).

14.7 Modelling of CYP450 Protein of *H. pluvialis*

In homology modelling, identifying an acceptable structure to predict the target protein model is the very first step in drug discovery. Due to the lack of a predicted structure for CYP450 in microalgae, AlphaFold programme was used to build the target protein structure (Fig. 14.1a). For the purpose of releasing the unstable atomic contact and tuning the irrational local structural conformations, the energies of the established bonds of the generated 3D models of CYP450 were minimized (Sureshkumar et al. 2018). The stereochemical characteristics of primary and sidechain amino acids (aa) of the CYP450 model protein of *H. pluvialis* was confirmed using a Procheck Ramachandran (RC) plot analysis. These findings support the secondary structure validation by demonstrating that 98.9% of amino acids present in the core region of RC plot, 0.4% of residues are in allowed regions, and 0.7% are in disallowed regions (Fig. 14.1b). The *H. pluvialis* envisaged CYP450 structure revealed, overall quality factors from ERRAT values of 92.02, which can be used as a benchmark for good models (Table 14.4).

The PDBSum webserver was generally used to pinpoint the CYP catalytic active site for the *Haematococcus pluvialis*. The 17 amino acids Ile271, Gly275, Thr278,

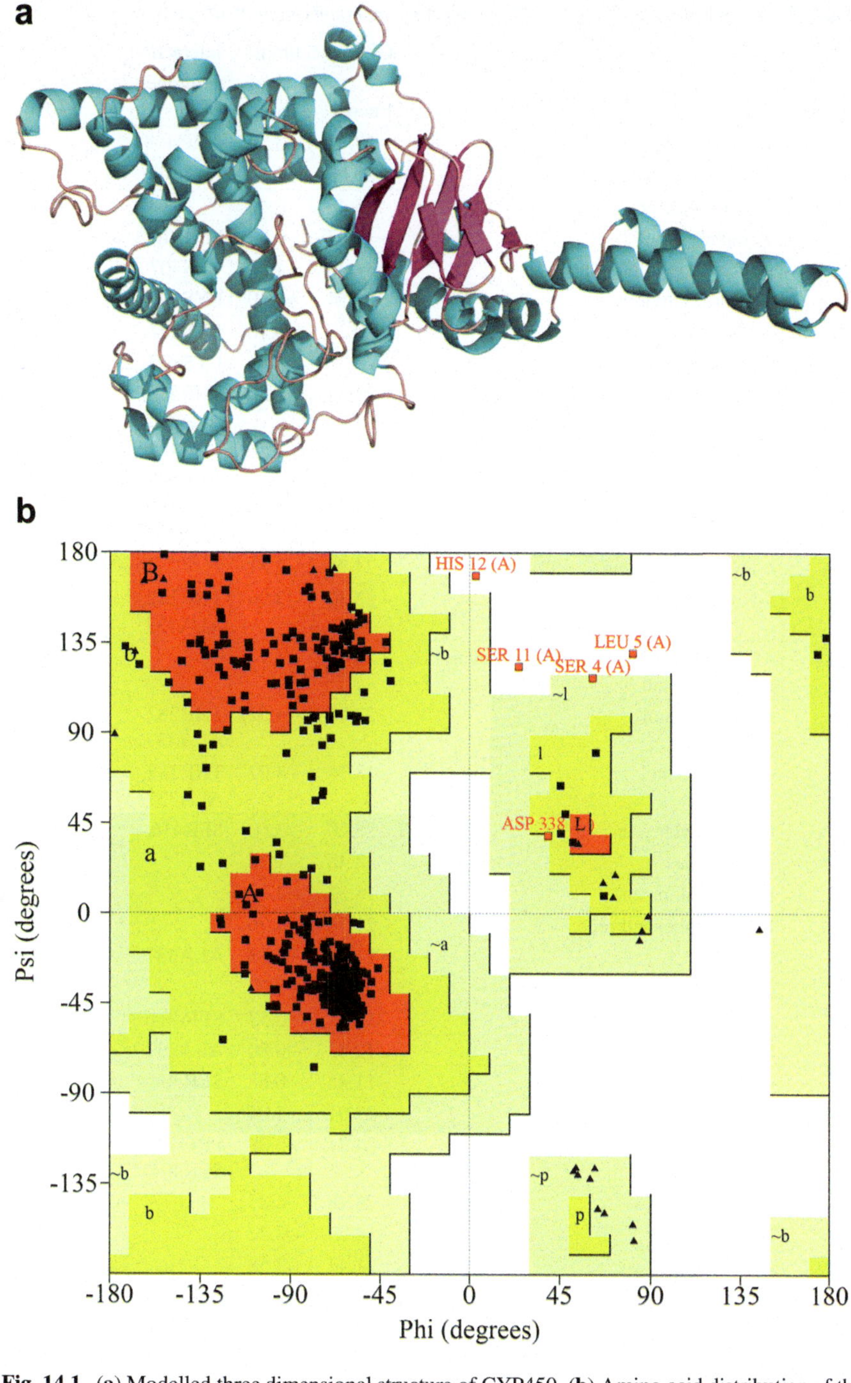

Fig. 14.1 (**a**) Modelled three dimensional structure of CYP450. (**b**) Amino acid distribution of the CYP protein of *H. pluvialis* in ramachandran plot

Table 14.4 Molecular docking of PAHs against *Haematococcus pluvialis* enzyme

S. No	Ligand name	Glide (kcal/mol)		Interaction	
		Energy	Score	Hydrogen bond	Pi-Pi
1.	1-Hydroxychrysene	−27.17	−3.74	ARG408	
2.	1-Hydroxypyrene	−25.08	−3.67	LYS407	
3.	1-Methylnaphthalene	−15.44	−3.43		PHE403
4.	1-Naphthol	−22.07	−4.63	SER416	
5.	2-Acetylaminofluorene	−31.40	−4.40	LYS407	
6.	2-Hydroxyfluorene	−24.54	−3.40	LYS407	
7.	2-Naphthol	−20.29	−3.16	LYS407	
8.	2-Phenanthrenol	−25.58	−3.45	LYS407	
9.	4-Octylphenol	−24.89	−3.52	THR282, ALA337	
10.	4-Butylphenol	−19.85	−4.41	THR282, ALA337	
11.	4-Heptylphenol	−24.42	−3.85	THR282, ALA337	
12.	4-Hydroxychrysene	−24.26	−3.81		
13.	4-Nonylphenol	−23.30	−3.96	THR282, ALA337	
14.	4-Oxo-1,2,3,4-tetrahydro naphthalene1-carboxylic acid	−20.46	−3.12		SER404
15.	4-Pentylphenol	−23.04	−4.52	THR282, ALA337	
16.	4-Propylphenol	−18.74	−4.22	THR282, ALA337	
17.	6-Hydroxychrysene	−28.20	−3.60	SER416	
18.	Anthracene	−18.48	−1.26		
19.	Benz(a)anthracene	−18.52	−2.22		
20.	Benzo(c)phenanthrene	−21.76	−1.76		
21.	Catechol	−21.56	−5.33	ALA337	PHE403
22.	Chrysene	−23.17	−3.02		
23.	Cyclohexaneacetic acid	−18.31	−2.52	SER404	
24.	Cyclohexanecarboxylic acid	−12.83	−0.80	ALA337	
25.	Cyclohexanepropionic acid	−11.48	−0.88	SER404	
26.	Dibenzo(a,l)pyrene	−29.01	−4.04		
27.	Hydroxyfluoranthene—1-carboxylic acid	−24.21	−3.38	PRO402, SER404	
28.	Methaqualone	−23.48	−4.27		
29.	Naphthalene	−14.23	−3.22		PHE403
30.	*n*-Hexadecane	−21.44	−2.58		
31.	Phenol	−18.03	−4.13	THR282, ALA337	

Ile279, Thr282, Val338, Leu341, Arg343, Pro402, Phe403, Ser404, Arg408, Lys409, Cys410, Ser412, and Ser416 make up the active site of *Haematococcus pluvialis*. The molecular docking experiments in Schrodinger software revealed the docking score, energy, represented as *gscore* and *genergy*, along with the hydrogen bond interactions were selected to evaluate the potential species of microalgae implicated in the deprivation of several kinds of polycyclic aromatic hydrocarbons. The contacts via Thr282, Ala337, Pro402, Ser404, Lys407, Arg408, and Ser416 through hydrogen bonding interactions, as well as Phe403 with -π interactions. Catechol, a single PAH, the lowest docking score of 5.33 (kcal/mol) among the *H. pluvialis* CYP, which was traced by the phenol molecule and their substructures. The enduring polycyclic aromatic hydrocarbon compounds had glide scores between 4 and 1 kcal/mol, which is below the permitted range for a docking interaction investigation. With *H. pluvialis* CYP protein, other polycyclic aromatic hydrocarbon compounds, including those top-ranked 3-ring and 2-ring structures, displayed gscores of 4.40 and 4.63 kcal/mol, respectively. However, the higher molecular weight polycyclic aromatic hydrocarbon compounds were unable to establish binding with the *H. pluvialis* CYP protein and could not even manage to establish a sole contact with the biding site amino acids.

14.8 Conclusion

Based on the computational analysis, the expression patterns of multiple genes are altered under various induction settings, as well as differentially expressed genes involved in cellular metabolism of *H. pluvialis* were completely analysis. Comparative de novo transcriptome assembly revealed a distinct picture in gene ontology of transcripts that were differentially expressed across wild and mutant strains of *H. pluvialis*. Important therapeutic targets are the genes and proteins involved in the mechanisms governing cell growth and stress responses. Studying the signal transduction regulatory mechanisms and cellular processes in *H. pluvialis* may benefit from the gene ontology enrichment analysis of DEG transcripts labelled as hypothetical or uncharacterized proteins. The majority of DEGs were involved in metabolic activity, catalytic activity, single-organism processes, single-organism cellular processes, and single-organism metabolic processes during the early phases of stress are the potential therapeutic targets. Understanding the complete genomic, proteomic, and transcriptomic data were exploited in different ways to explore the inclusive therapeutic knowledge.

Reference

Bairoch A, Apweiler R (2000) The SWISS-PROT protein sequence database and its supplement TrEMBL in 2000. Nucleic Acids Res 28(1):45–48

Benson DA, Cavanaugh M, Clark K, Karsch-Mizrachi I, Ostell J, Pruitt KD, Sayers EW (2018) GenBank. Nucleic Acids Res 46(D1):D41–D47

Blin K, Shaw S, Kloosterman AM, Charlop-Powers Z, van Wezel GP, Medema MH, Weber T (2021) antiSMASH 6.0: improving cluster detection and comparison capabilities. Nucleic Acids Res 49(W1):W29–W35

Choubey SK, Jeyaraman J (2016) A mechanistic approach to explore novel HDAC1 inhibitor using pharmacophore modeling, 3D-QSAR analysis, molecular docking, density functional and molecular dynamics simulation study. J Mol Graph Model 70:54–69

Cimermancic P, Medema MH, Claesen J, Kurita K, Wieland Brown LC, Mavrommatis K, Pati A, Godfrey PA, Koehrsen M, Clardy J, Birren BW, Takano E, Sali A, Linington RG, Fischbach MA (2014) Insights into secondary metabolism from a global analysis of prokaryotic biosynthetic gene clusters. Cell 158(2):412–421

Dhankhar J, Kadian SS, Sharma A (2012) Astaxanthin: a potential carotenoid. Int J Pharm Sci Res 3(5):1246

Gao Z, Li Y, Wu G, Li G, Sun H, Deng S, Shen Y, Chen G, Zhang R, Meng C, Zhang X (2015) Transcriptome analysis in Haematococcus pluvialis: astaxanthin induction by salicylic acid (SA) and jasmonic acid (JA). PLoS One 10(10):e0140609

Guerin M, Huntley ME, Olaizola M (2003) Haematococcus astaxanthin: applications for human health and nutrition. Trends Biotechnol 21(5):210–216

Guttula SV, Rao AA, Sridhar GR, Chakravarthy MS (2011) Protein ligand interaction analysis an insilico potential drug target identification in diabetes mellitus and nephropathy. J Bioinform Seq Anal 2(5):95–99

Gwak Y, Hwang YS, Wang B, Kim M, Jeong J, Lee CG, Hu Q, Han D, Jin E (2014) Comparative analyses of lipidomes and transcriptomes reveal a concerted action of multiple defensive systems against photooxidative stress in Haematococcus pluvialis. J Exp Bot 65(15):4317–4334

Hannigan GD, Prihoda D, Palicka A, Soukup J, Klempir O, Rampula L, Durcak J, Wurst M, Kotowski J, Chang D, Wang R, Piizzi G, Temesi G, Hazuda DJ, Woelk CH, Bitton DA (2019) A deep learning genome-mining strategy for biosynthetic gene cluster prediction. Nucleic Acids Res 47(18):e110

He B, Hou L, Dong M, Shi J, Huang X, Ding Y, Cong X, Zhang F, Zhang X, Zang X (2018) Transcriptome analysis in Haematococcus pluvialis: astaxanthin induction by high light with acetate and Fe^2. Int J Mol Sci 19(1):175

Higuera-Ciapara I, Félix-Valenzuela L, Goycoolea FM (2006) Astaxanthin: a review of its chemistry and applications. Crit Rev Food Sci Nutr 46(2):185–196

Iftime D, Kulik A, Härtner T, Rohrer S, Niedermeyer THJ, Stegmann E, Weber T, Wohlleben W (2016) Identification and activation of novel biosynthetic gene clusters by genome mining in the kirromycin producer Streptomyces collinus Tü 365. J Ind Microbiol Biotechnol 43:277–291

Jeyakanthan J, Inagaki E, Kuroishi C, Tahirov TH (2005) Structure of PIN-domain protein PH0500 from Pyrococcus horikoshii. Acta Crystallogr Sect F Struct Biol Cryst Commun 61(5):463–468

Jeyakanthan J, Pandian CJ, Jeyaseelan JM, Rajamanikandan S, Prabhu D, Ravi GR, Nachiappan M, Kingsley J (2023) An integrative approach to explore potent therapeutic protein targets in multidrug-resistant nosocomial pathogen Acinetobacter Baumannii. In: Therapeutic protein targets for drug discovery and clinical evaluation: bio-crystallography and drug design, pp 159–193

Kanehisa M, Araki M, Goto S, Hattori M, Hirakawa M, Itoh M, Katayama T, Kawashima S, Okuda S, Tokimatsu T, Yamanishi Y (2008) KEGG for linking genomes to life and the environment. Nucleic Acids Res 36(Database issue):D480–D484

Kaur N, Khokhar M, Jain V, Bharatam PV, Sandhir R, Tewari R (2013) Identification of druggable targets for Acinetobacter baumannii via subtractive genomics and plausible inhibitors for MurA and MurB. Appl Biochem Biotechnol 171(2):417–436

Kim S, Kim MJ, Jung MG, Lee S, Baek YS, Kang SH, Choi HG (2013) De novo transcriptome analysis of an Arctic microalga, Chlamydomonas sp. Genes Genom 35(2):215–223

Lee JH, Kim JW, Lee PC (2020) Genome mining reveals two missing CrtP and AldH enzymes in the C30 carotenoid biosynthesis pathway in Planococcus faecalis AJ003^T. Molecules 25(24): 5892

Ma T, Zhou Y, Li X, Zhu F, Cheng Y, Liu Y, Deng Z, Liu T (2016) Genome mining of astaxanthin biosynthetic genes from Sphingomonas sp. ATCC 55669 for heterologous overproduction in Escherichia coli. Biotechnol J 11:228–237

Mistry J, Chuguransky S, Williams L, Qureshi M, Salazar GA, Sonnhammer ELL, Tosatto SCE, Paladin L, Raj S, Richardson LJ, Finn RD, Bateman A (2021) Pfam: the protein families database in 2021. Nucleic Acids Res 49(D1):D412–D419

Myronovskyi M, Tokovenko B, Brötz E, Rückert C, Kalinowski J, Luzhetskyy A (2014) Genome rearrangements of Streptomyces albus J1074 lead to the carotenoid gene cluster activation. Appl Microbiol Biotechnol 98:795–806

Niero H, da Silva MAC, de Felicio R, Trivella DBB, de Souza LAO (2021) Carotenoids produced by the deep-sea bacterium Erythrobacter citreus LAMA 915: detection and proposal of their biosynthetic pathway. Folia Microbiol 66:441–456

Niwa T, Tsuruoka T, Inouye S, Naito Y, Koeda T, Niida T (1972) A new potent β glucuronidase inhibitor, D-glucaro-δ-lactam derived from nojirimycin. J Biochem 72(1):207–211

Pashkow FJ, Watumull DG, Campbell CL (2008) Astaxanthin: a novel potential treatment for oxidative stress and inflammation in cardiovascular disease. Am J Cardiol 101(10):S58–S68

Prabhu D, Rajamanikandan S, Anusha SB, Chowdary SC, Veerapandiyan M, Jeyakanthan J (2020a) In silico functional annotation and characterization of hypothetical proteins from Serratia marcescens FGI94. Biol Bull Russ Acad Sci 47:319–331

Prabhu D, Rajamanikandan S, Saritha P, Jeyakanthan J (2020b) Evolutionary significance and functional characterization of streptomycin adenylyltransferase from Serratia marcescens. J Biomol Struct Dyn 38(15):4418–4431

Prabhu D, Rajamanikandan S, Sureshan M, Jeyakanthan J, Saraboji K (2021) Modelling studies reveal the importance of the C-terminal inter motif loop of NSP1 as a promising target site for drug discovery and screening of potential phytochemicals to combat SARS-CoV-2. J Mol Graph Model 106:107920

Praveen K, Yellamma K (2017) Insilco studies on astaxanthin derivatives against tau protein-a novel approach to design anti-Alzheimers drug targets. Int J Pharm Sci Res 8(11):226–231

Rajamanikandan S, Srinivasan P (2017) Exploring the selectivity of auto-inducer complex with LuxR using molecular docking, mutational studies and molecular dynamics simulations. J Mol Struct 1131:281–293

Rismani-Yazdi H, Haznedaroglu BZ, Bibby K, Peccia J (2011) Transcriptome sequencing and annotation of the microalgae Dunaliella tertiolecta: pathway description and gene discovery for production of next-generation biofuels. BMC Genomics 14(12):148

Rodriguez-Amaya DB (2016) Natural food pigments and colorants. Curr Opin Food Sci 7:20–26

Safarov N, Agalarov R, Gasanov R (2018) Molecular docking of astaxanthin to β-glucuronidase. Adv Biol Earth Sci 3(3):176–182

Safarova G, Safarov N, Gasanov R (2016) Molecular docking of astaxanthin to monoamine oxidase. Adv Biol Earth Sci 1:45–50

Skinnider MA, Merwin NJ, Johnston CW, Magarvey NA (2017) PRISM 3: expanded prediction of natural product chemical structures from microbial genomes. Nucleic Acids Res 45(W1):W49–W54

Solanki V, Tiwari V (2018) Subtractive proteomics to identify novel drug targets and reverse vaccinology for the development of chimeric vaccine against Acinetobacter baumannii. Sci Rep 8:9044

Steven R, Humaira Z, Natanael Y, Dwivany FM, Trinugroho JP, Dwijayanti A, Kristianti T, Tallei TE, Emran TB, Jeon H, Alhumaydhi FA, Radjasa OK, Kim B (2022) Marine microbial-derived resource exploration: uncovering the hidden potential of marine carotenoids. Mar Drugs 20(6): 352

Suganya V, Anuradha V (2019) In silico molecular docking of astaxanthin and sorafenib with different apoptotic proteins involved in hepatocellular carcinoma. Biocatal Agric Biotechnol 19: 101076

Surekha K, Nachiappan M, Prabhu D, Choubey SK, Biswal J, Jeyakanthan J (2017) Identification of potential inhibitors for oncogenic target of dihydroorotate dehydrogenase using in silico approaches. J Mol Struct 100(1127):675–688

Sureshkumar P, Thomas J, Poornima V (2018) Structural insights on bioremediation of polycyclic aromatic hydrocarbons using microalgae: a modelling-based computational study. Environ Monit Assess 190:92

Takemura M, Takagi C, Aikawa M, Araki K, Choi S-K, Itaya M, Shindo K, Misawa N (2021) Heterologous production of novel and rare C30-carotenoids using Planococcus carotenoid biosynthesis genes. Microb Cell Fact 20:194

Wallace JL (2000) How do NSAIDs cause ulcer disease? Best Pract Res Clin Gastroenterol 14(1): 147–115

Wang L, Feng Z, Wang X, Wang X, Zhang X (2010) DEGseq: an R package for identifying differentially expressed genes from RNA-seq data. Bioinformatics 26(1):136–138

Xu X, Tian L, Xu J, Xie C, Jiang L, Huang H (2018) Analysis and expression of the carotenoid biosynthesis genes from Deinococcus wulumuqiensis R12 in engineered Escherichia coli. AMB Express 8:94

Yang Q, Zhang Z, Zhu X, Ruan H, Fu Y (2009) Therapeutic effect of astaxanthin on acetic acid-induced gastric ulcer in rats. Acta Pharm Sin 44(5):558–560

Young MD, Wakefield MJ, Smyth GK, Oshlack A (2010) Gene ontology analysis for RNA-seq: accounting for selection bias. Genome Biol 11(2):R14. https://doi.org/10.1186/gb-2010-11-2-r14

Zheng M, Tian J, Yang G, Zheng L, Chen G, Chen J, Wang B (2013) Transcriptome sequencing, annotation and expression analysis of Nannochloropsis sp. at different growth phases. Gene 523(2):117–121

Part V
Biotechnological and Industrial Applications

Chapter 15
Algal Polymers, Proteins, and Pigments for Industrial Applications

G. Uma, S. Jeraldine Nisha, D. Rameshbabu, T. Citarasu, and G. Immanuel

15.1 Introduction

Algae are photoautotrophic prokaryotic (cyanobacteria) or eukaryotic organisms and can also be considered as heterotrophs or mixotrophs that can convert nitrogen and phosphorus from the environment into biomass using light, CO_2, and water. The resulting biomass can then be fractionated using the appropriate technology to create various bioproducts (Ometto et al. 2014; Li-Beisson et al. 2019). Algae can be large multicellular or unicellular organisms with a variety of reproductive strategies (Brodie et al. 2017). Macroalgae and microalgae are two different forms of algae and the microalgae (microphytes) include diatoms as well as green algae (Chlorophyta), blue-green algae (Cyanobacteria), yellow-green algae (Ochrophyta or Xanthophyta), and golden algae (Ochrophyta or Chrysophyta) and diatoms (Bacillaryophyta). The macroalgae or simple algae includes red algae (Rhodophyta), green algae (Chlorophyta), and brown algae (Ochrophyta) (Lee et al. 2020). Macroalgae primarily inhabit the marine environment. They can either be found naturally or farmed on a large scale near the ocean (Chia et al. 2018).

Algae depend on the natural nutrients in the sea to grow and thrive (Sudhakar et al. 2019). Most of the principal metabolites of algae are composed of lipids, proteins, carbohydrates, and water. Chlorophyll, cytochromes, nucleotides, and other chemicals that are intermediates in other metabolic processes are also examples of primary metabolites (Andreeva et al. 2021). The biological makeup of microalgae makes them appropriate for producing a variety of compounds. The ratio of primary metabolites depends on the kind of algae and the growing environment (Sudhakar et al. 2019). In addition to primary metabolites, secondary metabolites—chemicals not involved in the primary metabolism and maybe specific to one or more algae

G. Uma (✉) · S. Jeraldine Nisha · D. Rameshbabu · T. Citarasu · G. Immanuel
Centre for Marine Science & Technology, Manonmaniam Sundaranar University, Rajakkamangalam, Tamil Nadu, India

R. Raja et al. (eds.), *Haematococcus*, https://doi.org/10.1007/978-981-99-2901-6_15

species—are also found in macroalgae. The algae generate a large number of secondary metabolites throughout their life cycle, and they represent a substantial natural source of these bioactive compounds (Tanna and Mishra 2018). In addition to that, macroalgae are a fantastic source of proteins, lipids, and carbohydrates, they also include a wide range of vitamins and macro and microelements (Yuan et al. 2018). On the other hand, the location and hour of collecting can alter the composition of the above metabolites and micro and macronutrients of algae (Praiboon et al. 2018). Considering the above importance of algae with particular reference to algal metabolites, the present review article is discussed almost the algal polymers as well as algal nutritional sources for industrial purposes.

15.2 Algal Polymers

15.2.1 Algal Polysaccharides

In marine algae, polysaccharides are widely distributed; mucopolysaccharides, storage polysaccharides, and cell wall-structuring polysaccharides are some of them. The main polysaccharides in red seaweeds are xylan, carrageenan, fucoidan, water-soluble sulfated galactan, porphyran, and mucopolysaccharides. Ulvans, sulphuric acid polysaccharides, xylan, and sulfated galactans are the predominant polysaccharides found in green algal seaweeds. Sulphur-containing polysaccharides include sargassan, fucoidan, laminarin, and alginic acid. Sulfated polysaccharides have biological effects such as immune-inflammatory, antiviral, and anticancer properties (Vavilala and D'Souza 2015). The following are the important algal polysaccharides having wide range of industrial applications.

15.2.2 Cellulose

Algae have an amorphous matrix of layers that make up their cell wall skeleton, and a fibrillar component is embedded within this matrix. The main component of the fibrillar component is cellulose, a polymer of 1,4-linked D-glucose. High cellulose content has been observed in several macroalgal genera, including *Ceramium* (18.5%), *Chaetomorpha* (36.5–41%), *Chondria* (16.4%), *Cladophora* (20–45%), *Corallina* (15.2%), *Fucus* (13.5%), *Gelidiella* (11.3–13.6%), *Gracilaria* (10.5%), *Griffithsia* (22%), *Halidrys* (14%), *Hypnea* (11.4%), *Laminaria* (1.1–20%), *Rhizoclonium* (38.6%), and *Ulva* (1.8–19%) species (Ross et al. 2021; Zanchetta et al. 2021). In 2020, the market for nanocellulose was found to be the worth of $500 million USD (Zanchetta et al. 2021). Tissue engineering and other significant biomedical applications have made use of cellulose scaffolds made from *Ulva* sp. and *Cladophora* sp. (Bar-Shai et al. 2021). The use of macroalgal cellulose as a growth medium is one of its additional indirect uses (Hoover et al. 2011).

15.2.3 Starch

Starch and fructosans are high molecular weight nonstructural carbohydrates found in algae, whereas sugars, glycosides, and polyols are low molecular weight derivatives. Many macroalgal species make starch, a granular polymer made of glucose that is stored in the body (Cazon et al. 2017). With limited digestibility and a higher capacity for hydration than rice starch, the green seaweed *Ulva ohnoi* could produce up to 50.37% native starch when grown in seawater under nutrient-starved conditions (Prabhu et al. 2019). Films, injection-molded parts, and thermoplastic materials are made with resins based on starch. The fermentation of starch results in the production of ethanol and lactic acid, which are then processed and polymerized to produce PLA, polyhydroxy alkanoates (PHA), and copolymers (Devadas et al. 2021). Furthermore, starch can be combined with other biopolymers like PLA and PHA and synthetic degradable polymers to create fully biodegradable polymer composite materials, which are covered later (Rai et al. 2021). These polymers are less suitable for commercial applications since they are currently thought to be less resilient and offer less water resistance than traditional plastics.

15.2.4 Agar

Agarose and agaropectin are two polysaccharides that make up the mixture known as agar and have interchangeable structural and functional features (Vavilala and D'Souza 2015). Red algae, also known as Rhodophyceae, produce a heterogenous polysaccharide called agar, which is based on sulfated galactan. Agar supports flexibility and environmental adaptation in algae, which benefits them. The most extensively used sources for the commercial manufacture of agar are *Gelidium* sp. and *Gracilaria* sp., which are typically processed using the freeze-thaw procedure, although there are more than a dozen different extraction methods (Kartik et al. 2021). Commercially, it serves as a gelling agent and a substitute for gelatin made from animal sources. As a hydrocolloid, it has a variety of uses in the food, pharmaceutical, and cosmetic industries, with the food industry accounting for over 90% of its overall production. By 2025, the market for agar products will be anticipated to be the worth around $300 million USD (Lee et al. 2017). The agricultural industry also uses agar as a static carrier for nutrient/fertilizer and growth-promoting compounds. Agar-based edible films have been promoted as a sustainable alternative for food packaging, but due to their inherently poor thermomechanical characteristics, such as thermal fragility and brittleness, reinforcements have been used to enhance their functional properties for such applications (Mostafavi and Zaeim 2020).

15.2.5 Alginate

A polymer called alginate is produced from the cell walls of brown algae (Phaeophyceae). Alginic acid is the name given to the linear polyuronic acid type. Alginate is a polyelectrolyte and a biopolymer (Vavilala and D'Souza 2015). The family of linear copolymers known as alginate now includes blocks of 1,4-linked mannuronate (M) and guluronate (G) residues. More than 200 distinct alginates are now being produced, with M and G contents and block lengths varying among the alginates recovered from various sources (Lee and Mooney 2012). Since the early twentieth century, alginates made by algae have been isolated. They are heterogeneous in composition and exist in the cell walls of brown algae as a salt form of alginic acid (Chudasama et al. 2021). The main genera that produce alginate are *Laminaria*, *Macrocystis*, *Ascophyllum*, *Ecklonia*, and *Lessonia* species, with an estimated yearly production of 30,000 metric tons. Alginates are employed in many different industries, including the beverage, cosmetic, paper, pharmaceutical, printing, and textile industries, as gelling, stabilizing, and thickening agents. Although, alginates are largely used as gelling agents in the food business (Rehm and Moradali 2018), it has recently been researched for biomedical uses like drug delivery and wound dressing hydrogels after functionalization, such as when gold-nanoparticle-encapsulated alginic acid hybrid nanospheres were used as optical-contrast reagents and as biocompatible carriers for loading and delivering anticancer agents (Fernando et al. 2019).

15.2.6 Laminarin

Laminarin, which can make up to 35% of the dry weight of algal cells, is the primary storage polysaccharide of brown seaweeds (Paulien et al. 2014). The linear polysaccharide is made up of 25–50 glucose units. Depending on the reducing end, there are two different types of laminarin chains: M and G. Unlike G chains, which have glucose as the reducing end, M chains finish with 1-O-substituted D-mannitol (Cybulska et al. 2016). It is a storage glucan that was taken out of *Laminaria* and *Saccharina* species, which are popular dietary fibre sources in various nations. Laminarin also has potent biofunctional properties, including anti-apoptotic and anti-tumour properties (Zargarzadeh et al. 2020). Laminarin contains both water-soluble and water-insoluble forms based on branching. Algal tissues make up 0.03–35% of the dry weight of laminarin, and they are extracted using a multi-step procedure that includes acid treatment and dialysis (Kadam et al. 2015).

15.2.7 Fucoidans

Fucoidans are a class of polysaccharide (fucans) which mostly consist of sulfated L-fucose with less than 10% of other monosaccharides. It is a ramified sulfated polysaccharide (Vavilala and D'Souza 2015). Compared to high molecular fucoidan (HMF), low molecular fucoidan (LMF) is more bioavailable (Cybulska et al. 2016). Brown seaweeds produce fucoidans, which are 1,3, and 1,4 linked-L-fucopyranose residual bonds. The value of the expanding worldwide fucoidan market is presently $30 million USD. Fucoidan can be obtained from *S. muticum* and *S. whighti* which is also a strong source of natural antioxidants (Immanuel et al. 2012; Caxiano et al. 2022). 6.89% of the fucoidan produced by *Eckloniamaxima* had anti-diabetic effects (Zayed and Ulber 2019). Fucoidans and their reduced molecular weight derivatives could also be used in cell therapy and regenerative medicine. Fucoidan contains sulphate groups, which contribute to its bioactivities, including its antimicrobial, anticoagulant and antioxidant effects (Immanuel et al. 2012; Fernando et al. 2019).

15.2.8 Carrgeenan

Carrageenan makes up 30–75% of the dry weight of red seaweed and is one of the main components of the cell wall. They are linear hydrophilic polysaccharides that alternate 3- and 4-linked D-galactopyranose (G-units) and 3- and 6-linked 3,6-anhydro-D-galactopyranose (DA-units) units with an extra sulphate group (Campo et al. 2009). Carrageenan can be divided into three major economically significant families, namely kappa (κ), iota (ι), and lambda (λ), based on the quantity and positioning of sulphate groups in the repeating galactose units. Carrageenan's sulphate groups are primarily responsible for its chemical reactivity, and its diverse structural makeup explains its widespread use (Vavilala and D'Souza 2015; Campo et al. 2009). Carrageenan is produced by the red algae genera *Chondrus*, *Eucheuma*, *Gigartina*, *Iridaea*, and *Kappaphycus* species, in particular *E. denticulatum* and *K. alvarezii* providing 88% of the raw material for carrageenan production (Campbell and Hotchkiss 2017). Globally, the carrageenan market is anticipated to grow from an estimated 931.6 million USD in 2020 to 1.2 billion USD by 2025 (Chudasama et al. 2021). Carrageenan is used in anti-tumour immunity, encapsulating ingredients in food additives, anticoagulants, and antiviral compounds, among other things. Comparing κ-carrageenan and λ-carrageenan, the latter demonstrates a better suppression of tumour HeLa cells due to the higher degree of sulfation in the former (Suganya et al. 2016; Khotimchenko et al. 2020).

15.2.9 Ulvan

Green algae belonging to the genus *Ulva*, such as *U. flexuosa* and *U. gigantina*, are the source of ulvans. Ulvan comprise 8–29% of the dry weight of algae and make up the majority of the cell walls. With just trace amounts of mannose, arabinose, and galactose, ulvan is primarily composed of glucose, rhamnose, xylose, glucuronic acid, and sulphate. The two most common varieties of ulvan that have been identified are the water-soluble and the insoluble cellulose-like materials (Vavilala and D'Souza 2015). In the algae's fibrillar wall and intercellular space, where it is more concentrated, ulvan is distributed rather uniformly across the frond (Alves et al. 2013). Hybrid biomaterials such as nanofibers, membranes, hydrogels, 3D structures, and particles have been made using ulvan. These materials have demonstrated a range of biomedical applications, including the delivery of medication, the dressing of wounds, and the engineering of bone tissue (Tziveleka et al. 2019).

15.3 Polymers from Cyanobacteria

Natural polymers (carbohydrates, proteins, etc.) and other small molecules (sugar, disaccharides, and fatty acids) can all be used as sources for the production of bioplastics. From around two million tonnes in 2014 to about 6.7 million tonnes in 2018, there was an increase in the production capacity of bioplastics, which were mostly made from starch and poly (lactic acid) (PLA)-based polymers (Thielen 2014; European Bioplastics, Bioplastics Market Data 2020). Nowadays, bioplastics compete with food supplies, because they are made from terrestrial crops such as maize and potatoes (Rahman and Miller 2017). Additionally, producing bioplastic from agricultural crops consumes a lot of land, water, and nutrients (European Bioplastics, Bioplastics Market Data 2020). According to a number of studies, this type of bioplastic synthesis is not long-term viable and also there have been numerous issues with the large-scale commercialization of bio-based pharmaceuticals over the past 10 years (Di Gregorio 2009). As microalgae doesn't compete with food sources, can grow on waste resources, and can accumulate significant levels of lipids, microalgae may be a preferable biomass source for the manufacturing of bioplastics (Hempel et al. 2011; Rahman and Miller 2017). Additionally, the manufacture of bioplastic from microalgae can be more environmentally friendly and contribute to both the circular economy and the bioeconomy. Bioplastics can be used in pharmaceuticals, cosmetics, and food packaging (Mohan et al. 2019).

Studies on the synthesis of bioplastics from microalgal sources can be divided into two categories. The term "bioplastic" can be used to describe composite materials created by mixing microalgal biomass, bio- or petroleum-based polymers, and the additives. Thermal-mechanical techniques, like compression moulding, are used to create these items. The alternative method of biopolymer production relies on the intracellular cultivation of biopolymers like starch and polyhydroxybutyrates

(PHBs) in microalgal cells. Then, these products can be removed and put through additional processing to make bioplastic, however, the microalgal cells are not directly used in this situation.

For the manufacture of biofuel, microalgae can take the place of a variety of other sources, and they can also be employed as additives in the food and cosmetic industries as well as in pharmaceutical formulations. A growing number of people are interested in microalgae because of its potential for the bioeconomy (Oh et al. 2018). Lipids (7–23%), carbohydrates (5–23%), and proteins (6–52%) make up the bulk of microalgae (Chandra and Mohan 2014). Additionally, microalgae contain copper (18–102 mg kg^{-1}), iron (1395–11,101 mg kg^{-1}), manganese (45–454 mg kg^{-1}), selenium (0–0.5 mg kg^{-1}), and zinc (28–64 mg kg^{-1}) (Tibbetts et al. 2015).

15.3.1 Polyhydroxyalkanoates (PHA)

Polyhydroxyalkanoates (PHA), a class of polyesters produced by cyanobacteria and reported from algae, has piqued the interest of researchers since 1966. PHA possesses chemical characteristics similar to those of synthetic polymers (Singh and Mallick 2017). PHA is known to build up as granules in the cytoplasm of cyanobacteria and be isolated from specific microalgal species (Garcia et al. 2021; Madadi et al. 2021). The most typical variety of PHA is PHB. In phosphorus-limited conditions, the thermophilic cyanobacterium *Synechococcus* sp. MA19's intracellular PHB level produced 55% DW PHB (Nishioka et al. 2001). Numerous cyanobacterial strains, including *Calothrix* sp., *Oscillatoria* sp., *Microcystis* sp., *Nostoc* sp., *Spirulina* sp., and *Synechocystis* sp., have been found to contain PHA in concentrations ranging from 1 to 55%. *Oscillatoriaokeni* TISTR 8549 spontaneously produced 42% of the functionally superior poly(3-hydroxybutyrate-*co*-3-hydroxyvalerate) under nitrogen deficiency circumstances (Madadi et al. 2021).

The idea of producing bacterial PHA from microalgal biomass is gaining more interest worldwide. In order to limit the use of single-use plastics, more research is required on the sorts of polymers that can be produced from microalgal biomass. Microalgae provide photosynthetic biomass for bacterial PHA synthesis as a substitute renewable feedstock that can be grown with limited natural resources and has the added advantages of phycoremediation in addition to CO_2 sequestration in the current post-COVID economy's shift towards sustainable energy (Tan et al. 2022). PHA is a potential substance with numerous uses, including as a bioplastic, a packaging material, an implant material, a drug carrier, a starting point for the manufacture of drugs, and a source of biofuels. PHA is a biodegradable material that is thought to be environmentally friendly and has been used as a bioplastic for packaging. PHA was created by ICI in the 1990s under the trade name Biopol and used to make shampoo bottles. However, because of the high production costs, large-scale industrial production and application are still not very competitive (Wang and Chen 2017).

15.3.2 Polyhydroxybutyrate (PHB)

Phaeodactylumtricornutum can produce bioplastic PHB at up to 10.6% of the dry weight of the algal biomass (Hempel et al. 2011). Fourteen to 17% of the dry weight of PHB bioplastic is produced by *Chlorogloea fritschii* (Monshupanee et al. 2016). Forty-three percent of the dry weight PHB is produced in *Nostocmuscorum* (Sharma and Mallick 2005). PHB can also be produced from *Synechococcus* sp. MA19 (Costa et al. 2018), *Synechocystis* sp. PCC6803 (Shrivastav et al. 2010), *Calothrix scytonemicola* (Kaewbai-Ngam et al. 2016), *Aulosira fertilissima* (Samantaray and Mallick 2012), *Anabaena cylindrical* (Lama et al. 1996), and *Spirulina maxima* (De Philippis et al. 1992).

PHB is most likely to be widely used in food packaging, whether it is for fresh or long-term storage. Examples of bioplastics' usage in agriculture to maintain ideal conditions and safeguard the crop from physical and biological threats include grow sacks, protection nets, and greenhouse films (Sidek et al. 2019). PHB-based wound dressings can be loaded with the medication, and the material's fibrous composition makes it easier for the medication to release into the wound. Cancer cells have been shown to adhere to PHB sheets and contact angle techniques can be used to spot these characteristics. This method is less painful than biopsy, which is commonly used in medical research (Abd El-malek et al. 2020). It is possible to use PHB with a high molecular weight as a medication carrier (Narancic et al. 2020). PHB can be used as an efficient antibacterial agent in aquaculture because of its non-adhesive characteristics, which stop pathogens from building biofilms and hence inhibit infection. PHB is an antifouling material that can be mixed with other metals and applied to the hull to deter unwanted marine life from congregating there (Kavitha et al. 2018).

15.3.3 Polyurethane (PU), Cellulose Acetate and Polylactic acid (PLA)

Commercially available biopolymer polyurethane is frequently used in foams, varnishes, elastomers, and adhesives (Ozcimen et al. 2017). They can be produced using oils, of which triglycerides from algae are a sustainable supply. The fatty acid content in the algae varies from species to species. *Chlorella* was used by Pawar et al. (2016) to oxidize polyols and generate them. Additionally, the algal oil-based polyurethane coatings demonstrated antibacterial and anticorrosive properties (Patil et al. 2018). Due to its dimensional stability, stiffness, and printability, cellulose acetate, a material used to make disposable jewellery, combs, cigarette filters, and sight frames, is also recognized in food packaging, where it was used to wrap baked goods (Stevens 2013). Polylactic acid is usually made from algal feedstocks. Medical implants, computer and mobile phone casings, as well as other packaging materials, including cups, tins, and bottles, all make use of polylactic acid mixes (Reddy et al. 2013).

15.4 Algae-Polymer Biocomposites and Applications

The utilization of microalgae at various concentrations to create hybrid composites with petroleum polymers was the subject of multiple investigations. These composites can be used in a wide variety of industries. Microalgae are utilized in composite materials as both fillers and reinforcing fibres. Blends made from microalgae replace some petroleum-based plastic, lowering the cost and carbon impact of polymers (Bulota and Budtova 2015). Because it increases the volume of the composite material, reduces density, and also boosts tensile strength, chlorella was discovered to be appropriate filler for PVC (Zhang et al. 2000).

Defatted algal biomass can be mixed with polyurethane (PU) in various mass ratios to create composite films. These films, when compared to the neat PU-polyethylene glycol film, showed improved antioxidant activity and bulk hydrophilicity at 70% biomass content (Saha et al. 2021). *Nannochloropsis*-containing biocomposite has increased matrix stiffness and oxygen permeability (Fabra et al. 2018). Using 1-allyl-3-methylimidazolium chloride, cellulose/microalgae composites with microalgae contents ranging from 19.14 to 78.83 wt% were created (AmimCl). The composite films demonstrated strong mechanical characteristics, thermal stability, and a low hydrolytic degradation rate (Yan et al. 2016). Containers, car parts, and personal care items can be made from the thermoplastic blend of algae and starch that is created (Shi et al. 2012). The packaging, biomedical, and agricultural sectors are predicted to use the articles and blends from PCL and PHB (Barghini et al. 2010). Due to their antibacterial properties, silver nanoparticles made from marine macroalgal polysaccharides were immobilized on cotton fibre, which has biomedical uses (El-Rafiea et al. 2013). Similar to this, the PU composite foam can be created by alkalinizing biomass made up of *Cladophora* sp. and *Microspora* sp. The resulting composite has a lower loss modulus, a higher transition temperature, and enhanced stiffness between 25 and 50 °C. The composite's capacity to absorb shock may also be increased (Jonjaroena et al. 2020).

15.5 Algal Proteins

A good supply of protein is thought to be found in seaweed and microalgae, it is known that some types of seaweed and microalgae have protein levels that are comparable to those of conventional protein sources such as meat, eggs, soybeans, and milk (Gouveia et al. 2008). In terms of productivity and nutritional value, using algae to produce protein provides a number of advantages over using traditional high-protein crops. When compared to terrestrial crops such as soybean, pulse legumes, and wheat (0.6–1.2, 1–2, and 1.1 tons/ha/year, respectively), seaweed and microalgae offer greater protein yields per unit area (2.5–7.5 tons/ha/year and 4–15 tons/ha/year, respectively) (Van Krimpen et al. 2013). Some indigenous people have used seaweed and microalgae as a source of human sustenance for thousands of years (Chakdar et al. 2012). Seaweed and microalgae contain a lot of protein, as much as or even more than some plant sources. This is one of the main reasons why they are so popular (Becker 2007).

Algae are frequently on par with other protein sources, including soybean and egg, and their EAA composition complies with FAO standards; therefore, they are typically regarded as a valid source of protein (Fleurence 1999). Because marine algae are not widely consumed, there is a dearth of in vivo studies on the ideal digestion of algae, which makes it difficult to compare the quality of algal protein with different species of algae and with that of other protein sources. However, the scarcest amino acids in the majority of algal species are tryptophan and lysine. Additionally, methionine, cysteine, and lysine are frequently limited in brown algal species, but leucine and isoleucine are frequently found in low percentages in red algal species (Misurcova et al. 2014). Many types of seaweed often contain low quantities of cysteine, which is frequently undetectable (Kakinuma et al. 2001).

Many seaweed species have relatively high concentrations of aspartic acid and glutamic acid, which are largely responsible for the unique umami flavour that is associated with seaweed (MacArtain et al. 2007). For instance, it has been noted that the two amino acids aspartic acid and glutamic acids account for 22–44% of the total amino acids in *Fucus* sp. and 26–32% in *Ulva* sp., respectively. *Pyropia palmata* and *P. tenera* are two examples of red seaweed species (Rhodophyta) that have been found to have up to 33 and 47% dw, respectively. Similar to this, other microalgal species, including *Spirulina* sp., have been reported to contain considerably higher quantities, up to 63% dw (Tokusoglu and Uunal 2003). As much as 70% of the dry weight of some species of microalgae can be made up of proteins. *Arthrospira*, *Chlorella*, *Aphanizomenon*, and *Nostoc* are well-known, protein-rich microalgal species (Caporgno and Mathys 2018). All of the essential amino acids are present in microalgae proteins, which typically have a balanced total amino acid profile (TAA) (EAA). The FAO and WHO stated that the amino acid profiles of proteins isolated from *Arthrospira*sp. are compatible with those that are advised for human consumption (Habib et al. 2008).

15.6 Applications of Algal Proteins

15.6.1 Nutritional Applications

Many Asian cultures frequently employ *Pyropiatenera* (nori) as a sushi wrapper (Holdt and Kraan 2011). The amino acids aspartic acid and glutamic acid, which have a distinctive and intriguing flavour and are particularly abundant in many kinds of seaweed, are responsible for the discovery of the umami flavour (MacArtain et al.

2007). The brown seaweed *Laminaria japonica* (kombu), which has been proven to particularly appeal to the umami taste experience, is where the flavour enhancer monosodium glutamate was initially isolated (Marcus 2007). Up to 10% of *U. pinnatifida* (wakame) in pasta shows acceptable sensory qualities and antioxidant activity (Prabhasankar et al. 2009). When consumed with a meal, bread supplemented with 4% *A. nodosum* can dramatically lower energy intake in overweight people (Hall et al. 2012). Similar amount of renin-inhibitory peptides from *P. palmata* hydrolysates were also used to make bread, which similarly had good sensory qualities and was reported to have retained its bioactive components after baking (Fitzgerald et al. 2014).

The most common forms of microalgae used as dietary supplements are powder, pills, and tablets (Pulz and Gross 2004). They have, however, also been added to a variety of functional meals, including noodles, bread, biscuits, beverages, candies, and beer (Liang et al. 2004). AlgaVia® (www.algavia.com), which creates protein- and lipid-rich algal flour from *Chlorella protothecoides*, is one of many companies set up to sell algal goods. The most popular microalgae taken is *Spirulina*, which has a high-protein content and additional nutritional advantages such as anti-hypertension, renal protection, anti-hyperlipidemia, and anti-hyperglycemia (Spolaore et al. 2006). *Spirulina* is a great source of protein, but it is also high in B vitamins, phycobiliproteins that scavenge free radicals, and γ-linoleic acid (GLA), which lowers cholesterol (Sajilata et al. 2008). Due to its nutrient-dense qualities, the World Health Organization (WHO) has designated it as a "super food," and the National Aeronautics and Space Administration (NASA) has even sent it into space (Khan et al. 2005). To illustrate this, *Spirulina* has 5100% more iron than spinach, 3100% more beta-carotene than carrots, and 180% more calcium than milk (Capelli and Cysewski 2010).

The primary component of *Chlorella* that is good for human health is having β-1,3-glucan, an active immunostimulant, free-radical scavenger, and blood lipid reduction (Yamaguchi 1996), and it is a popular microalga (Yaakob et al. 2014). Additionally, *Chlorella* has high level of phosphorus (1761.5 mg/100 g dw), polyunsaturated fatty acids (PUFAs) (39% of total lipids), and proteins (48% dw) (Tokusoglu and Uunal 2003).

15.6.2 Industrial Applications

Two classes of bioactive algal proteins, such as lectins and phycobiliproteins, have been used in numerous industrial processes.

(a) **Lectins:** Lectins are glycoproteins renowned for their aggregation and highly specific binding with carbohydrates without beginning alteration through related enzymatic activity (Hayes 2013). Lectins are most frequently isolated from macroalgal sources. Apoptosis induction, cancer metastasis, host-pathogen interactions, cell–cell communication, and antiviral activity are just a few of the biological processes that lectins are involved in (Ziolkowska and Wlodawer 2006). Lectins are employed in blood grouping, antiviral (including HIV-1), cancer biomarkers, and targets for medication administration due to their capacity to bind carbohydrates with high specificity (Naeem et al. 2007). Compared to other plant-derived lectins, lectins generated from algae have not undergone the same amount of characterization. Algal lectins have been found to have a variety of biological actions, including those that are mitogenic, cytotoxic, antibacterial, anti-nociceptive, anti-inflammatory, antiviral (HIV-1), and platelet aggregation inhibition (Harnedy and Fitz Gerald 2011).

(b) **Phycobiliproteins:** Red algae can contain considerable amount of phycobiliproteins, particularly phycoerythrin, with values of 1.2% (total dw) observed for *P. palmatum* (Wang et al. 2001). Fluorescence labelling, flow cytometry, fluorescent microscopy, and fluorescent immunohistochemistry technique could be employed using phycobiliproteins (Aneiros and Garateix 2004). Phycocyanin, a phycobiliprotein, a blue pigment used in chewing gum, popsicles, confectionery, soft drinks, dairy products, and wasabi, as well as in cosmetics like lipstick and eyeliner (Spolaore et al. 2006). Numerous patents on the health-promoting bioactivities of phycobiliproteins, including their antioxidative, anti-inflammatory, antiviral, anti-tumour, neuroprotective, and hepatoprotective properties, have already been evolved for nutraceutical uses (Sekar and Chandramohan 2008).

15.7 As Feed for Animals

Algae's high-protein content makes it advantageous for usage as feed for pets, farm animals, and aquaculture. Animal feed accounts for 30% of estimated global algal production, with 50% of biomass from *Spirulina* being used as a feed supplement in particular diets due to its superior nutritional profile (Becker 2004). The health, production, and worth of chickens can all be increased by adding microalgae as a protein source to their meal. Several species, including *Chlorella* sp., *Arthrospira* sp., *Porphyridium* sp., and *Haematococcus* sp., have been used to illustrate this (Holman and Malau-Aduli 2013). *Spirulina* supplementation has been linked to greater viability, better general health, and lower plasma levels of cholesterol, triglycerides, and fatty acids in chickens (Mariey et al. 2014). *Spirulina* in dietary supplements somewhat slowed development rates, improved fertility rates, and intensified the colour of egg yolks (Ross and Dominy 1990). The yield of breast muscle in broiler chickens was improved by adding 3% *U. lactuca* (Abudabos et al. 2013). The red seaweed *Polysiphonia* sp. increased the nutritional profile and pellet binding of duck feed (El-Deek and Brikaa 2009). Red seaweeds *Chondrus crispus* and *Sarcodiotheca gaudichaudii* have also been shown to function well as prebiotics to enhance the gastrointestinal health, productivity, and quality of chicken eggs (Kulshreshtha et al. 2017). *Arthrospiramaxima*, *A. platensis*, and *C. vulgaris* proteins have been observed to be appropriate for replacing up to 33% of soy proteins in pig feed without causing any negative effects (Gatrell et al. 2014). Incorporating *Spirulina* into meals instead of pellets was found to increase average daily weight gain in pigs (Grinstead et al. 2000), improve pig fertility by boosting sperm motility and storage viability, whereas reduce average daily weight gain when added to pellets.

For the artificial reproduction of several aquaculture species, particularly mollusks, microalgae play an important role (Becker 2004). Other than serving as a source of food for zooplanktons, microalgae are crucial in aquaculture because they increase rearing medium quality, stabilize pH, and promote better survival and growth than artificial diets (Lio-Po et al. 2005). Salmonoids, shrimp, lobsters, and crayfish consume carotenoid pigments such as astaxanthin produced from *Haematococcus pluvalis* to produce their distinctive pink meat (Muller-Feuga 2000). The colour pattern and intensity of ornamental fish such as tetras, cichlids, gouramis, damos, goldfish, and koi, are also greatly improved by the addition of astaxanthin at a concentration of 30 ppm (Lorenz and Cysewski 2000). Red seaweed has been proposed as a good feed additive and protein source. According to Lozano et al. (2016), adding 10% *Gracilariachilensis* to the diet of Atlantic salmon (*Salmonsalar*) increased specific growth rate considerably and increased antiviral activity against the infectious salmon anaemia (ISA) virus. Similar to this, adding *P. palmata* to *S. salar* has been shown to enhance hepatic function and have a favourable impact on body lipid levels (Wan et al. 2016).

15.8 Bioactive Peptides

Although they are less well defined than bioactive peptides from other sources, it has been discovered that bioactive peptides from algae also exhibit bioactive capabilities. *Saccharina longicruris* protein hydrolysates include antimicrobial peptides that considerably slow down the rate of multiplication of *Staphylococcus aureus* (Beaulieu et al. 2015). In skin fibroblasts, the hexapeptide Glu-Asp-Arg-Leu-Lys-Pro isolated from *Ulva* sp. was shown to have mitogenic activity (Ennamany et al. 1998). Seaweed and microalgae like *Undaria pinnatifida*, *Eckloniacava*, *Porphyraye zoensis*, *Hizikia fusiformis*, *Palmaria palmate*, *Chlorella vulgaris*, *C. ellipsoidea*, *Arthrospira platensis*, and *Nannochloropsis oculata*, have all been found to contain bioactive peptides that inhibit the angiotensin-I-converting enzyme (Fan et al. 2014). *Porphyra columbina* protein hydrolysates have been shown to have immunosuppressive, anti-hypertensive, and antioxidant properties (Cian et al. 2012). Microalgae have been used to make anti-inflammatory peptides, such as Chlorella 11-peptide from *C. pyrenoidosa* and peptides from *A. maxima* with anti-atherosclerosis bioactivity (Vo et al. 2013). The bioactivities of algal peptides have been linked to hepatoprotection, immunomodulation, protection against ultraviolet (UV) radiation, prevention of osteoporosis, and anticoagulant effects (Kang et al. 2013).

The bioactive substances from microalgae can be utilized in cosmetics as antioxidants, free-radical scavengers, stress relievers, immune system boosters, odour maskers, makeup pigments, sunscreen protectors, and antiageing agents. Active compounds derived from microalgae have a variety of functions, such as blemish avoidance, skin damage repair, seborrhea improvement, inflammation process inhibition, speeding up the healing process, and skin moisture maintenance. Numerous skin care products for individuals are derived from microalgae. For instance, the business Soliance employs the complete *Arthrospira* species, and personal skin care products contain the *Chlorella* species-derived peptide sequence LVMH. Alguronic acid is another ingredient in the antiageing skin care products made by Solazyme. Additionally, Soliance incorporates the algae *Skeletonema costatum* into its moisturizing skin care products and *Dysmorphococcus globosus* into its anti-inflammatory products (Enzing et al. 2014). Microalgae are also of greatest interest in the field of thalasso therapy due to their distinct cellular composition and amount of PUFAs, including DHA, EPA, vitamins, and folic acid. Thalassotherapy is a contemporary medical practice that uses marine organisms and seaweed, such as microalgae, mud, sand, or plankton, for therapeutic and preventive purposes (Lee et al. 2017).

15.9 Algal Pigments

Pigments, which are complexes that capture visible light, are widely distributed in algae. Algal pigments can be divided into three groups: carotenoids, phycobilins, and chlorophylls. Astaxanthin, lutein, fucoxanthin, canthaxanthin, zeaxanthin, and β-cryptoxanthin are common carotenoids that have uses as antioxidants, anti-inflammatory agents, immunopreventive agents, and anticancer agents. In addition to chlorophyll, the red algae also produce phycoerythrin, a red protein pigment (Rossano et al. 2003). *Irishmoss*, *Gracillaria* sp., *Porphyra* sp., and other red algae species are among those that are utilized in cosmetics. Chlorophyll, a photosynthetic pigment is produced by green algae, additionally, it has anti-inflammatory properties. *Chlorella vulgaris*, *Ulva lactuca*, and other species of green algae are among those utilized in cosmetics. β-carotene produced by *Dunaliella salina* is used as a food additive and supplement because it is a precursor to vitamin A (Surabhi et al. 2018). An additional pigment called fucoxanthin can be found in the chloroplasts of brown algae. Tyrosinase inhibitory properties serve to minimize or control skin pigmentation. It also has anti-inflammatory properties and aids in the slowing of skin ageing by promoting collagen synthesis (a structural protein that tends to disperse with age). Additionally, the pigment keeps the skin wet and the skin cells functioning properly. *Isochrysis* species, *Postelsiapa maeformis*, *Laminaria digilata*, and other species of brown algae are among those utilized in cosmetics (Shimoda et al. 2010).

15.10 Chlorophylls

Chlorophylls are green, photosynthetic pigments found in higher plants, microorganisms, and algae. It is a fat-soluble substance having bioactive qualities like antioxidant and antimutagenic actions (Lanfer-Marquez et al. 2005). Lee et al. (2021) studied that the red seaweed *Grateloupia elliptica's* chlorophyll and chlorophyll derivatives had the ability to reduce obesity, and chlorophyll inhibited lipid formation by reducing the expression of adipogenic proteins at the intracellular level in 3T3-L1 adipocytes in in vitro experiments without being harmful to cells. Red seaweeds' chlorophyll a (CAS 479-61-8) is being investigated as a pigment or colourant for the textile, food (which is already approved by the EFSA under code E140), animal feed, skin care, and cosmetic industries (Freitas et al. 2022).*Chlorella* contains two primary forms of chlorophylls (a and b), which are their most prevalent pigments and for which they can accumulate significant levels (up to 4.5% of dry weight). The term "emerald food" refers to *Chlorella* because of its high chlorophyll concentration. However, even at lower levels, it also includes additional pigments like lutein, beta-carotene, c-astaxanthin, and astaxanthin (Khanra et al. 2018).

15.11 Phycobilins

Phycobilins are used in the industry in the form of a water-soluble protein, and their commercial applications rely on broad knowledge and technological foundations, which are still in short supply. As a result, up to this point, the vast majority of unique applications involving the usage of phycobilins have only been documented in patents (Mysliwa-Kurdziel and Solymosi 2016). Phycobiliproteins are protein-based pigments that capture light. The two most popular microalgal sources for the production of phycobiliproteins are the cyanobacteria species *Spirulina* (*Arthrospira*) sp. (phycocyanin) and *Porphyridium* sp. (phycoerythrin) (Dineshbabu et al. 2019). Also the principal phycobiliprotein present in *Spriulina* contains the naturally occurring blue pigment known as phycocyanin (De Morais et al. 2016). The Food and Drug Administration approved the use of this pigment as a colour additive for food matrices in 2013 (FDA 2020). It is being utilized as a natural colourant in foods like beverages and confectionary goods. Furthermore, it has a wide application in different industries like in the cosmetics industry (for example, lipstick and eyeliners), in the pharmaceutical industry due to its medicinal benefits (antioxidant, anti-inflammatory, and anticancer activity), and as fluorescent markers (Panutai and Iamtham 2019).The commercially available blue colourant Linablue® *Spirulina* extract is sold as a food product decoration component, and a number of businesses advertise using the microalgae-based natural colours in cosmetic items, such as phycocyanin from *Arthrospira* sp. and phycoerythrin from *Porphyridium* sp., and R-Phycoerythrin as a consumable for laboratories is used in biotechnological or biomedical research (from *Neopyro piatenera* or *Neogastroclonium subarticulatum*) (Freitas et al. 2022).

15.12 Carotenoids

A class of substances known as carotenoids has hues ranging from red and brown to orange and yellow. Both α- and β-carotene have pro-vitamin A activity as well as anti-inflammatory properties. However, carotenes also possess additional biological properties, including antioxidant, anticancer, antiageing and cosmeceutical properties. As a result, the food, cosmetics, and pharmaceutical industries have a lot of potential when it comes to this class of pigments (Freitas et al. 2022). The most abundant source of β-carotene in microalgae is the halotolerant green microalga *Dunaliella salina*. Up to 10%, on a dry biomass basis, of β-carotene may be accumulated in *D. salina*. The European Commission has authorized the food colour beta-carotene from *Dunaliella salina*. Due to their function as active components, carotenoids can be used in a variety of industries, including animal feed, cosmetics, pharmaceuticals, and food colouring additives [e.g., β-carotene (EC 160)] (e.g., aquaculture of salmon). Additionally, carotenoids are known to be powerful antioxidants with health-promoting functionalities that include lowering triglycerides and raising HDL (high-density lipoprotein) cholesterol, as well as preventing cancer (Tanaka et al. 2012).

Lutein and zeaxanthin, which are structural isomers, are non-provitamin carotenoids known as xanthophylls. The green microalgae *D. salina* or *Chloroidium ellipsoideum* are the main sources of zeaxanthin, while *Corallina officinalis*, a red seaweed, has also been found to be an excellent source of this chemical of interest. Additionally, the red seaweeds *Gracilaria corticata* and *Grateloupia filicina* showed intriguing amount of lutein and zeaxanthin in their xanthophylls. Zeaxanthin is regarded as a tool against tumour and cancer development due to its anti-inflammatory capabilities, and its use in chemotherapy may be advantageous for patients. As a photoprotective agent, this chemical can also be used to treat eye conditions like cataracts or slow macular degeneration. From a different angle, lutein is already extensively used in sectors including food, pharmaceuticals, and cosmetics due to its colour and biotechnological applications. In fact, numerous studies have demonstrated that lutein possesses anticancer qualities and guards against cataracts and age-related macular degeneration. For instance, oral lutein supplementation was reported to lesser effects of ultraviolet (UV) irradiation by minimizing the hyperproliferative rebound and the first inflammatory reactions brought on by UV rays (Freitas et al. 2022).

15.13 Astaxanthin

Astaxanthin (3,3′-dihydroxy-carotene-4,4′-dione) is a fatty-soluble orange-red pigment that belongs to the same carotenoid family as β-carotene. A green, freshwater unicellular microalga called *Haematococcus pluvialis* has a high astaxanthin content. Common uses for this microalga include nutraceuticals, medicines, aquaculture, food items, and cosmetics (Paniagua-Michel 2015). Clinical trials have expanded and commercial production has increased due to the potential for astaxanthin to be used as a nutritional supplement in treatment or prevention plans for a variety of health issues brought on by oxidative stress, UV-light photooxidation or inflammation, cancers, and other pathological conditions (Freitas et al. 2022).

15.14 Phycoerythrin, Phycocyanin and Allophycocyanin

Based on their taxonomic position and the growth conditions of algae, the production of the pigments phycocerthrin, phycocyanin, and allophycocyanin varies. Phycoerythrin production by *Microcoleus autumnalis* (PACC 5505, PACC 5522, and PACC 5527) is very prospective. Similar to this, one strain of *Leptolyngbya boryana* (CCALA 084) and one strain of *Microcoleus autumnalis* (PACC 5522) are ideal for biotechnological production of phycocyanins as well as allophycocyanins (Basheva et al. 2018). The red pigment, phycoerythrin, gets its name from the Greek word "erythros," which means "red." A photosynthetic pigment called phycoerythrin is found in cryptomonads, red microalgae, and red macroalgae (Freitas et al. 2022).

The name of the blue pigment, phycocyanin, is partially derived from the English word "cyan," which itself comes from the Greek word "kyanos." The intense cobalt blue hue of phytocyanin is one that nearly no other natural pigment can match (Mysliwa-Kurdziel and Solymosi 2016). The research also indicates that phycocyanin, particularly when prepared to be used as a nutraceutical agent, will lead the market over the coming years, while phycocyanin in powder form or phycocyanin of a food grade got lead the market in 2020. The blue-green pigment allophycocyanin gets its name from the Greek words "allos" (other) and "kyanos" (blue). It seems like there has not been much research done on allophycocyanin yet. Allophycocyanin is distinctive since nearly no other natural pigment can reproduce its blue-green hue (Mysliwa-Kurdziel and Solymosi 2016), which likely raises its worth as a rare component (Freitas et al. 2022).

15.15 Conclusion

Algal biopolymers provide a greener alternative to synthetic polymers. It is time to do extensive research of algae as potential replacements for commonplace and one-time use plastics. In order to meet the growing demand for eco-friendly algal polymers for a world with net-zero carbon emissions, sustained bioprospecting for newer species/strains of algae with high precursor content, application of omics technologies, innovative screening methods for novel polymers and process optimization to increase productivity, green production processes, and addressing specific, polymer-based, supply-side and demand-side limitations are crucial. Additionally, it has been claimed that algae exhibit in vitro protein digestibility that is comparable to that of frequently ingested vegetables and high-grade proteins in terms of EAA concentration.

Seaweed proteins have been recommended as a healthy dietary supplement for enhancing animal growth, meat quality, and value. Additionally, cultivable finfish and shell fish species in particular have long been raised in aquaculture using microalgae as their primary feed. Algal pigments are increasingly in demand as food additives, feed, mineral supplements, and dyes because they have a substantially higher bioactive potential than other biological pigments. Additionally, it has gained popularity recently for use in medicines, cosmeceuticals, and nutraceuticals. These pigments have a wider range of uses as anti-inflammatory and anticancer medicines because of their great antioxidative potential.

References

Abd El-Malek F, Khairy H, Farag A, Omar S (2020) The sustainability of microbial bioplastics, production and applications. Int J Biol Macromol 157:319–328

Abudabos AM, Okab AB, Aljumaah RS, Samara EM, Abdoun KA, Al-Haidary AA (2013) Nutritional value of green seaweed (*Ulvalactuca*) for broiler chickens. Ital J Anim Sci 12:e28

Alves A, Sousa RA, Reis RL (2013) A practical perspective on ulvan extracted from green algae. J Appl Phycol 25(2):407–424

Andreeva A, Budenkova E, Babich O, Sukhikh S, Dolganyuk V, Michaud P, Ivanova S (2021) Influence of carbohydrate additives on the growth rate of microalgae biomass with an increased carbohydrate content. Mar Drugs 19:381

Aneiros A, Garateix A (2004) Bioactive peptides from marine sources: pharmacological properties and isolation procedures. J Chromatogr B 803:41–53

Barghini A, Ivanova VI, Imam SH, Chiellini E (2010) Poly-(ε-caprolactone) (PCL) and poly (hydroxy-butyrate) (PHB) blends containing seaweed fibers: morphology and thermal-mechanical properties. J Polym Sci A Polym Chem 48(23):5282–5288

Bar-Shai N, Sharabani-Yosef O, Zollmann M, Lesman A, Golberg A (2021) Seaweed cellulose scaffolds derived from green macroalgae for tissue engineering. Sci Rep 11(1):11843

Basheva D, Moten D, Stoyanov P, Belkinova D, Mladenov R, Teneva I (2018) Content of phycoerythrin, phycocyanin, alophycocyanin and phycoerythrocyanin in some cyanobacterial strains: applications. Eng Life Sci 18(11):861–866

Beaulieu L, Bondu S, Doiron K, Rioux LE, Turgeon SL (2015) Characterization of antibacterial activity from protein hydrolysates of the macroalga *Saccharina longicruris* and identification of peptides implied in bioactivity. J Funct Foods 17:685–697

Becker W (2004) Microalgae in human and animal nutrition. In: Richmond A (ed) Handbook of microalgal culture: biotechnology and applied phycology. Wiley-Blackwell, Hoboken, p 312

Becker E (2007) Micro-algae as a source of protein. Biotechnol Adv 25:207–210

Brodie J, Chan CX, De Clerck O, Cock JM, Coelho SM, Gachon C, Grossman AR, Mock T, Raven JA, Smith AG (2017) The algal revolution. Trends Plant Sci 22:726–738

Bulota M, Budtova T (2015) PLA/algae composites: morphology and mechanical properties. Compos A Appl Sci Manuf 73:109–115

Campbell R, Hotchkiss S (2017) Carrageenan industry market overview. In: Hurtado AQ, Critchley AT, Neish IC (eds) Tropical seaweed farming trends, problems and opportunities: focus on *Kappaphycus* and *Eucheuma* of commerce. Springer, Cham, pp 193–205

Campo VL, Kawano DF, da Silva DB Jr, Carvalho I (2009) Carrageenans: biological properties, chemical modifications and structural analysis - a review. Carbohydr Polym 77(2):167–180

Capelli B, Cysewski GR (2010) Potential health benefits of Spirulina microalgae. Forum Nutr 9: 19–26

Caporgno MP, Mathys A (2018) Trends in microalgae incorporation into innovative food products with potential health benefits. Front Nutr 5:58

Caxiano IN, Mello PA, Alijo PHR, Teixeira LV, Cano RF, Maia JGSS, Bastos JBV, Pavao MSG (2022) Continuous design and economic analysis of a *Sargassum muticum* biorefinery process. Bioresour Technol 343:126152

Cazon P, Velazquez G, Ramirez JA, Vazquez M (2017) Polysaccharide-based films and coatings for food packaging: a review. Food Hydrocoll 68:136–148

Chakdar H, Jadhav SD, Dhar DW, Pabbi S (2012) Potential applications of blue green algae. J Sci Ind Res 71:13–20

Chandra R, Mohan SV (2014) Enhanced bio-hydrogenesis by co-culturing photosynthetic bacteria with acidogenic process: augmented dark-photo fermentative hybrid system to regulate volatile fatty acid inhibition. Int J Hydrog Energy 39:7604–7615

Chia SR, Ong HC, Chew KW, Show PL, Phang SM, Ling TC, Nagarajan D, Lee DJ, Chang JS (2018) Sustainable approaches for algae utilisation in bioenergy production. Renew Energy 129: 838–852

Chudasama NA, Sequeira RA, Moradiya K, Prasad K (2021) Seaweed polysaccharide based products and materials: an assessment on their production from a sustainability point of view. Molecules 26(9):2608

Cian RE, Martinez-Augustin O, Drago SR (2012) Bioactive properties of peptides obtained by enzymatic hydrolysis from protein byproducts of *Porphyra columbina*. Food Res Int 49:364–372

Costa SS, Miranda AL, Andrade BB, de Jesus AD, Souza CO, de Morais MG (2018) Influence of nitrogen on growth, biomass composition, production, and properties of polyhydroxyalkanoates (PHAs) by microalgae. Int J Biol Macromol 116:552–562

Cybulska J, Halaj M, Cepak V, Lukavsky J, Capek P (2016) Nanostructure features of microalgae biopolymer. Starch-Starke 68:629–636

De Morais MG, Morais D, Greque E, da Silva Vaz B, Gonçalves CF, Lisboa C, Vieira Costa JA (2016) Nanoencapsulation of the bioactive compounds of *Spirulina* with a microalgal biopolymer coating. J Nanosci Nanotechnol 16:81–91

De Philippis R, Sili C, Vincenzini M (1992) Glycogen and poly-β-hydroxybutyrate synthesis in *Spirulina maxima*. Microbiology 138(8):1623–1628

Devadas VV, Khoo KS, Chia WY, Chew KW, Munawaroh HSH, Lam M-K, Lim JW, Ho Y-C, Lee KT, Show PL (2021) Algae biopolymer towards sustainable circular economy. Bioresour Technol 325:124702

Di Gregorio BE (2009) Biobased performance bioplastic: Mirel. Chem Biol 16(1):1–2

Dineshbabu G, Goswami G, Kumar R, Sinha A, Das D (2019) Microalgae–nutritious, sustainable aqua- and animal feed source. J Funct Foods 62:103545

El-Deek A, Brikaa MA (2009) Nutritional and biological evaluation of marine seaweed as a feedstuff and as a pellet binder in poultry diet. Int J Poult Sci 8:875–881

El-Rafiea HM, El-Rafieb MH, Zahran MK (2013) Green synthesis of silver nanoparticles using polysaccharides extracted from marine macro algae. Carbohydr Polym 96(2):403–410

Ennamany R, Saboureau D, Mekideche N, Creppy E (1998) Secma 1®, a mitogenic hexapeptide from *Ulva algeae* modulates the production of proteoglycans and glycosaminoglycans in human foreskin fibroblast. Hum Exp Toxicol 17:18–22

Enzing CM, Ploeg M, Barbosa MJ, Sijtsma L (2014) Microalgae-based products for the food and feed sector: an outlook for Europe. Petten, Joint Research Centre, pp 19–37

European Bioplastics, Bioplastics Market Data, 2020. https://www.european-bioplastics.org/market/

Fabra MJ, Martinez-Sanz M, Gomez-Mascaraque LG, Gavara R, Lopez-Rubio A (2018) Structural and physicochemical characterization of thermoplastic corn starch films containing microalgae. Carbohydr Polym 186:184–191

Fan X, Bai L, Zhu L, Yang L, Zhang X (2014) Marine algae-derived bioactive peptides for human nutrition and health. J Agric Food Chem 62:9211–9222

Fernando IPS, Kim D, Nah JW, Jeon YJ (2019) Advances in functionalizing fucoidans and alginates (bio)polymers by structural modifications: a review. Chem Eng J 355:33–48

Fitzgerald C, Gallagher E, Doran L, Auty M, Prieto J, Hayes M (2014) Increasing the health benefits of bread: assessment of the physical and sensory qualities of bread formulated using a renin inhibitory *Palmaria palmata* protein hydrolysate. LWT Food Sci Technol 56:398–405

Fleurence J (1999) Seaweed proteins: biochemical, nutritional aspects and potential uses. Trends Food Sci Technol 10:25–28

Food and Drug Administration (2020) Listing of color additives exempt certification. https://www.accessdata.fda.gov/scripts/cdrh/cfdocs/cfcfr/CFRSearch.cfm?fr=73.530

Freitas MV, Pacheco D, Cotas J, Mouga T, Afonso C, Pereira L (2022) Red seaweed pigments from a biotechnological perspective. Phycology 2:1–29

Garcia G, Sosa-Hernandez JE, Rodas-Zuluaga LI, Castillo-Zacarias C, Iqbal H, Parra-Saldívar R (2021) Accumulation of PHA in the microalgae *Scenedesmus* sp. under nutrient-deficient conditions. Polymers 13(1):131

Gatrell S, Lum K, Kim J, Lei X (2014) Nonruminant nutrition symposium: potential of defatted microalgae from the biofuel industry as an ingredient to replace corn and soybean meal in swine and poultry diets. J Anim Sci 92:1306–1314

Gouveia L, Batista AP, Sousa I, Raymundo A, Bandarra N (2008) Microalgae in novel food products. In: Konstantinos N, Papadopoulos PP (eds) Food chemistry research development. Nova Science, New York, pp 75–112

Grinstead G, Tokach M, Dritz S, Goodband R, Nelssen J (2000) Effects of *Spirulina platensis* on growth performance of weanling pigs. Anim Feed Sci Technol 83:237–247

Habib MAB, Parvin M, Huntington TC, Hasan MR (2008) A review on culture, production and use of *Spirulina* as food for humans and feeds for domestic animals and fish, FAO Fisheries and Aquaculture Circlar No. 1034

Hall A, Fairclough A, Mahadevan K, Paxman J (2012) *Ascophyllum nodosum* enriched bread reduces subsequent energy intake with no effect on post-prandial glucose and cholesterol in healthy, overweight males. A pilot study. Appetite 58:379–386

Harnedy PA, Fitz Gerald RJ (2011) Bioactive proteins, peptides, and amino acids from macroalgae. J Phycol 47:218–232

Hayes M (2013) Biological activities of proteins and marine-derived peptides from byproducts and seaweeds. In: Kim S-K (ed) Marine proteins and peptides: biological activities and applications. Wiley, Chichester, pp 139–165

Hempel F, Bozarth AS, Lindenkamp N, Klingl A, Zauner S, Linne U, Steinbüchel A, Maier UG (2011) Microalgae as bioreactors for bioplastic production. Microb Cell Factories 10:81

Holdt SL, Kraan S (2011) Bioactive compounds in seaweed: functional food applications and legislation. J Appl Phycol 23:543–597

Holman B, Malau-Aduli A (2013) *Spirulina* as a livestock supplement and animal feed. J Anim Physiol Anim Nutr 97:615–623

Hoover SW, Marner WD, Brownson AK, Lennen RM, Wittkopp TM, Yoshitani J, Zulkifly S, Graham LE, Chaston SD, McMahon KD, Pfleger BF (2011) Bacterial production of free fatty acids from freshwater macroalgal cellulose. Appl Microbiol Biotechnol 91(2):435–446

Immanuel G, Sivagnana Velmurugan M, Marudupandi T, Radhakrishnan S, Palavesam A (2012) The effect of fucoidan from brown seaweed *Sargassum wightii* on WSSV resistance and immune activity in shrimp *Penaeus monodon* (Fab). Fish Shell Fish Immunol 32:551–564

Jonjaroena V, Ummartyotin S, Chittapun S (2020) Algal cellulose as a reinforcement in rigid polyurethane foam. Algal Res 51:102057

Kadam SU, Tiwari BK, O'Donnell CP (2015) Extraction, structure and biofunctional activities of laminarin from brown algae. Int J Food Sci Technol 50(1):24–31

Kaewbai-Ngam A, Incharoensakdi A, Monshupanee T (2016) Increased accumulation of polyhydroxybutyrate in divergent cyanobacteria under nutrient-deprived photoautotrophy: an efficient conversion of solar energy and carbon dioxide to polyhydroxybutyrate by *Calothrix scytonemicola* TISTR 8095. Bioresour Technol 212:342–347

Kakinuma M, Park CS, Amano H (2001) Distribution of free L-cysteine and glutathione in seaweeds. Fish Sci 67(1):194–196

Kang KH, Qian ZJ, Ryu B, Karadeniz F, Kim D, Kim SK (2013) Hepatic fibrosis inhibitory effect of peptides isolated from *Navicula incerta* on TGF-1 induced activation of LX-2 human hepatic stellate cells. Prev Nutr Food Sci 18:124–132

Kartik A, Akhil D, Lakshmia D, Gopinath KP, Arun J, Sivaramakrishnan R, Pugazhendhi A (2021) A critical review on production of biopolymers from algae biomass and their applications. Bioresour Technol 329:124868

Kavitha G, Rengasamy R, Inbakandan D (2018) Polyhydroxybutyrate production from marine source and its application. Int J Biol Macromol 111:102–108

Khan Z, Bhadouria P, Bisen P (2005) Nutritional and therapeutic potential of *Spirulina*. Curr Pharm Biotechnol 6:373–379

Khanra S, Mondal M, Halder G, Tiwari ON, Gayen K, Bhowmick TK (2018) Downstream processing of microalgae for pigments, protein and carbohydrate in industrial application: a review. Food Bioprod Process 110:60–84

Khotimchenko M, Tiasto V, Kalitnik A, Begun M, Khotimchenko R, Leonteva E, Bryukhovetskiy I, Khotimchenko Y (2020) Antitumor potential of carrageenans from marine red algae. Carbohydr Polym 246:116568

Kulshreshtha G, Rathgeber B, Mac Isaac J, Boulianne M, Brigitte L, Stratton G, Thomas NA, Critchley AT, Hafting J, Prithiviraj B (2017) Feed supplementation with red seaweeds, *Chondrus crispus* and *Sarcodiotheca gaudichaudii*, reduce *Salmonella enteritidis* in laying hens. Front Microbiol 10(8):567

Lama L, Nicolaus B, Calandrelli V, Manca MC, Romano I, Gambacorta A (1996) Effect of growth conditions on endo and exopolymer biosynthesis in *Anabaena cylindrica*. Phytochemistry 42(3):655–659

Lanfer-Marquez UM, Barros RMC, Sinnecker P (2005) Antioxidant activity of chlorophylls and their derivatives. Food Res Int 38:885–891

Lee KY, Mooney DJ (2012) Alginate: properties and biomedical applications. Prog Polym Sci 37(1):106–126

Lee WK, Lim YY, Leow ATC, Namasivayam P, Ong Abdullah J, Ho CL (2017) Biosynthesis of agar in red seaweeds: a review. Carbohydr Polym 164:23–30

Lee XJ, Ong HC, Gan YY, Chen WH, Mahlia TMI (2020) State of art review on conventional and advanced pyrolysis of macroalgae and microalgae for biochar, bio-oil and bio-syngas production. Energy Convers Manag 210:112707

Lee HG, Lu YA, Je JG, Jayawardena TU, Kang MC, Lee SH, Kim TH, Lee DS, Lee JM, Yim MJ (2021) Effects of ethanol extracts from *Grateloupia elliptica*, a red seaweed, and its chlorophyll derivative on 3T3-L1 adipocytes: suppression of lipid accumulation through downregulation of adipogenic protein expression. Mar Drugs 19:91

Liang S, Liu X, Chen F, Chen Z (2004) Current microalgal health food R & D activities in China. In: Ang PO Jr (ed) Asian Pacific phycology in the 21st century: prospects and challenges. Springer, Heidelberg, pp 45–48

Li-Beisson Y, Thelen JJ, Fedosejevs E, Harwood JL (2019) The lipid biochemistry of eukaryotic algae. Prog Lipid Res 74:31–68

Lio-Po GD, Leano EM, Penaranda MMD, Villa-Franco AU, Sombito CD, Guanzon NG (2005) Anti-luminous vibrio factors associated with the 'green water' grow-out culture of the tiger shrimp *Penaeus monodon*. Aquaculture 250:1–7

Lorenz RT, Cysewski GR (2000) Commercial potential for *Haematococcus* microalgae as a natural source of astaxanthin. Trends Biotechnol 18:160–167

Lozano I, Wacyk JM, Carrasco J, Cortez-San Martín MA (2016) Red macroalgae *Pyropia columbina* and *Gracilaria chilensis*: sustainable feed additive in the *Salmon salar* diet and the

evaluation of potential antiviral activity against infectious salmon anemia virus. J Appl Phycol 28:1343–1351
MacArtain P, Gill CI, Brooks M, Campbell R, Rowland IR (2007) Nutritional value of edible seaweeds. Nutr Rev 65:535–543
Madadi R, Maljaee H, Serafim LS, Ventura SPM (2021) Microalgae as contributors to produce biopolymers. Mar Drugs 19(8):466
Marcus JB (2007) Enhancing umami taste in foods. In: Taylor A, Hort J (eds) Modifying flavour in food. CRC Press, Cambridge, pp 202–220
Mariey Y, Samak H, Abou-Khashba H, Sayed M, Abou-Zeid A (2014) Effect of using *Spirulina platensis* algae as a feed additives for poultry diets: 2 productive performance of broiler. Egypt Poult Sci 34:245–258
Misurcova L, Bunka F, Ambrozova JV, Machu L, Samek D, Kracmar S (2014) Amino acid composition of algal products and its contribution to RDI. Food Chem 151:120–125
Mohan SV, Hemalatha M, Chakraborty D, Chatterjee S, Ranadheer P, Kona R (2019) Algal biorefinery models with self-sustainable closed loop approach: trends and prospective for blue-bioeconomy. Bioresour Technol 295:122128
Monshupanee T, Nimdach P, Incharoensakdi A (2016) Two-stage (photoautotrophy and heterotrophy) cultivation enables efficient production of bioplastic poly-3-hydroxybutyrate in auto-sedimenting cyanobacterium. Sci Rep 6:37121
Mostafavi FS, Zaeim D (2020) Agar-based edible films for food packaging applications - a review. Int J Biol Macromol 159:1165–1176
Muller-Feuga A (2000) The role of microalgae in aquaculture: situation and trends. J Appl Phycol 12:527–534
Mysliwa-Kurdziel B, Solymosi K (2016) Phycobilins and phycobiliproteins used in food industry and medicine. Mini Rev Med Chem 17(13):1173–1193
Naeem A, Saleemuddin M, Hasan Khan R (2007) Glycoprotein targeting and other applications of lectins in biotechnology. Curr Protein Pept Sci 8:261–271
Narancic T, Cerrone F, Beagan N, O'Connor KE (2020) Recent advances in bioplastics: application and biodegradation. Polymers 12(4):920
Nishioka M, Nakai K, Miyake M, Asada Y, Taya M (2001) Production of poly-β-hydroxybutyrate by thermophilic cyanobacterium, *Synechococcus* sp. MA19, under phosphate-limited conditions. Biotechnol Lett 23(14):1095–1099
Oh YK, Hwang KR, Kim C, Kim JR, Lee JS (2018) Recent developments and key barriers to advanced biofuels. Bioresour Technol 257:320–333
Ometto F, Quiroga G, Psenicka P, Whitton R, Jefferson B, Villa R (2014) Impacts of microalgae pre-treatments for improved anaerobic digestion: thermal treatment, thermal hydrolysis, ultrasound and enzymatic hydrolysis. Water Res 65:350–361
Ozcimen D, Inan B, Morkoc O, Efe A (2017) A review on algal biopolymers. J Chem Eng Res Updates 4:7–14
Paniagua-Michel J (2015) Handbook of marine microalgae. Microalgal nutraceuticals. Academic Press, Cambridge, pp 255–267
Panutai W, Iamtham S (2019) Extraction, purification and antioxidant activity of phycobiliprotein from *Arthrospira platensis*. Process Biochem 82:189–198
Patil CK, Jirimali HD, Paradeshi JS, Chaudhari BL, Alagi PK, Hong SC (2018) Synthesis of biobased polyols using algae oil for multifunctional polyurethane coatings. Green Mater 6(4): 165–177
Paulien FHH, Martijn MH, Harriette L (2014) Bos green building blocks for bio-based plastics. Biofuels Bioprod Biorefin 8:306–324
Pawar MS, Kadam AS, Dawane BS, Yemul OS (2016) Synthesis and characterization of rigid polyurethane foams from algae oil using biobased chain extenders. Polym Bull 73(3):727–741
Prabhasankar P, Ganesan P, Bhaskar N, Hirose A, Stephen N, Gowda LR, Hosokawa M, Miyashita K (2009) Edible Japanese seaweed, wakame (*Undaria pinnatifida*) as an ingredient in pasta: chemical, functional and structural evaluation. Food Chem 115:501–508

Prabhu M, Chemodanov A, Gottlieb R, Kazir M, Nahor O, Gozin M, Israel A, Livney YD, Golberg A (2019) Starch from the sea: the green macroalga *Ulvaohnoi* as a potential source for sustainable starch production in the marine biorefinery. Algal Res 37:215–227

Praiboon J, Palakas S, Noiraksa T, Miyashita K (2018) Seasonal variation in nutritional composition and anti-proliferative activity of brown seaweed, *Sargassum oligocystum*. J Appl Phycol 30:101–111

Pulz O, Gross W (2004) Valuable products from biotechnology of microalgae. Appl Microbiol Biotechnol 65:635–648

Rahman A, Miller CD (2017) Microalgae as a source of bioplastics. In: Rastogi RP, Madamwar D, Pandey A, Rastogi RP, Madamwar D, Pandey A (eds) Algal green chemistry: recent Progress in biotechnology. Elsevier, Amsterdam, pp 121–138

Rai P, Mehrotra S, Priya S, Gnansounou E, Sharma SK (2021) Recent advances in the sustainable design and applications of biodegradable polymers. Bioresour Technol 325:124739

Reddy RL, Reddy VS, Gupta GA (2013) Study of bio-plastics as green and sustainable alternative to plastics. Int J Emerg Technol Adv Eng 3(5):76–81

Rehm BH, Moradali MF (2018) Alginates and their biomedical applications. Springer, Singapore

Ross E, Dominy W (1990) The nutritional value of dehydrated, blue-green algae (*Spirulina plantensis*) for poultry. Poult Sci 69:794–800

Ross IL, Shah S, Hankamer B, Amiralian N (2021) Microalgal nanocellulose-opportunities for a circular bioeconomy. Trends Plant Sci 26(9):924–939

Rossano R, Ungaro N, D'Ambrosio A, Liuzzi GM, Riccio P (2003) Extracting and purifying R-phycoerythrin from Mediterranean red algae *Corallina elongate* Ellis & Solander. J Biotechnol 101:289–293

Saha P, Aloui H, Yun JH, Kim HS, Kim BS (2021) Development of a novel composite film based on polyurethane and defatted *Chlorella* biomass: physical and functional characterization. J Appl Polym Sci 138(14):50152

Sajilata M, Singhal R, Kamat M (2008) Fractionation of lipids and purification of γ-linolenic acid (GLA) from *Spirulina platensis*. Food Chem 109:580–586

Samantaray S, Mallick N (2012) Production and characterization of poly-β- hydroxybutyrate (PHB) polymer from *Aulosira fertilissima*. J Appl Phycol 24(4):803–814

Sekar S, Chandramohan M (2008) Phycobiliproteins as a commodity: trends in applied research, patents and commercialization. J Appl Phycol 20:113–136

Sharma L, Mallick N (2005) Accumulation of poly-β-hydroxybutyrate in *Nostoc muscorum*: regulation by pH, light-dark cycles, N and P status and carbon sources. Bioresour Technol 96(11):1304–1310

Shi B, Wideman G, Wang JH (2012) A new approach of BioCO_2 fixation by thermoplastic processing of microalgae. J Polym Environ 20(1):124–131

Shimoda H, Tanaka J, Shan SJ, Maoka T (2010) Anti-pigmentary activity of fucoxanthin and its influence on skin mRNA expression of melanogenic molecules. J Pharm Pharmacol 62:1137–1145

Shrivastav A, Mishra SK, Mishra S (2010) Polyhydroxyalkanoate (PHA) synthesis by *Spirulina subsalsa* from Gujarat coast of India. Int J Biol Macromol 46(2):255–260

Sidek IS, Draman SF, Abdullah SR, Anuar N (2019) Current development on bioplastics and its future prospects: an introductory review. INWASCON Technol Mag 1:3–8

Singh AK, Mallick N (2017) Advances in cyanobacterial polyhydroxyalkanoates production. FEMS Microbiol Lett 364(20). https://doi.org/10.1093/femsle/fnx189

Spolaore P, Joannis-Cassan C, Duran E, Isambert A (2006) Commercial applications of microalgae. J Biosci Bioeng 101:87–96

Stevens CV (2013) Bio-based plastics: materials and applications. Wiley, New York

Sudhakar MP, Kumar BR, Mathimani T, Arunkumar K (2019) A review on bioenergy and bioactive compounds from microalgae and macroalgae-sustainable energy perspective. J Clean Prod 228: 1320–1333

Suganya AM, Sanjivkumar M, Navin Chandran M, Palavesam A, Immanuel G (2016) Pharmacological importance of sulphated polysaccharide carrageenan from red seaweed *Kappaphycus alvarezii* in comparison with commercial carrageenan. Biomed Pharmacother 84:1300–1312

Surabhi J, Roshani K, Vivek NU (2018) Applications of algae in cosmetics: an overview. Int J Innov Res Sci Eng Technol 7(2):1269–1278

Tan FHP, Nadir N, Sudesh K (2022) Microalgal biomass as feedstock for bacterial production of PHA: advances and future prospects. Front Bioeng Biotechnol 10:879476

Tanaka T, Shnimizu M, Moriwaki H (2012) Cancer chemoprevention by carotenoids. Molecules 17:3202–3242

Tanna B, Mishra A (2018) Metabolites unravel nutraceutical potential of edible seaweeds: an emerging source of functional food. Compr Rev Food Sci Food Saf 17:1613–1624

Thielen M (2014) Bioplastics: plants and crops raw materials products. Fachagentur Nachwachsende Rohsto e eV (FNR) Agency for Renewable Resources, Gulzow

Tibbetts SM, Milley JE, Lall SP (2015) Chemical composition and nutritional properties of freshwater and marine microalgal biomass cultured in photobioreactors. J Appl Phycol 27: 1109–1119

Tokusoglu O, Uunal M (2003) Biomass nutrient profiles of three microalgae: *Spirulina platensis, Chlorella vulgaris*, and *Isochrisis galbana*. J Food Sci 68:1144–1148

Tziveleka LA, Ioannou E, Roussis V (2019) Ulvan, a bioactive marine sulphated polysaccharide as a key constituent of hybrid biomaterials: a review. Carbohydr Polym 218:355–370

Van Krimpen M, Bikker P, Van der Meer I, Van der Peet-Schwering C, Vereijken J (2013) Cultivation, processing and nutritional aspects for pigs and poultry of European protein sources as alternatives for imported soybean products. Lelystad, Wageningen UR Livestock Research, p 48

Vavilala SL, D'Souza JS (2015) Marine algae extracts: processes, products, and applications. In: Kim S-K, Chojnacka K (eds) Marine algae extracts. John Wiley & Sons, pp 411–452

Vo TS, Ryu B, Kim SK (2013) Purification of novel anti-inflammatory peptides from enzymatic hydrolysate of the edible microalgal *Spirulina maxima*. J Funct Foods 5:1336–1346

Wan AH, Soler-Vila A, O'Keeffe D, Casburn P, Fitzgerald R, Johnson MP (2016) The inclusion of *Palmaria palmata* macroalgae in Atlantic salmon (*Ulvalactuca*) diets: effects on growth, haematology, immunity and liver function. J Appl Phycol 28:3091–3100

Wang Y, Chen GQ (2017) Polyhydroxyalkanoates: sustainability, production, and industrialization. In: Tang C, Ryu CY (eds) Sustainable biopolymers from biomass. Wiley, New York, pp 11–33

Wang G, Sun H, Fan X, Tseng C (2001) Large-scale isolation and purification of R-phycoerythrin from red alga *Palmaria palmata* using the expanded bed adsorption method. Acta Bot Sin 44: 541–546

Yaakob Z, Ali E, Zainal A, Mohamad M, Takriff MS (2014) An overview: biomolecules from microalgae for animal feed and aquaculture. J Biol Res 21:1

Yamaguchi K (1996) Recent advances in microalgal bioscience in Japan, with special reference to utilization of biomass and metabolites: a review. J Appl Phycol 8:487–502

Yan C, Wang R, Wan J, Zhang Q, Xue S, Wu X, Zhang J, Zhang J, Lu Y, Cong W (2016) Cellulose/ microalgae composite films prepared in ionic liquids. Algal Res 20:135–141

Yuan Y, Zhang J, Fan J, Clark J, Shen P, Li Y, Zhang C (2018) Microwave assisted extraction of phenolic compounds from four economic brown macroalgae species and evaluation of their antioxidant activities and inhibitory effects on α-amylase, α-glucosidase, pancreatic lipase and tyrosinase. Food Res Int 113:288–297

Zanchetta E, Damergi E, Patel B, Borgmeyer T, Pick H, Pulgarin A, Ludwig C (2021) Algal cellulose, production and potential use in plastics: challenges and opportunities. Algal Res 56: 102288

Zargarzadeh M, Amaral AJR, Custodio CA, Mano JF (2020) Biomedical applications of laminarin. Carbohydr Polym 232:115774

Zayed A, Ulber R (2019) Fucoidan production: approval key challenges and opportunities. Carbohydr Polym 211:289–297

Zhang F, Kabeya H, Kitagawa R, Hirotsu T, Yamashita M, Otsuki T (2000) An exploratory research of PVC-Chlorella composite material (PCCM) as effective utilization of *Chlorella* biologically fixing CO_2. J Mater Sci 35(10):2603–2609

Ziolkowska NE, Wlodawer A (2006) Structural studies of algal lectins with anti-HIV activity. Acta Biochim Pol 53:617–626

Chapter 16
Valorising *Haematococcus* Biomass for Commercial Applications

Madhubalaji Chegu Krishnamurthi, Sarada Ravi, and Vikas Singh Chauhan

16.1 Introduction

Haematococcus pluvialis containing commercial products are widely used in the food and beverage industry and the cosmetics and health supplement sectors. *Haematococcus pluvialis* is a freshwater microalga having the potential ability for natural Astaxanthin production (Lorenz and Cysewski 2000). *Haematococcus* is packed with an abundance of essential fatty acids, amino acids, vitamins, minerals, and other nutrients that are essential for healthy growth and development. Furthermore, *Haematococcus* commercial products have become increasingly popular as a natural source of astaxanthin (Tripathi et al. 1999). Astaxanthin, an oxygen-rich red coloured keto carotenoid pigment is produced by microalgae as a defence mechanism against harsh environmental conditions and is known for its potent antioxidant activity (Vidhyavathi et al. 2008). For example, astaxanthin has ten times more antioxidant activity than other carotenoids and is more than 500 times more potent than alpha-tocopherol, a form of vitamin E (Ranga et al. 2009). Astaxanthin has significant effect on reducing inflammation, enhancing the immune system, and providing protection from UV damage (Ambati et al. 2014). Several in vitro studies have also shown it to be an effective quencher of reactive oxygen species and nitrogen species, including singlet oxygen and one- and two-electron oxidants. This powerful antioxidant effect has led to the widespread use of astaxanthin in foods, dietary supplements (nutraceuticals), cosmetics, and animal feed. In 1987, the U.S. Food and Drug Administration (US FDA) approved astaxanthin as a feed additive for the aquaculture industry (Brendler and Williamson 2019); it was also

M. C. Krishnamurthi · S. Ravi · V. S. Chauhan (✉)
Plant Cell Biotechnology Department, CSIR-Central Food Technological Research Institute, Mysuru, Karnataka, India

Academy of Scientific and Innovative Research (AcSIR), Ghaziabad, India
e-mail: vikas@cftri.res.in

R. Raja et al. (eds.), *Haematococcus*, https://doi.org/10.1007/978-981-99-2901-6_16

approved as a dietary supplement in 1999. *Haematococcus* commercial products are used in various ways to produce natural food colourings, flavourings, preservatives, and dietary supplements.

The astaxanthin in *Haematococcus* also gives the characteristic pink colour to salmon and shrimp.

16.2 *Haematococcus*-Based Products: International Market

The *Haematococcus pluvialis* market is expected to reach USD 148.1 million by 2027, growing at a CAGR of 13.2% during the forecast period, 2020–2027, as reported by Meticulous research (*Haematococcus pluvialis* Market by Size, Share, Forecasts, & Trends Analysis n.d.). Another report by Grand View Research, Inc. predicts that the global astaxanthin market size is set to skyrocket to USD 4.75 billion by 2028, with an estimated CAGR of 16.8% (Astaxanthin Market Size Worth $4.75 Billion By 2028 n.d.). The exponential increase is due to using natural astaxanthin in poultry and aquaculture, which is expected to create lucrative opportunities for manufacturers in the *Haematococcus pluvialis* market. However, the high production cost is expected to hinder the market growth of *Haematococcus pluvialis* to some extent. The *Haematococcus pluvialis* market is segmented based on application: nutraceuticals, food and beverages, pharmaceuticals, cosmetics, and animal feed. The nutraceuticals segment is expected to hold the dominant share of the overall *Haematococcus pluvialis* market by 2025 (Meticulous Research *Haematococcus pluvialis* Market by Size, Share, Forecasts, & Trends Analysis n.d.).

Haematococcus commercial products come in many forms, including dried powder or whole-cell form, lipid extract, flakes, capsule, tablet, paste, liquid extract, and oil. They are easy to use and can be added to foods, beverages, cosmetics (moisturisers, facial creams, and anti-ageing serums) and nutritional supplements. *Haematococcus pluvialis* extract is one of the most common forms of *Haematococcus* commercial products and is available in various concentrations, from 10 to 50%. This extract is rich in astaxanthin and other antioxidants, minerals, Vitamin A, Vitamin C, and Vitamin E. Capsules and tablets are the most popular form of *Haematococcus* commercial products. These products are easy to consume and can be taken daily for optimal health benefits of the natural astaxanthin. Capsules and tablets typically come in dosages ranging from 4 to 12 mg of astaxanthin and often include additional ingredients such as Vitamin E and other antioxidants. Capsules are a convenient way to get a high supplement dose without measuring out a powder or liquid extract.

Haematococcus available in powder form, is usually mixed into foods and drinks and often contains additional antioxidants and nutrients. Powdered *Haematococcus* is a great way to prepare a concentrated dose of astaxanthin and is often used in smoothies, energy bars, and other health foods. Astaxanthin paste is widely used in the food industry, as it can provide a natural colourant and boost food flavour. *Haematococcus pluvialis* extract can also be combined with other active ingredients, such as lycopene, omega-3 fatty acids, vitamin E, vitamin C, and beta-carotene, to create a powerful antioxidant blend. The most effective way to take astaxanthin is in a product designed to be absorbed by the body. Astaxanthin supplements are generally considered safe.

Synthetic astaxanthin is cheaper but less effective than natural astaxanthin (Capelli et al. 2013). Natural astaxanthin is more expensive but more potent and provides more health benefits. It is recommended to look for a product that contains natural astaxanthin from algae. In addition, plant-based astaxanthin is being investigated as a viable option for antioxidant formulations in the food, beverage, and pharmaceutical industries. In 2016, a team of researchers from the East Indian state of Odisha published a study on the extraction of astaxanthin from mangrove leaves (Banerjee et al. 2017).

16.3 In Vitro and In Vivo Studies

Astaxanthin has been shown to have a variety of health benefits in both in vitro and in vivo studies in addition to human studies. *Haematococcus* biomass potentially impacts various diseases, including cancer, hypertension, diabetes, cardiovascular diseases, gastrointestinal diseases, age related diseases, liver diseases, neurodegenerative diseases, lupus and arthritis and boosts immune system (Oslan et al. 2021; Mularczyk et al. 2020). In vitro and animal studies have shown that it can scavenge free radicals, protect cells and DNA from oxidative damage, and reduce inflammation. Astaxanthin is also gaining attention as a natural treatment for inflammation and joint pain and various skin conditions. It has been widely used to treat Alzheimer's, Parkinson's, liver dysfunction, high cholesterol, and macular degeneration (Alugoju et al. 2022). Astaxanthin could reduce the number of beta-amyloid plaques in the brains of rats with Alzheimer's disease (Alugoju et al. 2022; Alghazwi et al. 2019; Chen et al. 2021). Over the past few years, various astaxanthin drugs have entered the market.

Several in vitro and animal studies have shown that astaxanthin helps protect the brain against oxidative and inflammatory damage, which can help improve cognitive function and reduce the risk of age-related cognitive decline (Li et al. 2016). Human studies on the cognitive effects of astaxanthin are still in the early stages. However, the results of a clinical trial suggested that astaxanthin improves cognitive function in healthy adults (Sekikawa et al. 2020).

Astaxanthin improved exercise performance and reduced muscle damage in healthy adults. Additionally, astaxanthin supplementation has been shown to reduce symptoms of carpal tunnel syndrome (MacDermid et al. 2012). The antioxidant properties of Astaxanthin can help lower LDL (bad) cholesterol levels, reduce blood clotting, and prevent arterial plaque formation. Additionally, *Haematococcus* may help reduce inflammation throughout the body, including in arteries, which can help reduce the risk of heart attack and stroke.

Haematococcus has recently been studied for its potential to help manage healthy blood sugar levels. In a study found that astaxanthin supplementation was associated with decreased oxidative stress markers and inflammation in people with type 2 diabetes (Kanwugu et al. 2022). The antioxidants in astaxanthin can help reduce insulin sensitivity and improve glucose uptake in the cells, which can help reduce the risk of Type 2 diabetes (Li et al. 2018). These studies suggest that *Haematococcus* may be a promising supplement to improve the overall health, as it is rich in powerful antioxidants with various potential benefits. *Haematococcus* also helps improve mood, as it helps regulate the production of serotonin and dopamine.

16.4 *Haematococcus* in Various Industrial Sectors

Haematococcus biomass is a renewable source of sustainable biomass and is one of the most promising options used for various industrial applications. One of the key benefits of using *Haematococcus* biomass in the industrial sector is its high astaxanthin content. Furthermore, it can increase the sustainability of products, helping to reduce their environmental impact. It has been utilised in various sectors, such as aqua culture, cosmetics, food industry, livestock feed, energy, and healthcare (Fig. 16.1). In the industrial sector, the utilisation of *Haematococcus* biomass has seen significant growth. Its use in industrial processes has enabled manufacturers to reduce costs, improve efficiency, and create more sustainable products.

Haematococcus biomass can be used in feedstocks for animal, for production and purification of pharmaceuticals, and for production of bioplastics, paper, and textile. It is used as an ingredient in food products and as a supplement in nutritional and herbal products. The spent biomass of *Haematococcus* can also be used as a fertiliser, a source of protein in animal feed, a bio-pesticide, and a soil conditioner. Additionally, it is used as a coating material for packaging. As a result, *Haematococcus* biomass has become an essential part of the industrial sector, providing a variety of uses and applications.

16.5 *Haematococcus* in Aquaculture

Haematococcus-based commercial products are also used for aquaculture to enhance the colour of farmed fish. The natural astaxanthin can be added to the feed of fish, shrimp, and other aquatic animals to give them a reddish hue. This helps to improve the marketability of the fish and shrimp. *Haematococcus* improve the health of fish in several ways. *Haematococcus* provide a source of essential fatty acids, which are beneficial for maintaining the health of the fish's skin and scales (Shah et al. 2016). These fatty acids help to promote the growth of beneficial bacteria while also helping to reduce the levels of unhealthy bacteria in the water. Additionally, the amino acids in these products help strengthen the fish's immune system while aiding in the

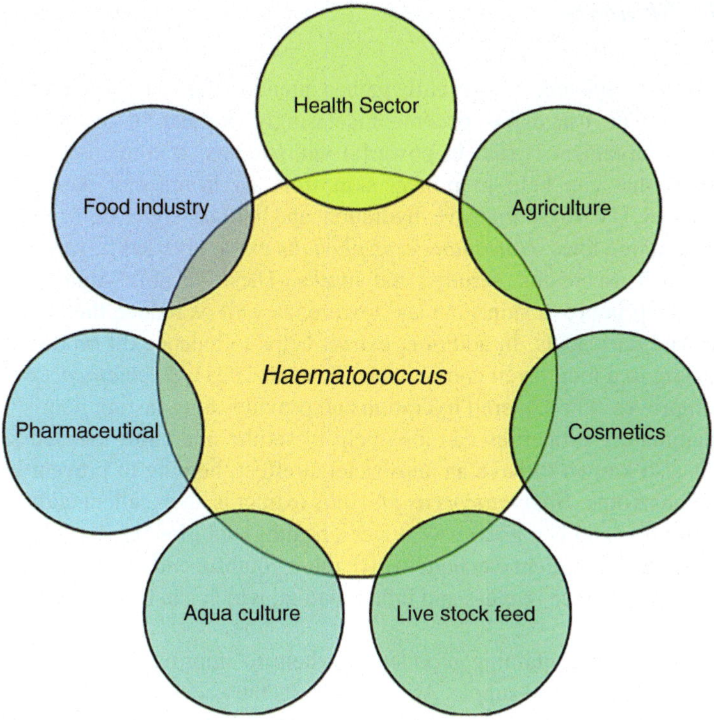

Fig. 16.1 *Haematococcus* biomass utilisation in various industrial applications

digestion of food. *Haematococcus* also contains several vitamins, minerals, and other nutrients. These nutrients are essential for the healthy growth and development of fish. For example, the high levels of Vitamin A found in these products help improve the fish's vision. These products also provide essential nutrients for the overall health of aquatic life. These products are also easy to use and can be added directly to the fish tank, allowing easy access to all the beneficial nutrients available.

16.6 *Haematococcus* in Cosmetics

Haematococcus pluvialis is gaining attention due to its many applications in cosmetics and other personal care products. *Haematococcus* has been used for centuries in Asian and European cultures for its medicinal and cosmetic uses. It is now gaining more popularity in the Western world as an ingredient in many cosmetics. Various products on the market contain *Haematococcus*, including creams, serums, masks, and shampoos. These products also contain other potent antioxidants, such as Vitamin C and green tea extract, to protect the skin from environmental damage. *Haematococcus* popularity is increasing due to its ability to produce astaxanthin, a powerful antioxidant in many skin care products. With its anti-ageing properties, astaxanthin is becoming a key ingredient in cosmetics.

16.6.1 Skincare

Haematococcus pluvialis has recently gained attention due to its potential benefits in skincare, and its usage has become increasingly popular in recent years. This microscopic organism is rich in powerful antioxidants. It contains various active compounds that can help to protect skin from environmental damage such as pollution and UV rays, improve hydration and reduce signs of ageing such as wrinkles and fine lines. *Haematococcus pluvialis* extract is used in various skincare products such as creams, serums, and masks. These versatile algae extracts are known to help improve skin tone and texture, as well as reduce the appearance of wrinkles and dark spots. In addition, extract helps to balance oil production in the skin, leading to a more even complexion (To et al. 2021). *Haematococcus pluvialis* extract improves skin's overall hydration and provides long-lasting nourishment. Its anti-inflammatory properties can also help to soothe and calm irritated skin. The extract is also known to have an anti-bacterial effect, helping to prevent blemishes and acne breakouts. *Haematococcus pluvialis* extract is an excellent addition to any skincare routine and is a great way to get a healthy and glowing complexion. With regular use, this ingredient can help to give more youthful and radiant complexion. It can also help to reduce redness and inflammation, which can be caused by exposure to the sun.

Haematococcus-containing products can help to improve hydration levels and keep skin feeling soft and supple. Additionally, its fatty acids can help to hydrate and soothe the skin, while its amino acids can help to repair damaged skin (Marinoa et al. 2020). *Haematococcus* powder can be mixed with carrier oil, such as jojoba oil, to create a mask or scrub. Many skincare products contain *Haematococcus* as a liquid or powder, which can be added to lotions, creams, serums, moisturisers, cleansers, and other products. Most products should be used twice a day, morning and night, and avoiding contact with the eyes is essential. Additionally, it is recommended to use products specifically formulated for the face, as these will be gentler on the skin.

16.6.2 Hair Care

Haematococcus pluvialis is gaining attention due to its high content of antioxidants and its ability to provide a range of hair treatments and protective benefits. The most notable benefit of *Haematococcus pluvialis* is its ability to differentiate between dry, damaged hair and healthy hair, making it an excellent choice for anyone looking to restore their hair's natural shine and lustre. This microalga also contains carotenoids, which help protect hair from environmental and free radical damage. In addition, *Haematococcus pluvialis* can reduce the appearance of split ends and make hair stronger, more manageable, and easier to style (Siahaan et al. 2021). With these benefits, it is no wonder why the use of this microalga in hair care products is increasing. A significant benefit of using *Haematococcus pluvialis* in hair care products is that it is a natural ingredient that is not likely to cause any irritation or side effects. This makes it an excellent choice for sensitive skin or syntactic allergies. In addition, *Haematococcus pluvialis* is a good source of essential fatty acids necessary for maintaining healthy hair and scalp. These essential fatty acids help to nourish the hair and scalp and can also help to reduce dandruff. Additionally, this microalga helps seal moisture, making it great for dry, brittle hair (Siahaan et al. 2021).

16.6.3 Environmental Benefits of Haematococcus *in Cosmetics*

Using natural ingredients in cosmetics has become increasingly popular in recent years, as many people are looking for ways to reduce their environmental impact. *Haematococcus* is known for its high concentrations of astaxanthin, an antioxidant beneficial for skin health. It provides a sustainable, natural ingredient that has excellent skin health benefits, as well as environmental benefits. *Haematococcus* can be cultivated in controlled environments, eliminating the risks associated with wild harvesting. It is a natural ingredient with no synthetic chemicals, making it better for the environment than many other ingredients used in cosmetics. It is a renewable resource that does not deplete natural resources like other ingredients that come from animals or plants. *Haematococcus* helps reduce water pollution, as it can be grown in tanks free from contaminants. Using *Haematococcus* in cosmetics is one way of reducing our environmental footprint and helping keep our planet healthy for future generations.

16.6.4 Safety and Regulations of Haematococcus *in Cosmetics*

To use *Haematococcus* in cosmetics, specific safety and regulation considerations need to be taken into account. *Haematococcus* is a type of algae, and as such, it is not subject to the same regulations as other cosmetics ingredients. As a result, it is essential to ensure that products that contain this ingredient are adequately tested and that the manufacturer's instructions regarding its use are followed. In terms of regulation, the European Union has established safety guidelines for using *Haematococcus* in the cosmetics (Leinemann and Mabilia 2019). These guidelines include using good manufacturing practices, using certified materials, and testing finished products to ensure that they are safe for use. In addition, the European Commission has issued a set of regulatory requirements for the labelling of cosmetics containing *Haematococcus*. Similar regulations will come in future throughout the globe.

16.7 *Haematococcus* in the Healthcare Sector

Haematococcus biomass is an increasingly popular component in the healthcare sector. *Haematococcus* biomass has been used in medical treatments and products to help treat a variety of medical problems and to aid in developing new medical therapies (Yamashita 2013). *Haematococcus* biomass contains various beneficial oils, lipids, and vitamins which are used to create nutritional supplements. These supplements can improve general health, treat specific medical conditions such as inflammation, and help promote wound healing. In addition, *Haematococcus* biomass is being studied as a potential source of nutrition for cancer patients (Nagaraj et al. 2012). *Haematococcus* biomass is also being used to develop new drug delivery systems. Recent advances in producing and formulating *Haematococcus* biomass have enabled the creation of a wide range of pharmaceutical products, including creams, ointments, and tablets. The use of *Haematococcus* biomass in the pharmaceutical industry has opened up new opportunities for personalised medicine and offered cost-effective treatments for rare and hard-to-treat ailments.

The research has shown that the carotenoids and other bioactive compounds found in *Haematococcus* are essential in promoting eye health and reducing the risk of vision-related diseases (Lin et al. 2020). *Haematococcus* extracts are also used in products designed to support eye health, such as eye drops and creams. This ingredient can help reduce eye fatigue, improve night vision, and reduce the risk of developing cataracts. The astaxanthin found in the biomass is especially beneficial for eyesight, as it helps protect the eyes from the sun's harmful UV rays. *Haematococcus* also contains lutein, an essential nutrient for eye health and vision. *Haematococcus* biomass are a promising and versatile resource with many potential applications in the healthcare sector.

One of the most prominent applications of *Haematococcus* biomass in the healthcare sector is its potential to fight cancer. Studies have shown that this microalga can produce compounds that can inhibit the growth of cancer cells (Nagaraj et al. 2012; Palozza et al. 2009). Furthermore, the antioxidants found in *Haematococcus* biomass may be able to protect the body from the damage caused by radiation therapy (Rao et al. 2013). Another potential benefit of *Haematococcus* biomass is its potential to help with weight loss. Studies have shown that the compounds found in *Haematococcus* biomass may help reduce the body's ability to store fat. In addition, studies have also suggested that the compounds in *Haematococcus* biomass may be able to reduce the signs of ageing. *Haematococcus* is also used to produce nutraceutical products, such as dietary supplements, and prebiotics. *Haematococcus* biomass has also been used in the development of medical devices, such as implantable electrodes and drug delivery systems, due to its biocompatibility. Furthermore, its antioxidant properties have been utilized to develop cell-based therapies for cancer and other diseases.

Moreover, the use of this biomass in the healthcare sector can also include the production of specific nanomaterials, such as nanoparticles and nanocomposites, to deliver drugs to target areas in the body (Jafari et al. 2021). These nanomaterials are more effective in drug delivery than traditional methods due to their small size and high surface area ratio. The potential of *Haematococcus* biomass in the healthcare sector is vast and exciting. The use of this renewable resource in the development of treatments and medical devices could open up new possibilities for the improvement of patient care. This algal biomass could become a valuable asset in the healthcare industry with further research.

16.8 *Haematococcus* in the Pharmaceutical

Haematococcus biomass is highly versatile material increasingly used for many applications in the pharmaceutical industry. It can produce various pharmaceuticals, from anti-inflammatory drugs to skin care products. This makes it an attractive option for pharmaceutical companies looking to diversify their product offerings. One of the most exciting applications of *Haematococcus* biomass in the pharmaceutical industry is its use in producing drugs for treating various conditions and other chronic diseases. Furthermore, high antioxidant content of *Haematococcus* makes it an excellent choice for developing drugs to help combat the damaging effects of free radicals in the body. *Haematococcus* biomass its antioxidant properties and high content of vitamins and minerals makes it an ideal choice for producing dietary supplements. It can create various products, such as vitamin pills, energy drinks, and other food supplements.

Some pharmaceutical products that use *Haematococcus* biomass include dietary supplements, skin creams, eye drops, and anti-ageing creams (Zakaria et al. 2021). *Haematococcus* makes it an effective treatment for various skin conditions such as eczema and psoriasis (To et al. 2021). The astaxanthin can help protect cells from oxidative damage and can even help reduce the risk of developing certain diseases for example cardiovascular diseases by reducing cholesterol levels. *Haematococcus* biomass is an excellent source of carotenoids and one of the pharmaceutical industry's most promising applications of *Haematococcus* biomass is producing carotenoid-based drugs. Carotenoids are a class of natural substances that have been shown to have a range of therapeutic effects, including antioxidant activity, anti-inflammatory effects, and immunomodulatory activity. Carotenoid-based drugs are being developed for various diseases, including cancer, cardiovascular diseases, and neurological disorders. Many anti-cancer drugs are derived from proteins abundant in *Haematococcus* biomass. In addition, *Haematococcus* biomass is a rich source of fatty acids, which can be used to produce anti-cancer drugs (Rao et al. 2013).

16.9 *Haematococcus* in Food Industry

Food safety has become a significant concern with the rapid growth of the food industry. *Haematococcus* is increasingly used as an additive in food production due to its valuable health benefits. Recently, it has gained attention as a "superfood" due to its potent health benefits. *Haematococcus* is increasingly used in various food products, from smoothies, shakes, salad dressings, sauces, and juices to dietary supplements and energy bars. Its high protein content makes it an excellent substitute for animal-based proteins, and its natural flavour makes it a great addition to any meal. It is also increasingly popular in vegan and vegetarian diets, providing a complete source of protein and essential nutrients. When biomass is added to smoothies, its unique texture and flavour make it a great addition to these drinks. Its antioxidant capabilities make it a great way to boost body's energy and nutrition. It can also be used in baking or added to salads or other dishes for a nutritious boost. It is also gaining popularity as a nutritional supplement due to its high antioxidant content and health benefits. As interest in *Haematococcus* continues to grow, its potential as a beneficial food ingredient will also grow.

Haematococcus is also rich in pigments that can be used to add colour to food products. These pigments are also known for their antioxidant and anti-inflammatory properties, making them even more beneficial. They are commonly used in the food industry as natural food colourings and can also be used as a supplement to enhance the nutritional content of food products (Tri et al. 2021). Haematococcus-derived astaxanthin powder is used as a natural food colourant to enhance the aesthetics of food products, such as cakes, cookies, bread and ice cream giving them a unique and vibrant hue (Gouveia et al. 2006). Therefore, it can be used to enhance the appearance of food products. This is especially beneficial for vegan and vegetarian products, as it can make them look more attractive without using artificial colourings. *Haematococcus* is also used to make natural food dyes, which are more vibrant and longer-lasting than artificial dyes.

Its high protein content makes it a valuable source of nutrition, while its high antioxidant content makes it an ideal natural preservative. *Haematococcus* is also known to possess anti-microbial properties, which can help reduce the risk of contamination in food products. This can help increase their shelf life and ensure they remain safe to consume.

Haematococcus has various uses in different cuisines, from being a source of flavourings and food colouring to being used as a health supplement. The algae have a slightly sweet, earthy flavour, which makes them ideal for adding unique flavours to different dishes. Its unique flavour and health benefits make it an excellent addition to any dish.

Overall, the incorporation of *Haematococcus* into food products can offer many benefits, such as improving nutrition, enhancing the appearance and prolonging the shelf life of food products. It is an excellent option for the food industry and should be considered when looking to improve food quality. *Haematococcus* can also be used to increase the nutritional value of food products. It is a rich source of Omega 3 fatty acids and other essential micronutrients, such as vitamins A and E (Yu et al. 2022). These micronutrients can help improve consumers' overall health and well-being. *Haematococcus* is becoming increasingly popular in the food industry due to its health benefits and potential culinary applications. *Haematococcus* is used as a seasoning or added to sauces or dressings. Alternatively, *Haematococcus* in various ready-made products such as energy bars and protein powders. Adding

Haematococcus to cooking is a great way to increase the nutritional value of meals. It contains vitamins and minerals, such as iron, potassium, magnesium, zinc, and vitamin C, which can help boost the immune system and improve overall health.

16.10 *Haematococcus* in Agriculture

Agriculture and farming are an integral part of modern society. It is the basis for the production of food that sustains us. *Haematococcus* can be used in agriculture in various ways, from being used as a soil amendment to being used as a feed ingredient. As a soil amendment, *Haematococcus* helps to increase soil fertility and improve the structure of the soil. It also increases water-holding capacity, which helps to reduce runoff and soil erosion (Shah 2019; Chettri et al. 2022). The use of *Haematococcus* in agriculture is becoming increasingly popular, as it is a natural and sustainable way to increase crop yield and improve soil fertility. They can also help to reduce runoff and soil erosion and improve the health and performance of animals, with the increasing popularity of sustainable and natural agricultural practices.

16.10.1 **Haematococcus** *as a Natural Fertiliser*

Haematococcus is a type of single-celled microalga that can be used as a natural fertiliser for agricultural plants (Sani et al. 2022). This alga is a rich source of nitrogen, phosphorus, and potassium, making it an excellent choice for boosting crop yields and increasing soil health. It contains many other beneficial nutrients, including proteins, lipids, carbohydrates, vitamins, and minerals. Using *Haematococcus* as a natural fertiliser can provide significant benefits to farmers, including increased yields and improved soil fertility (Das et al. 2019). Using *Haematococcus* as a natural fertiliser can also help reduce the need for chemical fertilisers and pesticides, which can have long-term environmental impacts.

16.10.2 **Haematococcus** *for Livestock Feed*

Haematococcus has been gaining popularity as a natural feed supplement for livestock in recent years (Saadaoui et al. 2021). These microscopic green algae produce an astounding array of health-promoting compounds, including vitamins, minerals, and antioxidants. In addition, *Haematococcus* are rich in essential fatty acids, high-quality protein, and other valuable nutrients that can benefit both production and performance of animals. As a feed ingredient, *Haematococcus* helps to boost animal health and performance (Nishshanka et al. 2022). It is an excellent source of the beneficial antioxidant astaxanthin and helps to improve the immune system, as well as to help to improve the growth rate of animals. Astaxanthin is known to help animals combat oxidative stress, maintain healthy organ function, and improve overall skin and coat health. When it comes to livestock feed, *Haematococcus* has several advantages over other supplements. It is more cost-effective than many commercial alternatives, and animals more readily digest it. Additionally, *Haematococcus* does not contain hormones or antibiotics, so there is no risk of adverse side effects for the animals. The algae are also high in essential minerals such as calcium, magnesium, and zinc, which can help to promote healthy bones and joints in animals.

Haematococcus is an excellent source of omega-3 fatty acids, which are essential for a healthy immune system. Omega-3s are also believed to support a healthy heart and may even reduce inflammation and help to control weight in some animals. These fatty acids are not found in other feed supplements so *Haematococcus* can provide a unique and beneficial addition to a livestock diet (Valente et al. 2021). *Haematococcus* is easy to add to existing feed mixes, as it is typically sold as a powder or paste. This means that farmers and ranchers do not need to change their current setup to take advantage of the nutritional benefits of these algae. Therefore, *Haematococcus* is an attractive and cost-effective option for boosting the health and performance of livestock animals. *Haematococcus* is also increasingly being used as a feed ingredient in aquaculture and animal feed. This is due to its high nutritional content. In addition, its antioxidant and anti-inflammatory properties can also help reduce inflammation and improve the health of the animal (Dineshbabu et al. 2019). Common *Haematococcus*-containing products for animal feed include Tetravite-T, Aquarian-1000, and Naturals-Plus. These products were developed to provide animals with optimal nutrition and health benefits.

16.10.3 **Haematococcus** ***in Pest Control***

Haematococcus pluvialis is gaining increased attention in the agricultural industry due to its potential to be used in pest control fields (Sonawane et al. 2022). It has been studied in relation to agricultural pest control because of its high levels of astaxanthin, an antioxidant which can be used in plant protection. It has been found that *Haematococcus* can control various pests, such as aphids, whiteflies, and thrips. In addition, astaxanthin has been shown to reduce the growth of fungi and other plant pathogens that can cause crop diseases. Astaxanthin from *Haematococcus* has also been effective in controlling insect pests. Studies have found that when sprayed onto crops, *Haematococcus* reduces the number of insects on the plants by up to 72%. This is especially beneficial for organic farming, allowing pest control without chemical pesticides. *Haematococcus* subcritical water extraction is reported to contain propanoic acid and butanoic acid, which have potential pesticidal activity (Rodríguez-Meizoso et al. 2010).

Overall, *Haematococcus* is an effective and sustainable option for agricultural pest control. Its high levels of astaxanthin make it an ideal choice for organic farmers looking for a natural alternative to chemical pesticides.

16.10.4 **Haematococcus** *in Soil Health and Conservation*

Haematococcus is known to bring numerous benefits to agricultural production. Its presence in soils can help improve the soil's health and promote conservation. As a result, these algae can be an essential component in agriculture, helping to increase yields and reduce soil erosion. *Haematococcus* is a primary producer, meaning that it can produce organic matter from light and carbon dioxide. The presence of this alga in the soil can help to increase the amount of available organic matter, improving the soil's fertility. It can also help to improve the soil's water-holding capacity, reducing the need for extra irrigation. In addition, this algae's presence can help increase the activity of beneficial microorganisms in the soil, improving its biological activity. *Haematococcus* can also help to reduce soil erosion. Its presence in the soil can help bind together particles, reducing the amount of sediment carried away by runoff. This can help improve the soil's overall stability and reduce the amount of soil loss due to erosion. In addition, the presence of this algae can help to improve the soil's ability to absorb and retain nutrients, reducing the amount of nutrient runoff (Shah 2019). Overall, *Haematococcus* can be a valuable component in agricultural production. Its presence in the soil can help to improve its fertility, water-holding capacity, and stability. In addition, it can help to reduce soil erosion and nutrient runoff. As a result, this type of algae can be an essential component in agriculture, helping to increase yields and promote soil conservation.

16.10.5 Cost Benefits of **Haematococcus** *in Agriculture*

Understanding the cost benefits of utilising *Haematococcus* in agriculture is becoming increasingly important. The most cost-effective way to use *Haematococcus* in agriculture is as a natural fertiliser. *Haematococcus* is rich in nitrogen, phosphorus, and potassium, essential nutrients for plant growth and development. It can help improve soil fertility, reduce the need for chemical fertilisers, and lower overall costs. *Haematococcus* can also be used as an insect repellent, helping to reduce the need for expensive chemical insecticides.

Haematococcus can also be used as a soil supplement, adding essential minerals and trace elements. Trace elements such as zinc, copper, and manganese are vital for healthy plant growth and can be lacking in some soils. By adding *Haematococcus* to the soil, farmers can ensure that their crops get the necessary minerals and trace elements. *Haematococcus* can also help to improve the water retention capacity of the soil. This means that farmers can use less water for irrigation and reduce their overall water usage. This can help to reduce costs associated with irrigation, as well as help conserve water resources. Overall, understanding the cost benefits of utilizing *Haematococcus* in agriculture is essential for optimising production and improving yields by using it as a natural fertiliser, insect repellent, soil supplement, and water.

16.11 Challenges in the Production and Processing of *Haematococcus*

The main challenge facing the industry is ensuring the safety and efficacy of the products. As such, it is essential to ensure that all *Haematococcus* biomass used in pharmaceutical products meets strict safety standards. Another challenge faced by the industry is the formulation of *Haematococcus* biomass. This complex process requires careful consideration of solubility, stability, and concentration (Jannel et al. 2020). To ensure that the *Haematococcus* biomass is accurately formulated, it is essential to use a high-quality *Haematococcus* biomass supplier experienced in pharmaceutical formulation. *Haematococcus* biomass used in the production of pharmaceutical products is safe for use in humans. To do this, manufacturers must follow strict safety protocols and use only high-quality *Haematococcus* biomass that has been tested for contamination. One of the main challenges is the cost. There are only a few sources of *Haematococcus* biomass, and the cost of production is high (Zhang et al. 2020). This is due to the large number of resources required to cultivate biomass, such as water, temperature, and other environmental factors. Additionally, the cost of harvesting and processing biomass can be high.

Another challenge is quality control. As *Haematococcus* biomass is used in medical applications, it is essential that the product is of the highest quality. This means it needs to be free of contaminants, such as bacteria and other impurities (Yu et al. 2022). However, harvesting and processing biomass can introduce these contaminants, making it challenging to maintain a high-quality standard. Despite these challenges, there is still great potential for using *Haematococcus* biomass in the healthcare sector, as it is rich in essential nutrients and antioxidants.

16.12 International Brands of *Haematococcus*

Some leading astaxanthin-based product manufacturers include AlgaeLife, BioAstin, and AstaReal. AlgaeLife is based in South Korea and offers a range of astaxanthin products, including supplements, skincare products, and food and beverage products. The company has a strong presence in Asia and is expanding into Europe and the USA. BioAstin is another leading manufacturer of astaxanthin-based products. The company, based in the USA, offers a wide range of supplements, skincare products, and food and beverage products. BioAstin has a strong presence in the USA, Europe, and Asia. AstaReal is a Swedish company that is the world's leading astaxanthin manufacturer. The other companies producing *Haematococcus*-based products worldwide are mentioned in Table 16.1.

Table 16.1 Some companies producing *Haematococcus*-based products

S. No	Company	Product name	Details	Website
1.	CyanoTech Corporation, USA	BioAstin	SCE oleoresin, Softgels, water dispersible powder, and dietary supplement	https://www.cyanotech.com/astaxanthin/
2	Mera Pharmaceutical Inc., USA	AstaFactor®	Dietary supplement of *Haematococcus,* soft gels of astaxanthin	www.astafactor.com
4	Algaltechnologies Ltd., Israel	AstaPure™	*Haematococcus* biomass, astaxanthin oleoresin and beadlets	www.algatech.com
5	Valensa International, USA	Zanthin®	Haematococcus extract, softgel, beadlets	www.usnutra.com
6	Stazen Inc., USA	Stazen	Haematococcus-containing dietary supplement, crushed and dried biomass	www.stazen.com
7	BioReal (Sweden)AB	AstaXin®	*Haematococcus*-containing dietary supplement	www.bioreal.se
8	BioReal (Sweden)AB	AstaCarox®	Haematococcus in crushed and dried form	www.bioreal.se
9	BioReal (Sweden)AB	AstaEquus®	Feed supplement for horses	www.bioreal.se
10	BioReal (Sweden)AB	AstaREAL®	Supercritical fluid-oil extract derived from crushed algae	www.bioreal.se
11	BioReal (Sweden)AB	Novaasta®	*Haematococcus*-containing feed supplement for animals	www.bioreal.se
12	Cyanotech Corporation, USA	Naturose™	Haematococcus meal, pigmentation source for fish and animals	www.cyanotech.com
13	Jingzhou Natural Astaxanthin Inc., China	NaturAsta™	Haematococcus dry biomass, softgel	www.asta.cn
14	Britannia Heal products Ltd. UK	Britaxan®	Astaxanthin complex with other carotenoid containing capsule	www.britannia-health.co.uk
15	Supreme Biotechnologies NZ Ltd. New Zealand	AstaSupreme®	Haematococcus biomass, astaxanthin oleoresin and softgel	www.supremebiotech.com
16	Atacama Bio Natural, Chile	Supreme Astaoil™	Oleoresin for food, nutraceutical and cosmetic products	www.atacamabionatural.com
17	Atacama Bio Natural, Chile	Supreme AstaPowder™	Biomass powder for animal feed and supplement	www.atacamabionatural.com

(continued)

Table 16.1 (continued)

S. No	Company	Product name	Details	Website
18	Parry Nutraceutical ltd., India	Zanthin®	Astaxanthin oleoresin, beadlets, and softgel	www.eidparry.com
19	AlgaTech International SDN BHD, Malaysia	Astaxanthin Premia-Ex	Haematococcus biomass, softgel	www.algaetech.com
20	Fuji Chemical Industry Co. Ltd., Japan, Sweden, USA	AstaREAL®	Astaxanthin oleoresin, water dispersible and soluble powder	www.fujichemical.co.jp
21	Kunming Biogenic Co. Ltd. China	AstaBio®	*Haematococcus* biomass, beadlets, softgel	www.bgenic.com
22	Beijing Ginko Group (BGG) Biological Technology Co. Ltd. China	AstaZain®	Oil, powder and beadlets	www.gingkogroup.com.cn
23	Wefirst Biotechnology Co. Ltd. China	AstaFirst™	Haematococcus powder, oleoresin, and softgel	www.astawefirst.com

16.13 Conclusion

In conclusion, *Haematococcus* is a valuable alga that can be used to create a variety of commercial products. Its unique properties give it multiple applications, making it a valuable and versatile resource. *Haematococcus* is commercially cultivated to produce astaxanthin-rich products for food, nutraceuticals, and cosmetic applications. Nutritious and delicious products made from *Haematococcus* algae are becoming increasingly popular in the food industry. This powerful microalga is loaded with essential fatty acids, vitamins, and minerals, which are essential to human health. It is a rich source of proteins and antioxidants that help to reduce inflammation and improve overall health. In the next decade, *Haematococcus* production is expected to increase exponentially and have potential novel applications in pharmaceutical, agriculture, cosmetics, and food industries.

References

Alghazwi M, Smid S, Musgrave I, Zhang W (2019) *In vitro* studies of the neuroprotective activities of astaxanthin and fucoxanthin against amyloid beta (Aβ1-42) toxicity and aggregation. Neurochem Int 124:215–224

Alugoju P, Krishna Swamy VKD, Anthikapalli NVA, Tencomnao T (2022) Health benefits of astaxanthin against age-related diseases of multiple organs: a comprehensive review. Crit Rev Food Sci Nutr:1–66

Ambati RR, Siew Moi P, Ravi S, Aswathanarayana RG (2014) Astaxanthin: sources, extraction, stability, biological activities and its commercial applications-a review. Mar Drugs 12(1): 128–152

Astaxanthin Market Size Worth $4.75 Billion By 2028 (n.d.). https://www.grandviewresearch.com/press-release/astaxanthin-market. Accessed 2 Dec 2022

Banerjee K, Chowdhury GR, Mitra A (2017) Astaxanthin level of dominant mangrove floral species in Indian Sundarbans. MOJ Biorg Org Chem 1(2):00008

Brendler T, Williamson EM (2019) Astaxanthin: how much is too much? A safety review. Phytother Res 33(12):3090–3111

Capelli B, Bagchi D, Cysewski GR (2013) Synthetic astaxanthin is significantly inferior to algal-based astaxanthin as an antioxidant and may not be suitable as a human nutraceutical supplement. Forum Nutr 12(4):145–152

Chen M-H, Wang T-J, Chen L-J, Jiang M-Y, Wang Y-J, Tseng G-F, Chen J-R (2021) The effects of astaxanthin treatment on a rat model of Alzheimer's disease. Brain Res Bull 172:151–163

Chettri D, Verma AK, Verma AK (2022) Microbes and their application in the food and agriculture industry. In: Industrial microbiology and biotechnology. Springer, Cham, pp 625–642

Das P, Khan S, Chaudhary AK, AbdulQuadir M, Thaher MI, Al-Jabri H (2019) Potential applications of algae-based bio-fertilizer. In: Biofertilizers for sustainable agriculture and environment. Springer, Cham, pp 41–65

Dineshbabu G, Goswami G, Kumar R, Sinha A, Das D (2019) Microalgae–nutritious, sustainable aqua-and animal feed source. J Funct Foods 62:103545

Gouveia L, Raymundo A, Batista AP, Sousa I, Empis J (2006) Chlorella vulgaris and Haematococcus pluvialis biomass as colouring and antioxidant in food emulsions. Eur Food Res Technol 222:362–367

Haematococcus pluvialis Market by Size, Share, Forecasts, & Trends Analysis (n.d.). https://www.meticulousresearch.com/product/haematococcus-pluvialis-market-5142. Accessed 2 Dec 2022

Jafari Z, Bigham A, Sadeghi S, Dehdashti SM, Rabiee N, Abedivash A, Bagherzadeh M, Nasseri B, Karimi-Maleh H, Sharifi E (2021) Nanotechnology-abetted astaxanthin formulations in multimodel therapeutic and biomedical applications. J Med Chem 65(1):2–36

Jannel S, Caro Y, Bermudes M, Petit T (2020) Novel insights into the biotechnological production of Haematococcus pluvialis-derived astaxanthin: advances and key challenges to allow its industrial use as novel food ingredient. J Mar Sci Eng 8(10):789

Kanwugu ON, Glukhareva TV, Danilova IG, Kovaleva EG (2022) Natural antioxidants in diabetes treatment and management: prospects of astaxanthin. Crit Rev Food Sci Nutr 62(18):5005–5028

Leinemann F, Mabilia V (2019) European Union Legislation and policies relevant for algae. In: Grand challenges in algae biotechnology. Springer, New York, pp 577–591

Li X, Qi Z, Zhao L, Yu Z (2016) Astaxanthin reduces type 2 diabetic-associated cognitive decline in rats via activation of PI3K/Akt and attenuation of oxidative stress. Mol Med Rep 13(1):973–979

Li Y, Zhang B, He Q, Liu R (2018) The effects of *Haematococcus pluvialis* astaxanthin on Streptozotocin-induced diabetes in rats. Int J Food Eng 4(2):157

Lin W-N, Kapupara K, Wen Y-T, Chen Y-H, Pan I-H, Tsai R-K (2020) *Haematococcus pluvialis*-derived astaxanthin is a potential neuroprotective agent against optic nerve ischemia. Mar Drugs 18(2):85

Lorenz RT, Cysewski GR (2000) Commercial potential for *Haematococcus* microalgae as a natural source of astaxanthin. Trends Biotechnol 18(4):160–167

MacDermid JC, Vincent JI, Gan BS, Grewal R (2012) A blinded placebo-controlled randomized trial on the use of astaxanthin as an adjunct to splinting in the treatment of carpal tunnel syndrome. Hand 7(1):1–9

Marinoa T, Iovinea A, Casellab P, Martinoc M, Chianesea S, Laroccac V, Musmarra D, Molino A (2020) From *Haematococcus pluvialis* microalgae a powerful antioxidant for cosmetic applications. Chem Eng 79:271–276

Mularczyk M, Michalak I, Marycz K (2020) Astaxanthin and other nutrients from *Haematococcus pluvialis*-multifunctional applications. Mar Drugs 18(9):459

Nagaraj S, Rajaram MG, Arulmurugan P, Baskaraboopathy A, Karuppasamy K, Jayappriyan KR, Sundararaj R, Rengasamy R (2012) Antiproliferative potential of astaxanthin-rich alga

Haematococcus pluvialis Flotow on human hepatic cancer (HepG2) cell line. Biomed Prev Nutr 2(3):149–153

Nishshanka GKSH, Liyanaarachchi VC, Nimarshana PHV, Ariyadasa TU, Chang J-S (2022) *Haematococcus pluvialis*: a potential feedstock for multiple-product biorefining. J Clean Prod 344:131103

Oslan SNH, Tan JS, Oslan SN, Matanjun P, Mokhtar RAM, Shapawi R, Huda N (2021) *Haematococcus pluvialis* as a potential source of astaxanthin with diverse applications in industrial sectors: current research and future directions. Molecules 26(21):6470

Palozza P, Torelli C, Boninsegna A, Simone R, Catalano A, Mele MC, Picci N (2009) Growth-inhibitory effects of the astaxanthin-rich alga *Haematococcus pluvialis* in human colon cancer cells. Cancer Lett 283(1):108–117

Ranga R, Sarada AR, Baskaran V, Ravishankar GA (2009) Identification of carotenoids from green alga *Haematococcus pluvialis* by HPLC and LC-MS (APCI) and their antioxidant properties. J Microbiol Biotechnol 19(11):1333–1341

Rao AR, Sindhuja HN, Dharmesh SM, Sankar KU, Sarada R, Ravishankar GA (2013) Effective inhibition of skin cancer, tyrosinase, and antioxidative properties by astaxanthin and astaxanthin esters from the green alga *Haematococcus pluvialis*. J Agric Food Chem 61(16):3842–3851

Rodríguez-Meizoso I, Jaime L, Santoyo S, Señoráns FJ, Cifuentes A, Ibáñez E (2010) Subcritical water extraction and characterization of bioactive compounds from *Haematococcus pluvialis* microalga. J Pharm Biomed Anal 51(2):456–463

Saadaoui I, Rasheed R, Aguilar A, Cherif M, Al Jabri H, Sayadi S, Manning SR (2021) Microalgal-based feed: promising alternative feedstocks for livestock and poultry production. J Anim Sci Biotechnol 12(1):1–15

Sani I, Kurnia YW, Sinthya HC, Anthony R, Situmorang EC, Utomo C, Liwang T (2022) Exploring the potency of microalgae-based biofertilizer and its impact on oil palm seedlings growth. AGRIVITA J Agric Sci 44(1):139

Sekikawa T, Kizawa Y, Li Y, Takara T (2020) Cognitive function improvement with astaxanthin and tocotrienol intake: a randomized, double-blind, placebo-controlled study. J Clin Biochem Nutr 67:19–116

Shah MMR (2019) Astaxanthin production by microalgae *Haematococcus pluvialis* through wastewater treatment: waste to resource. In: Application of microalgae in wastewater treatment. Springer, New York, pp 17–39

Shah MMR, Liang Y, Cheng JJ, Daroch M (2016) Astaxanthin-producing green microalga *Haematococcus pluvialis*: from single cell to high value commercial products. Front Plant Sci 7:531

Siahaan EA, Pangestuti R, Pratama IS, Putra Y, Kim S-K (2021) Beneficial effects of astaxanthin in cosmeceuticals with focus on emerging market trends. In: Global perspectives on astaxanthin. Elsevier, Amsterdam, pp 557–568

Sonawane H, Shelke D, Chambhare M, Dixit N, Math S, Sen S, Borah SN, Islam NF, Joshi SJ, Yousaf B (2022) Fungi-derived agriculturally important nanoparticles and their application in crop stress management–prospects and environmental risks. Environ Res 212:113543

To QM, Tran ND, Vo TTT, Huynh TT, Tran DQ, Ta TNA, Lai BD, Nguyen DH, Le LT (2021) Determining the ability of astaxanthin from *Haematococcus pluvialis* on the protection of skin in the mouse model. J Appl Biol Biotechnol 9(4):8–2

Tri DT, Dung TTP, Uk DVQ, Mind NH, Cong NT, Nguyen Umbrella B (2021) Research using Haematoccus pluvialis powder in biofilm has been astaxanthin purpose to make additional eating for Japan koi fish. Sci Mag 18(9):1660

Tripathi U, Sarada R, Rao SR, Ravishankar GA (1999) Production of astaxanthin in *Haematococcus pluvialis* cultured in various media. Bioresour Technol 68(2):197–199

Valente LM, Cabrita AR, Maia MR, Valente IM, Engrola S, Fonseca AJ, Ribeiro DM, Lordelo M, Martins CF, e Cunha, L. F. (2021) Microalgae as feed ingredients for livestock production and aquaculture. In: Microalgae. Elsevier, Amsterdam, pp 239–312

Vidhyavathi R, Venkatachalam L, Sarada R, Ravishankar GA (2008) Regulation of carotenoid biosynthetic genes expression and carotenoid accumulation in the green alga *Haematococcus pluvialis* under nutrient stress conditions. J Exp Bot 59(6):1409–1418

Yamashita E (2013) Astaxanthin as a medical food. Funct Foods Health Dis 3(7):254–258

Yu BS, Lee SY, Sim SJ (2022) Effective contamination control strategies facilitating axenic cultivation of Haematococcus pluvialis: risks and challenges. Bioresour Technol 344:126289

Zakaria NNA, Zamzurie NA, Harith ZT (2021) Evaluation of sunscreen cream incorporated with astaxanthin from Haematococcus pluvialis in different storage conditions. IOP Conf Ser: Earth Environ Sci 756(1):012078. IOP Publishing

Zhang C, Chen X, Too H-P (2020) Microbial astaxanthin biosynthesis: recent achievements, challenges, and commercialization outlook. Appl Microbiol Biotechnol 104:5725–5737

Chapter 17
Biotechnological Applications of *Haematococcus*: Future Perspectives

Çağla Yarkent and Suphi Ş. Öncel

17.1 Introduction

Microalgae naturally synthesize primer and seconder metabolites which have different biological properties (Gürlek et al. 2020a). Their polysaccharides and lipids (hydrocarbons, polyunsaturated fatty acids, etc.) are evaluated in pharmaceutical, nutraceutical, and cosmetic applications (Yarkent et al. 2020). Proteins are considered as huge family consisting of enzymes, amino acids, and polypeptides and they have been gained attention by pharmaceutical and nutraceutical industries. Moreover, their carotenoids/chlorophylls have an important place in pharmaceutical applications (Gürlek et al. 2020b). Thus, microalgae are regarded as natural and sustainable sources of bioactive compounds for the production of vaccines, antibiotics, agrochemicals, and cosmetics (Mularczyk et al. 2020). As valuable natural sources, microalgae present important advantages compared to terrestrial plants regards to their higher photosynthetic efficiency, and growth rate compared (Saini and Keum 2019). They can be produced on marginal or unproductive land by using less water and nutrients; therefore, they do not compete with conventional crops (Di Sanzo et al. 2018). In addition, they can live in various waters from salty to brackish and even waste waters which gave them the ability to be cultivated with alternative sources resulting in economic advantage (Villaro et al. 2021).

Among different microalgae species, *Haematococcus* has gained attention due to its biochemical composition and the important effects of its metabolites. The interest in this context can be seen from the evaluation on Web of Science with the main keyword as "*Haematococcus*." The results showed that there is a remarkable increase in the investigation of *Haematococcus* culture over the last 5 years (Fig. 17.1). In the light of the research studies on Web of Science, the trend in

Ç. Yarkent · S. Ş. Öncel (✉)
Department of Bioengineering, Faculty of Engineering, University of Ege, Izmir, Turkey
e-mail: suphi.oncel@ege.edu.tr

R. Raja et al. (eds.), *Haematococcus*, https://doi.org/10.1007/978-981-99-2901-6_17

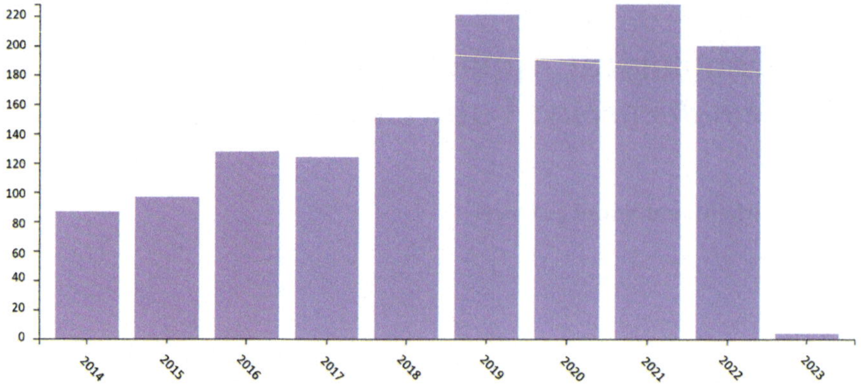

Fig. 17.1 The number of research studies on microalga *Haematococcus* over the years (Web of Science, December 2022)

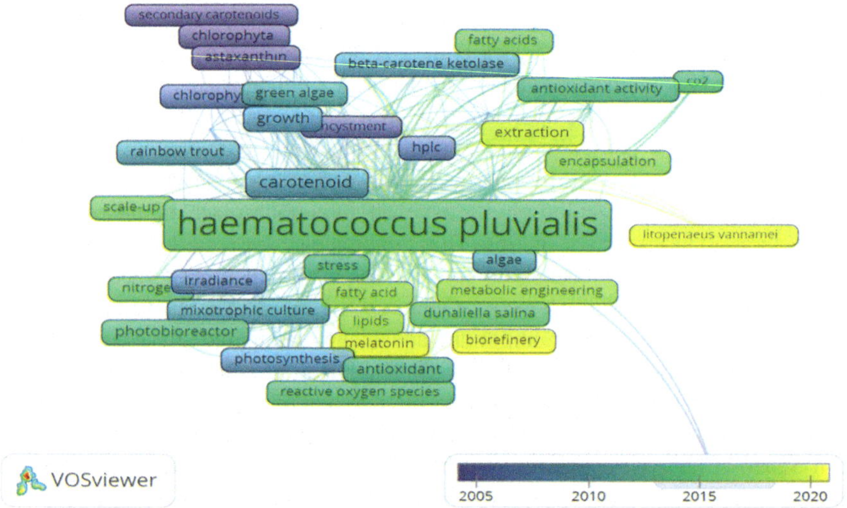

Fig. 17.2 The trend in the research studies on microalga *Haematococcus* over the years

Haematococcus was bibliometrically analyzed by VosViewer Software (Fig. 17.2). Considering a more focused analysis limited to recent years starting it 2005, the results showed that astaxanthin has still been the focus of attention, and thus this situation has led to a strong increase in the interest in *Haematococcus* since 2010. Noteworthy, the trend in evaluating the *Haematococcus* culture has focused on applications through biorefinery concept.

The purpose of this chapter is to emphasize the importance of *Haematococcus* in biotechnological applications. In this regard, the nature of *Haematococcus* cells and the cultivation strategies for *Haematococcus* were explained. Each process step followed from the cultivation of *Haematococcus* cells to obtain natural astaxanthin was examined. The impact of astaxanthin in different biological applications was emphasized and the mechanisms behind the effect of astaxanthin were highlighted. As a result, comprehensive recommendations were presented for future studies on *Haematococcus* cultivation, commercial astaxanthin production, utilization of microalgal astaxanthin as an end-product, and microalgal astaxanthin as a diet supplementary.

17.2 Astaxanthin as a Powerful Active Compound

Astaxanthin is a liposoluble keto-carotenoid (Brown et al. 2021; Nasri et al. 2021). Among the member of huge carotenoid family, this compound shows extraordinary antioxidant activity (approximately ten times higher) compared to other carotenoids (canthaxanthin, lutein, zeaxanthin, β-carotene) (Álvarez et al. 2020; Mularczyk et al. 2020; Ren et al. 2021). Also, this carotenoid is more effective than α-tocopherol (100 fold), vitamin E (550 fold), vitamin C (6000 fold), and coenzyme Q (800 fold) (Krestinin et al. 2020; Kumar et al. 2022; Marino et al. 2020; Yamashita 2021). In the literature, the photostability of three carotenoids in human dermal fibroblasts demonstrated as astaxanthin > cantaxanthin ≫ β-carotene. Only astaxanthin efficiently protects the skin from ultraviolet-A (UVA) radiation, while the others induce defense mechanisms (Yamashita 2021). Besides that, astaxanthin release remarkable antioxidant impact without showing any pro-oxidative features (Mularczyk et al. 2020; Nishida et al. 2022; Saini and Keum 2019). Furthermore, this compound has important biological functions as anti-inflammatory, and immunostimulator effects (Chen et al. 2021; Li et al. 2020; Villaro et al. 2021). Therefore, it has therapeutic impacts in the treatment of many ailments such as cancer, cardiologic, diabetes, eye, lung, metabolic, muscle, neural, skin diseases (Chen et al. 2020; Hasunuma et al. 2019; Kohandel et al. 2021; Yamashita 2021). Today, astaxanthin has a wide-range uses in pharmaceutical, nutraceutical, cosmetic, and food applications (Nasri et al. 2021; Sztretye et al. 2020; Zhu et al. 2022).

The market size of astaxanthin is reached up to 240 million $ in 2021 and market price is higher than 2000$/kg (Ren et al. 2021; Stachowiak and Szulc 2021). Additionally, it is estimated that it will raise to 3400 million $ by 2027 (Stachowiak and Szulc 2021). The global market demands for natural astaxanthin was 385 million $ in 2018 and now it is reached up to 814 million $ in 2022 (Álvarez et al. 2020; Tramontin et al. 2019). It is noteworthy that natural astaxanthin production cost is high and natural sources contain low astaxanthin content in their biomass composition. Therefore, this situation has led to a shift of attention to synthetic astaxanthin (Zhu et al. 2022). Today, the major part of the market belongs to chemically synthetized astaxanthin (>95%) (Wang et al. 2022). On the other hand, the astaxanthin obtained from *H. pluvialis* is only <1% of the global market (Mularczyk et al. 2020). Considering the production cost, the algal astaxanthin cost >6000€/kg, while the synthetic one could be obtained as 600€/kg (Marino et al. 2020). Recently, a study revealed that the cost of astaxanthin using *H. pluvialis* could be lowered by 610€/kg. This evidence encouraged the belief that microalgal processes could be a potential candidate to compete with synthetic astaxanthin (Villaro et al. 2021). In a different study, the astaxanthin production was carried out in two different platforms. Firstly, green *H. pluvialis* cells were cultivated in a photobioreactor; then red cells were incubated in a raceway pond. At the end, the production cost of astaxanthin from was estimated as 718$/kg. Additionally, the effect of labor cost was examined as it has an important place in the production cost. The results demonstrated that the cost of astaxanthin production can be raised 4.5 times higher and reached up from 120$/kg to 600$/kg if only photobioreactors are used (Saini and Keum 2019).

Recently, consumers are more aware of concerns on energy and water cycles in mother nature. Therefore, they tend to prefer more biological and eco-friendly compositions in their products instead of chemical ingredients (Marino et al. 2020; Yamashita 2021). Thus, the search for natural ingredient-included products is a top trend in the industry (Kumar et al. 2022). Although synthetic astaxanthin is considered as favorable alternative due to its high availability, low cost for production, and long lifespan, petrochemical reactants have an important place in process during its synthesis. These reactants cause negative environmental effects, and therefore synthetic astaxanthin production is considered not sustainable and renewable (Villaro et al. 2021). On the other hand, microalgae utilize carbon and release oxygen; therefore they minimize carbon footprint and alleviate greenhouse effects (Kumar et al. 2022). In their growth phase, these microorganisms could capture 1.8–2 kg CO_2 per kilogram of dry biomass (Molino et al. 2018). Moreover, they can reduce CO_2 emission while they produce important biocompounds. Therefore, microalgae can be considered as beneficial sources of natural products through a sustainable approach (Di Sanzo et al. 2018). There are three different enantiomers of astaxanthin due to configuration of asymmetric carbon atoms; 3S,3′S-, 3R,3′S-, and 3R,3′R- (Kumar et al. 2022).

The chemical structure of natural astaxanthin includes 3S,3′S-isomer, while the chemically synthesized alternative is an esterified and in the form of mixture of 3S,3′S-, 3R,3′S-, and 3R,3′R- (1:2:1) (Tramontin et al. 2019; Zhu et al. 2022). Compared with natural astaxanthin, the synthetic compound contains impurities, therefore it is considered inefficient and not safe for biological applications (Hasunuma et al. 2019; Ren et al. 2021). Moreover, natural astaxanthin is more desired for nutraceutical, pharmaceutical, and cosmetic applications due to health concerns about chemically synthesized astaxanthin (Saini and Keum 2019; Zhu et al. 2022). In comparison of the features of natural and synthetic astaxanthin, natural astaxanthin affects 50 times higher in quenching of singlet oxygen and 20 times higher in neutralizing free radicals (Kumar et al. 2022). The pigmentation efficiency of natural astaxanthin is greater than synthetic alternatives in marine animals (shrimps) (Villaro et al. 2021). In more detail, 3S,3′S was reported as the highest antioxidant form (Yamashita 2021). Unfortunately, it exists in a significantly low concentration in the synthetic astaxanthin complex (Yao et al. 2020).

Table 17.1 A list of natural astaxanthin sources (Hongo et al. 2020; Kumar et al. 2022; Nishida et al. 2022; Stachowiak and Szulc 2021; Tang et al. 2022; Villaro et al. 2021; Yamashita 2021; Zhu et al. 2022)

Natural astaxanthin sources	
Marine organisms	• Crab • Crayfish • Crustaceans • Lobster • Red snapper • Salmon • Sea bream • Shrimp • Trout
Plants	• *Adonis aestivalis* • *Adonis annua* • *Solanum lycopersicum* • *Daucus carota*
Bacteria species	• *Agrobacterium aurantiacum* • *Alcaligens* sp. • *Brevibacterium* sp. • *Mycobacterium lacticola* • *Paracoccus carotinifaciens* • *Paracoccus* sp.
Yeast species	• *Phaffia rhodozyma* (anamorph) • *Xanthophyllomyces dendrorhous* (teleomorph)
Microalgae species	• *Chlorella zofingiensis* • *Chlorococcus* sp. • *Chromochloris zofingiensis* • *Haematococcus pluvialis*

Astaxanthin is naturally found in some marine organisms, plants, and various microorganism as given in Table 17.1 (Hongo et al. 2020; Kumar et al. 2022; Nishida et al. 2022; Stachowiak and Szulc 2021; Tang et al. 2022; Villaro et al. 2021; Yamashita 2021; Zhu et al. 2022). As natural astaxanthin sources, *Adonis* species have low flower productivity; therefore, they are not cost-effective sources for pigment production. Also, their astaxanthin content was mentioned as 1% of dry weight (Stachowiak and Szulc 2021). On the other hand, plant-based productions are generally hampered due to geographical and seasonal variations (Naz et al. 2021). Among mentioned microorganisms, *Xanthophyllomyces dendrorhous* and *H. pluvialis* are main microbial platforms to produce astaxanthin commercially (Tang et al. 2022; Villaro et al. 2021). The total cost for obtaining astaxanthin from *X. dendrorhous* was demonstrated as 2500$/kg, while it was higher for *H. pluvialis* as 7000$/kg. However, the content of astaxanthin in red *X. dendrorhous* yeast is 0.01–0.03% of dry weight (Stachowiak and Szulc 2021). It was demonstrated that *X. dendrorhous* can accumulate lower amounts of astaxanthin while the astaxanthin composition of *H. pluvialis* can reach up to 5% of dry biomass (Villaro et al. 2021). From another perspective, the natural astaxanthin synthesized by bacteria and yeast are unesterified isomers as 3S,3′S and 3R,3′R, respectively, while *H. pluvialis* synthesize esterified astaxanthin (Kumar et al. 2022). In the literature, it was underlined that esterified astaxanthin is more resistant to temperature changes and photo-oxidation compared to free astaxanthin (Stachowiak and Szulc 2021). Therefore, the antioxidant activity of natural astaxanthin derived from *H. pluvialis* is higher than synthetic alternatives (55-fold) or yeast derived astaxanthin (Mussagy et al. 2022; Yamashita 2021). Thus, astaxanthin from *P. rhodozyma* is preferred as a colorant agent in lots of animal, and fish feeds, while astaxanthin obtained from *H. pluvialis* is considered for human consumption (Yamashita 2021).

Considering microalgae species, *Chlorella zofingiensis* has some advantages compared to *H. pluvialis*. It can grow faster and produce higher amounts of biomass. The highest astaxanthin accumulation (1 mg/g) is generally reached during heterotrophically cultivations. The most important advantage of this microalga is that astaxanthin can be accumulated in one step process (Villaro et al. 2021). In one step production, microalgae are grown within essential nutrients under favorable environment. As the culture grows over time, nutrient depletion and pH changes occur, thus this situation induces stress on the cells. Therefore, the cells trigger the pathway leading to astaxanthin biosynthesis to protect them. However, this spontaneous process can induce negative effects on algae growth. In comparison to *H. pluvialis*, the total carotenoid content of *C. zofingiensis* is 76.3% of diester and 18% of monoester, while this composition is different in *H. pluvialis* is reported as 35.5% of diester and 60.9% of monoester. The chemical structure of a compound attributes to its characteristics like polarity and antioxidant activity. In literature, it was emphasized that the diester form of astaxanthin is twice more powerful than free astaxanthin and it is also 1.6 times more effective than the monoester form. However, astaxanthin derived from *H. pluvialis* is believed to be 95% pure and this value is greater than other microalgal sources (Kumar et al. 2022). Therefore, *H. pluvialis* is preferred as the main microalga used in commercial astaxanthin production (Sztretye et al. 2020; Tang et al. 2022; Wang et al. 2022; Zhu et al. 2022).

17.3 Molecular Mechanisms Behind the Effects of Astaxanthin

The human body naturally expresses and activates endogenous antioxidant enzymes (Mason et al. 2020; Villaro et al. 2021). However, these mechanisms are insufficient to balance the disrupted cellular redox state as a result of genetic disorders or extreme conditions especially under harmful external effects or during endurance exercises (Li et al. 2020; Yamashita 2021). The increase in reactive oxygen species and free radicals evokes oxidative modifications to deoxyribonucleic acid (DNA), lipids, and proteins (Chou et al. 2020; Kohandel et al. 2021). Therefore, this situation accelerates injuries and fatigues in the human body (Sztretye et al. 2020; Villaro et al. 2021). This imbalance situation can be avoided through free radical scavenging mechanisms and/or via regulation of related enzymes (Chen et al. 2020; Mason et al. 2020). Therefore, antioxidant supplements are considered as an alternative way to moderate negative effects of reactive oxygen species (ROS) like lipid peroxidation (Hongo et al. 2020; Krestinin et al. 2020; Mascia et al. 2017). In this case, evaluating the biomarkers of antioxidant activity and/or oxidative stress are good option to mention the efficiency of an antioxidant (Bhattarai et al. 2021; Kato et al. 2020).

Astaxanthin has been mentioned as the most powerful carotenoid among the others. The main reason for this difference is related to the chemical structure of this compound (Yamashita 2021). It has an oxygenated hydrocarbon chain due to the presence of two oxygenated groups on each ring (Kumar et al. 2022; Saini and Keum 2019). In fact, astaxanthin exists in lipid droplets, cell membranes, or bound to protein in biological cell matrix (Nishida et al. 2022). This chemical property allows it to activate and suppress several molecules and pathways (Huang et al. 2021; Lu et al. 2021; Naz et al. 2021; Wang et al. 2022). In many studies, nuclear factor (erythroid-derived 2) related factor 2 (Nrf2) was emphasized as a key marker of endogenous and exogenous oxidative stress (Kohandel et al. 2021). The activity of Nrf2 is related to several pathways as Kelch-like ECH-associated protein 1 (Keap1)/Nrf2, epigenetics, and PI3K (phosphatidylinositol 3-kinase)/Akt (Protein Kinase B) (Lee et al. 2020; Nishida et al. 2022). (Keap1)/Nrf2 is responsible to stimulate antioxidant enzyme transcription such as catalase, heme oxygenase-1 (HO-1), peroxidase, and superoxide dismutase (Chen et al. 2020; Kumar et al. 2022; Wang et al. 2022).

Epigenetics cause demethylation of the CpG islands that regulate Nrf2 induction (Kohandel et al. 2021). PI3K/Akt has key role on pro-inflammatory and anti-inflammatory mechanisms that in both Nfr2-dependent and Nfr2-independent pathways (Bhattarai et al. 2021; Nishida et al. 2022). In the literature, it has been demonstrated that astaxanthin acts as a potent activator of Nrf2. For example, astaxanthin exerted cellular protection against oxidative stress in human hepatocellular carcinoma (HepG2)-luciferase cell line by inducing Nrf2-ARE (antioxidant response element) pathway (Kohandel et al. 2021). In addition, astaxanthin induces HO-1 expression by modulating Nfr2 to protect skin dermal proteins against the effect of excess matrix metalloproteinases production (Kumar et al. 2022; Saini and Keum 2019; Sun et al. 2021). Also, it was demonstrated that astaxanthin is capable of improving the capacity of rodent skeletal muscle, and increasing the endurance performance (Nishida et al. 2022). This compound represents the ability to maintain the tissue homeostasis in the retina of C57BL/6 mice with acute glaucoma. These data were obtained by following the stress-induced expression of Nrf2 (Li et al. 2020). Besides, the induction of Nfr2 suppresses nuclear factor kappa B (NF-kB) inflammation activity, therefore astaxanthin also induces anti-inflammatory effects in addition to its antioxidant impact (Chou et al. 2020; Kwak et al. 2021; Shokrimashhadi et al. 2021). Moreover, it was reported that astaxanthin enhanced exercise-induced fatty acid metabolism (Liu et al. 2021; Nishida et al. 2022; Nishida et al. 2020). Moreover, it represents this feature by the activation of 5-adenosine monophosphate-activated protein kinase (AMPK) pathway, independent of its antioxidant activity (Nishida et al. 2022). Noteworthy, the effect of astaxanthin extends beyond neutralizing the produced ROS. This compound represents a remarkable transcriptional impact, while regulating cellular redox state at the same time (García et al. 2020; Kato et al. 2020).

Mitochondria are considered as a major source of cellular ROS (Sztretye et al. 2020). These organelles play key roles in energy metabolism, maintaining a redox state, and regulating apoptosis (Nishida et al. 2020). Therefore, it is crucial to promote mitochondrial biogenesis to increase mitochondrial oxidative phosphorylation (Sun et al. 2021). In the case of mitochondria becoming dysfunctional that causes oxidative stress in the cells (Hongo et al. 2020; Krestinin et al. 2020). Today, lots of in vivo and in vitro studies have proven that astaxanthin has positive impact on mitochondrial integrity, especially under pathological conditions (Chen et al. 2020; García et al. 2020; Kwak et al. 2021; Sztretye et al. 2020). In a recent study, C2C12 mouse myoblast cell line was heat induced to occur mitochondrial oxidative stress, dysfunction, and apoptosis. The results revealed that astaxanthin decreased mitochondrial degradation. In a different study, this compound protected mitochondrial homeostasis in the hypothalamus of heat-induced male C57BL/6J mice (Chen et al. 2021). In the literature, it was demonstrated that astaxanthin treatment also increases mitochondrial DNA copy number and protein expression in male insulin-induced C57BL/6J mice (Bhattarai et al. 2021; Nishida et al. 2020). Additionally, it has shown neuroprotective effects in animal models (Chen et al. 2021; Huang et al. 2021; Li et al. 2020; Villaro et al. 2021). For example, a neuronal cell line SH-SY5Y was induced with *N*-Methyl-D-aspartate (NMDA) to stimulate excitotoxicity. As a result, it was demonstrated that astaxanthin reduced basal mitochondrial ROS production and alleviated the mitochondrial dysfunction in NMDA treated SH-SY5Y cells (García et al. 2020). In another study, isoproterenol was injected into Wistar rats to inducing oxidative stress to disrupt mitochondrial permeability transition pore. Then, the heart mitochondria were isolated from the animal and examined to understand the effects of astaxanthin. The results demonstrated that astaxanthin raised mitochondrial respiration (Krestinin et al. 2020). In order to evaluate the effect of astaxanthin on the soleus muscle, a tail-suspension model mouse was fed with astaxanthin as a diet. The results revealed that this carotenoid induced the downregulation of the mitochondrial respiratory chain complexes I and III and consequently improved its respiration capacity (Sun et al. 2021).

17.4 The Importance of Natural Astaxanthin Derived from *Haematococcus*

The annual biomass yield of *H. pluvialis* reaches up to 300 tons (Ren et al. 2021). The market value of astaxanthin derived from *H. pluvialis* is estimated at 6000–7150$/kg. Besides, the global astaxanthin market was 1.0 billion $ in 2019 (Kim et al. 2022). It is expected that this compound will have annual growth rate of 16.2% and its market value will reach 3.4 billion $ by 2027 (Kim et al. 2022; Kumar et al. 2022). The natural astaxanthin is priced at 5000$/kg (Kumar et al. 2022). Most of the natural astaxanthin that is currently on sale is produced in Asia, whereas there are promising facilitators in The United States (US), New Zealand, Chile, Sweden, Europe, and Israel. The companies that focus on natural astaxanthin production are divided into several countries; Greece, Netherlands, China (Villaro et al. 2021). Besides, the microalgal astaxanthin has an important place in China, the US, Japan, and Europe (Kumar et al. 2022; Wang et al. 2022; Yamashita 2021). More detailed, astaxanthin production from *H. pluvialis* has an important place in US, Japan, India (Stachowiak and Szulc 2021). The brands which have products containing astaxanthin produced by *H. pluvialis* can be mentioned as AstraPure, AstraReal, AstaFactor, AstaEquus, AstaZanthin, BioAstin, NatAxtin, NovaAsta, and Natural Astaxanthin (Kumar et al. 2022; Miyakawa 2021; Mularczyk et al. 2020; Mussagy et al. 2022; Stachowiak and Szulc 2021; Villaro et al. 2021). Among all mentioned companies, Beijing Gingko Group (China) is a leader in the global market of astaxanthin derived from *H. pluvialis* (Kumar et al. 2022).

Human body has not the ability to synthesize astaxanthin; therefore it is suggested to intake astaxanthin as dietary (Kato et al. 2020; Mascia et al. 2017). Astaxanthin-derived from microalgae has been approved by the United States Food and Drug Administration (FDA), the European Union, and the European Commission for human consumption (Marino et al. 2020; Ren et al. 2021; Villaro et al. 2021; Yao et al. 2020). Astaxanthin can be combined with antioxidant products to boost the beneficial effects of supplementation (Brown et al. 2021; Kwak et al. 2021). Today, natural astaxanthin is already involved in many commercial food supplements (1–8 mg astaxanthin) (Villaro et al. 2021). Until now, no notable adverse effects (like pro-oxidants) of astaxanthin have been mentioned (Álvarez et al. 2020; Kohandel et al. 2021; Liu et al. 2021). Recently, the ergogenic benefit of astaxanthin obtained from *H. pluvialis* was evaluated with 12 recreationally trained male cyclists during a 40 km cycling time trial for 7 days. The results proved that this natural compound enhanced fat oxidation (Brown et al. 2021).

The skin is constantly exposed to UV radiation and that causes an increase in ailments in the skin (Zakaria et al. 2021). In the literature, several studies demonstrated the beneficial effects of purified astaxanthin from *H. pluvialis* in vitro and in vivo cosmetic applications (Kumar et al. 2022; Radice et al. 2021). In vitro, human keratinocyte cell line (HaCaT) and human normal fibroblast cell line (Hs68) were exposed to UVB radiation (30 mJ/cm^2) for 90 s. Then, the cells were treated with astaxanthin for 24 h. As a result, it was reported that DNA damage was reduced and collagen secretion was promoted. In vivo, Wistar rats were exposed to UVB (300 mJ/cm^2) for 15 min, then astaxanthin treated inflammatory symptoms and apoptotic injuries. The results supported that this carotenoid repaired UVB-triggered skin injuries and aging by scavenging reactive oxygen species. Thus, astaxanthin enhanced collagen synthesis in vitro and in vivo (Chou et al. 2020).

On the other hand, prolonged UV exposure induces skin deterioration and causes skin aging (Zhou et al. 2021). In a small clinical study, astaxanthin extracted from *H. pluvialis* was examined to understand its impact in cosmetic applications. A cream with astaxanthin (0.7 mg/g) was topically supplemented on the skin and a visible wrinkle reduction was observed after 3 weeks. Moreover, 2 mg/day of astaxanthin was used as dietary supplement for 4 weeks. Skin characteristics were reported as increased moisture levels and reduced fine wrinkles. In a different study, it was revealed that an oral supplementation of astaxanthin (4 mg/day) improved the moisture content, the fine lines, and elasticity of the skin after 6 weeks. A combination of oral supplementation (6 mg/day) and topical treatment (2 mL, 78.9 μM solution) of algal astaxanthin were studied and the results appeared as the improvements in age spot size, elasticity, moisture content, skin texture, and skin wrinkles (Yamashita 2021). Skin properties vary according to changes in humidity, temperature, and UV radiation during the seasonal periods. Therefore, further studies are needed to evaluate the dose-response effect of oral astaxanthin intake and its effects on the skin based on environmental conditions (Zhou et al. 2021).

A sunscreen cream aims to absorb/reflect UV radiations to protect the skin. Each cosmetic product has a different complex matrix and its own physicochemical features (pH, viscosity, etc.). In the case of a natural component being applied into a cosmetic formulation, the new physiochemical properties should be investigated to understand the usability of this product. In a recent study, astaxanthin obtained from *H. pluvialis* was added into the sunscreen cream to develop astaxanthin sunscreen cream. The storage conditions (4C, 25C, 40C) and the percentages of astaxanthin in the formulation (0%, 0.5%, 1%) were evaluated to underline the effects of this compound on the product for 14 days. It was demonstrated that astaxanthin increased the sun protective factor (SPF) value at all storage conditions and the highest SPF was observed at 25 °C. In addition, this carotenoid decreased the pH value at 4 °C and 40 °C, while there was no change at 25 °C. Moreover, it was mentioned that astaxanthin affected the viscosity only at 40 °C. In summary, it was emphasized that astaxanthin can increase the efficiency of a sunscreen cream without changing its pH, and viscosity values (Zakaria et al. 2021).

The accumulated evidence has indicated that natural astaxanthin has antimicrobial capacity against pathogens (Kumar et al. 2022; Lee et al. 2020). In a study, in which its antimicrobial potential was demonstrated with inhibition effects on *Escherichia coli*, *Helicobacter pylori*, *Proteus mirabilis*, *Aliivibrio fischeri*, *Enterobacter aerogenes*, and *Listeria monocytogenes*. Besides, ethanolic extracts of astaxanthin-rich *H. pluvialis* promoted antiviral effect on herpes simplex virus type-1 in Vero cells (Villaro et al. 2021). Also, it was demonstrated that bioactive compounds can be evaluated as coating material to improve the efficiency of vaccines. For example, squalene is a microalgal compound and it is used as an adjuvant in COVID-19 vaccines (Yarkent and Oncel 2022). Therefore, astaxanthin can be considered a potential candidate as curative agents in viral treatments.

17.5 *Haematococcus* Cells

Haematococcus is a genus consist of freshwater unicellular microalgae that belongs to *Chlorophyta* phylum, *Chlorophyceae* class, *Volvocales* order, and *Haematococcoseae* family (Li et al. 2019; Luo et al. 2019; Radice et al. 2021). Among *Haematococcus* species, the most known is *Haematococcus pluvialis.* The life cycle of this microalga can be typically divided into two phases as green vegetative growth, and encystment stages during the photosynthetic cultivation (Kim et al. 2022). This period can be detailed as; (1) germination (~24 h), (2) green vegetative cells with two flagellate (32 cells), (3) green vegetative cells but non-motile and, (4) the beginning of astaxanthin accumulation, and (5) astaxanthin-enrich cysts (Mascia et al. 2017; Miyakawa 2021; Ren et al. 2021). The microalga culture requires different environmental conditions; thus the optimum conditions vary according to the demands of the cells at each stage (Yao et al. 2020).

The biomass composition of *H. pluvialis* alters according to the changes during its life cycle (Kim et al. 2022; Radice et al. 2021). Most *H. pluvialis* green cells represent a protein content of 29–45% in the green stage. Then, this biochemical composition decreases to 21–23% in the red stage. The carbohydrate content reaches a high concentration from 15–17% to 60–74% during the transition between the stages. The lipid content is stated as 20–25% in the green stage, 30–35% in the red stage(Mularczyk et al. 2020). *H. pluvialis* biomass contains polyunsaturated fatty acids (PUFAs) consist of different fractions as docosahexaenoic acid (DHA), eicosapentaenoic acid (EPA), arachidonic acid, palmitic acid, oleic acid, linoleic acid, and linolenic acid (Álvarez et al. 2020; Ren et al. 2021; Wang et al. 2022). Lutein is regarded as the main carotenoid (70–80%) in green cells. It is followed by β-carotene (16.70%), violaxanthin (12.5%), neoxanthin (8.3%), and chlorophyll (1.5–2%), respectively. The biochemical content of mature red cysts has been listed as: carbohydrates (36–40%), lipids (32–40%), proteins (17–25%), and carotenoids (2–5%) (Kim et al. 2022). The cyst cells characteristically contain ~82% astaxanthin of their total carotenoids (β-carotene, lutein, canthaxanthin, etc.) (Di Sanzo et al. 2018).

Haematococcus cells can be cultivated in different production platforms. Typically, these systems are divided into two groups: open ponds and closed photobioreactors (Fig. 17.3). They present different advantages and disadvantages (Saini and Keum 2019). As an example, in a recent study, *H. pluvialis* was cultivated in an open pond with depth: 12.0 cm; length: 76.0 m; surface area: 417.0 m^2, and width: 5.5 m. The results demonstrated that the highest biomass of 206 g/m^3 with photosynthetic efficiency of 4.4% was observed and the production rate was determined as 15.1 g/m^2/day (Ren et al. 2021). Although their construction cost is very low, the risk of contamination is a crucial point during the production period. Besides, the operation conditions should be decided to avoid turbulence or loss of water content. In addition, light fluctuation is still an outstanding barrier to present high efficient algal productions for open pond systems (Miyakawa 2021). On the other hand, closed bioreactors offer potency to produce *H. pluvialis* under

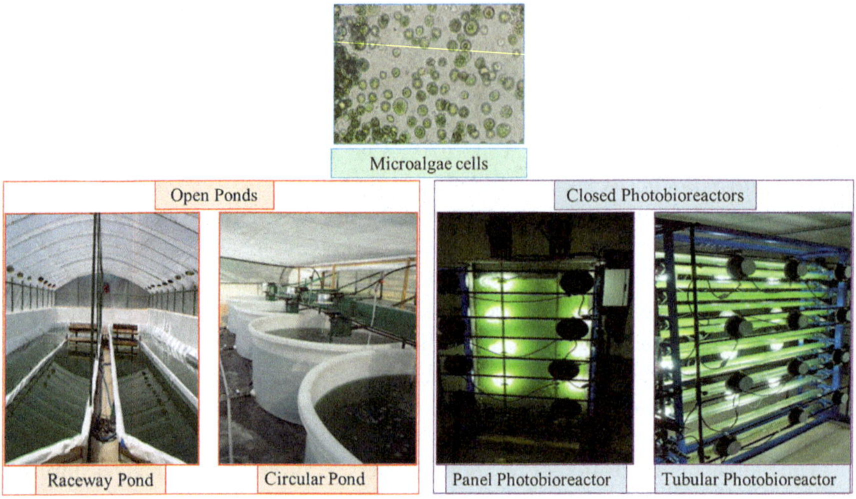

Fig. 17.3 Microalgae cultivation platforms for outdoor productions. (Courtesy of Authors)

controllable conditions (Hasunuma et al. 2019). At this point, low light efficiency and gas transfer limitations are considered as disadvantages that should be overcome (Wang et al. 2022). In one study, a 6 L bag-type bubble column photobioreactor was operated by applying adequate hydrodynamic mixing to avoid limitation effects of O_2 and CO_2 on *H. pluvialis* cultivation. A high biomass as 2.62 g/L and astaxanthin production as 78.37 mg/L were obtained at the end (Ren et al. 2021).

In the literature, it has been revealed that astaxanthin content in *H. pluvialis* can be reached up to 1.5–3% in open ponds, while this can be increased to >4% in a tubular photobioreactor (Miyakawa 2021). Comparing these two systems, the biomass productivity is stated as higher in the photobioreactor. However, keeping in mind the outdoor productions, the results vary based on the season changes and local climate features that affect light intensity and temperature during the production period (Villaro et al. 2021). In this regard, indoor systems are considered as a good option to minimize uncontrollable effects due to environment (Miyakawa 2021). On the other hand, because prolonged operation of intensive artificial light consumes a lot of energy, using sunlight is one and only economic strategy to achieve feasible energy management (Hasunuma et al. 2019; Kumar et al. 2022; Ren et al. 2021). Besides, the cost of using artificial light sources can be compensated with the price of natural astaxanthin (Miyakawa 2021).

In this context, it was revealed that *H. pluvialis* (even after astaxanthin extraction) presents notable biocomponents (lipids, carbohydrates, proteins, minerals, etc.) which can be considered as potent sources for biorefinery purposes (Djearamane et al. 2019; Kumar et al. 2022). Several studies have emphasized the potential of residual *H. pluvialis* biomass which can be used to produce food-grade oil, biodiesel, biohydrogen, ethanol, and methane (Kim et al. 2022; Ren et al. 2021). From another perspective, there is a strong research potential for the studies to increase the content of lipids and proteins. Additionally, simultaneously obtaining two or more metabolites from *H. pluvialis* at the same cultivation period is a good option to make the process economically profitable (Mularczyk et al. 2020). Hence, future studies may also focus on the production of these valuable metabolites. The production research can be shifted to the efficiency of a metabolite from *Haematococcus* by using different strategies like: (1) adding precursors; (2) operating stress conditions; (3) applying stressor chemicals; and (4) cultivation in wastewater or combined culture media (Saini and Keum 2019; Tramontin et al. 2019; Yamashita 2021).

17.6 *Haematococcus* Cells as Natural Astaxanthin Source

H. pluvialis is an important source for natural astaxanthin (3,3′-dihydroxy-β, β′--carotene-4,4′-dione) especially triggered when the cells are under stress conditions (Marino et al. 2020; Naz et al. 2021). Astaxanthin production is a process that consists of two consecutive stages (Hasunuma et al. 2019; Mascia et al. 2017). *H. pluvialis* cells exhibit altered morphological and chemical features in these stages (Luo et al. 2019; Ren et al. 2021). In the first stage, *H. pluvialis* is 8–20 μm in size and rapidly growth under favorable conditions (Radice et al. 2021; Yao et al. 2020). The morphological properties are seen as green round or oval motile cells with a distinct cell wall layer and two flagella (Miyakawa 2021). At this stage, the cells were demonstrated with chlorophyll content inside the cells (Luo et al. 2019). Then, the cells were transferred to a stress-inducing environment to trigger the activation of carotenogenesis (Mascia et al. 2017). This process causes a gradual change in the morphological properties of the cells. *Haematococcus* cells turn from the motile phase to the non-motile phase and they lose their flagella (Nasri et al. 2021). In encystment stage, the cell size of *H. pluvialis* is reached up to 20–50 μm and their color shift salters from green to reddish (Chou et al. 2020; Yao et al. 2020). In fact, it was demonstrated that the cell size could be reached up to 100 μm in some cases (Miyakawa 2021). The cells accumulate astaxanthin to protect themselves against the toxic effects of current conditions (Mascia et al. 2017). At this time, the microalga develop, the three-layered cell wall consists of algaenan biopolymers and cellulose-like polysaccharides (Nasri et al. 2021). This thick cell wall is typically in 1.8–2.2 μm, it protects the cell from enzymatic, chemical, and physical treatments (Yao et al. 2020).

Carotenoids are considered as indispensable components of photosystems due to their capacity for capturing light energy (Lu et al. 2021; Villaro et al. 2021). During the green phase, *H. pluvialis* contains more chlorophyll than carotenoids; on the contrary vice versa is valid at the red phase (Radice et al. 2021). In addition, lipid mechanism of the microalga alters to store neutral lipids during transition between these phases (Saini and Keum 2019). Therefore, the composition of algal cells differs according to the environmental conditions. However, considering the whole *H. pluvialis* cell, less than 1% of the total astaxanthin has bound to the photosystems, almost the entire astaxanthin is accumulated in the cytoplasm (Mascia et al. 2017). During the transition between the stages, *H. pluvialis* cells existing both motile and non-motile states which are defined as vegetative cells. Comparing the differences between these two cells, non-motile cells are more resistant against photo-oxidative stress, and begin to synthesize astaxanthin faster than motile cells. Indeed, it was revealed that most vegetative cells died under stress conditions that cause a remarkable decrease in overall astaxanthin productivity during two-stage production. Therefore, the selection of non-motile cells could enhance the stability of astaxanthin production. For non-motile cells, it was demonstrated that astaxanthin accumulation was correlated with the enlargement of the cell size. This information can be useful to separate the non-motile cells from the others (Li et al. 2019). In the literature, it was mentioned that there is a high correlation between biomass productivity and final astaxanthin concentration (Wang et al. 2022). In the case of microalga that accumulates a sufficient amount of biomass, it shows more adequate resistance against stress factors, and synthesizes more astaxanthin (Ren et al. 2021).

17.7 Astaxanthin Production from *Haematococcus* Cells

In each step of the entire production platform, the selection of techniques must be considered through a techno-economic approach (Mussagy et al. 2022). Considering the astaxanthin production from *H. pluvialis*, a techno-economic analysis was performed based on energy consumption of each operation. In order to obtain 0.5 kg/day on dry basis, the average cost of ~42.60€/day was needed to recover astaxanthin from algal biomass. The results revealed that the main part of the budget was spent for supercritical CO_2 extraction process (31.33€/day). Then, it was followed by drying process (7.43€/day) and the growth of microalgae (3.07€/day). The smallest parts of the cost were related to the nutrients (0.55€/day) and the mechanical pretreatment (0.24€/day) (Marino et al. 2020).

In the literature, it has been demonstrated that microalgae cultivation for obtaining astaxanthin is the part of 20–30% of the total production cost (Mularczyk et al. 2020). The total production cost depends on biomass production, yield of pigment production, metabolic activity of microalgae cells, isolation of pigment from cells, and the purification of pigment (Stachowiak and Szulc 2021). On the other hand, once the algal production is ended, harvesting and downstream processes should be followed to obtain astaxanthin from the biomass (Fig. 17.4). After the cultivation period, the astaxanthin-rich cells can be harvested by precipitation, centrifugation, filtration, etc. (Miyakawa 2021). This operation part typically accounts for 20–30% of the production cost (Kumar et al. 2022). The biomass can be spray dried, which is more cost-effective than freeze drying, and drum drying (Nasri et al. 2021). The main challenging issue is extraction of astaxanthin from cell biomass (Kim et al. 2022). For this purpose, the cells must be disrupted to increase astaxanthin bioavailability, then extraction process can be carried out by various methodologies (Di Sanzo et al. 2018; Villaro et al. 2021). Among different

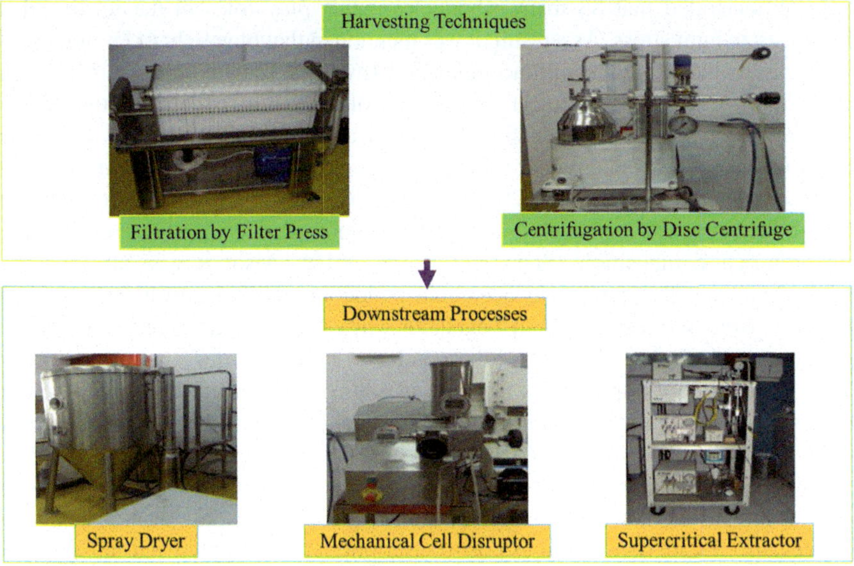

Fig. 17.4 Several harvesting and downstream strategies to obtain target compound from algal biomass. (Courtesy of Authors)

extraction techniques, the most suitable process should be selected with respect to the requirements for energy, cost, and its impact on the nature of the target product (Kim et al. 2022). At this point, temperature is considered as crucial parameter that has an effect on both extraction yield and degradation of the desired compound (Molino et al. 2018; Mularczyk et al. 2020). For example, supercritical CO_2 extraction process that combined with mechanical pretreatments is getting increasing attention to reach high quality products (Marino et al. 2020). However, the cost of downstream processes may account 20–90% of the total production cost (Kim et al. 2022).

17.8 Astaxanthin Accumulation in *Haematococcus* Cells

In microalgae cells, carotenoids have important biological functions in photosynthesis, photoprotection, coloration, and cell signaling (Saini and Keum 2019). In natural environments, microalgae have circadian rhythms; therefore a microalga culture reaches higher cell densities on the light-dark cycle (Ren et al. 2021). Besides, microalgae productions can be manipulated by inducing the culture under different stress conditions such as high illumination, high salinity, high temperature, nitrogen deficiency, and toxic compounds (Luo et al. 2019; Nasri et al. 2021; Wang et al. 2022). In the case of cells exposed to constant illumination as a stress factor, they accumulate more astaxanthin to survive (Chou et al. 2020). In the literature, it was demonstrated that 83 micro-ribonucleic acids play roles in the formation of responses to light stress. As a result, it was mentioned that blue light (425–500 nm) is more preferable to astaxanthin accumulation (by three-fold) compared to red light (640–700 nm) (Ren et al. 2021). In the light of this knowledge, recently, nuclear transformation vectors and genetic engineering of chloroplasts genomes have been investigated to make microalgal astaxanthin production competitive with synthetic astaxanthin (Tramontin et al. 2019). In microalgae productions, agro-industrial by-products can be considered as low-cost nutrients due to their carbon and nitrogen-rich composition (Saini and Keum 2019). Also, it was revealed that microalgae can also grow and accumulate biocompounds by utilizing nutrients/organics from wastes (fruit and vegetable wastes, dairy waste, and wastewater) (Villaro et al. 2021). Therefore, these microorganisms offer an opportunity to operate biorefinery applications through an eco-friendly perspective. As a related example, in one study, walnut shell extracts enhanced accumulation of astaxanthin in *H. pluvialis* culture by 77.57% (Ren et al. 2021).

In many studies, several chemicals have been mentioned with their effects on astaxanthin production (Villaro et al. 2021; Wang et al. 2022). For example, reactive oxygen species are considered as post-transcriptional factors involved in regulation of astaxanthin biosynthesis. Ethanol, ethylene, 1-aminocyclopro-pane-1-carboxylic acid, and 3-methyladenine have been mentioned as inducers of oxidative stress, thus they can boost astaxanthin accumulation. Additionally, butylated hydroxyanisole and melatonin were applied into the *H. pluvialis* culture and enhanced the astaxanthin concentration by 2.17-fold and 1.78-fold, respectively. Also, disodium 2-oxoglutarate, ammonium ferric acid, and butylated hydroxytoluene increased astaxanthin accumulation by 10.48%, 12.5%, and 10.71%, respectively (Ren et al. 2021). In another study, the effects of phytohormones on *H. pluvialis* culture were investigated. 1 μM rac-GR24, a synthetic analog of strigolactones, enhanced the astaxanthin biosynthesis from 0.63 to 5.32% of biomass (Wang et al. 2022). Recently, many nanoparticles (NPs) have been investigated as stress agents to understand their effects on microalgae productions (Djearamane et al. 2019). For example, titanium oxide, magnesium, iron, and zinc NPs were used for creating the cellular stress to enhance astaxanthin accumulation in *H. pluvialis* (Ren et al. 2021).

As an example, in a recent study, the zinc oxide (ZO) NPs was synthesized from peel extract of pomegranate (green method) and the particles were induced into the culture medium at the culture during the exponential growth phase. Among different ZO NPs concentrations (50–400 μg/mL), the maximum astaxanthin production was reached with 100 μg/mL, attained as 20 mg/g biomass after 15 days. At higher concentration than 200 μg/mL, the NPs showed toxic effects on cell viability (Nasri et al. 2021). In addition, a cytotoxic effect of NPs can cause a considerable decrease in biomass productivity and photosynthetic efficiency as well as surface and intra-cellular damage (Djearamane et al. 2019). This emphasized that high concentrations of metal-based NPs can be agglomerated with algal cells; therefore, this situation can decrease the light accessibility into the cells. The decrease in photosynthesis process might increase oxidative stress, and thus damage the cell integrity (Nasri et al. 2021). In this context, there is one study that explains the effect of a mechanical stress on

H. pluvialis cells. In this study, a microfluidic chip was designed to sort motile flagellated vegetative cells (15–20 μm) from non-motile astaxanthin-rich cells. Then, the selected cells were introduced into the micro-culture chambers in order to induce mechanical stress on them. As a result, the amount of astaxanthin was increased nine times higher compared to the control group (Yao et al. 2020). On a more commercialized scale, new mechanical operations can be designed to induce stress on the cells.

17.9 Astaxanthin Extraction from *Haematococcus* Cells

The most critical point of an extraction process is to obtain a successful recovery of astaxanthin without contamination of other cellular compounds or solvent residues (Álvarez et al. 2020; Molino et al. 2018; Mussagy et al. 2022). This process should be performed in a cost-effective manner through a sustainable approach (Di Sanzo et al. 2018). In this regard, it was emphasized that the astaxanthin recovery yield can be reached up to 40 g per kg of dry *H. pluvialis* (Luo et al. 2019). However, the effects of extraction procedures should be optimized to achieve high extraction yield. Extraction techniques can be divided to physical, and chemical treatments. Conventionally, astaxanthin can be extracted by using volatile organic solvents such as acetone, dichloromethane (DCM), dimethyl sulfoxide (DMSO), ethanol, ethyl acetate, hexane, and methanol (Miyakawa 2021). Among these chemicals, generally recognized as safe (GRAS) solvents like acetone and ethanol are more recommended due to health and safety concerns (Kim et al. 2022). In many applications, stir magnetic bar, Soxhlet, shaking, percolation, and maceration are preferred extraction techniques (Di Sanzo et al. 2018).

These processes provide basic information to understand the solubility of a target compound in a solvent system, critical temperature values, and suitable solvent/solute ratio. Accordingly, these techniques are useful for analytical, and experimental protocols. However, there are obvious limitations; such as the need for high energy, and long process periods. Moreover, these methods are less selective, efficient, and time-consuming for pilot or industrial operations (Mussagy et al. 2022). The most important point is that these techniques do not effectively disrupt the cell walls of *H. pluvialis* cells. Hence, the combination with two or more solvents is considered as potential strategy to enhance extraction efficiency. In a study, in which dried *H. pluvialis* was pretreated with dimethyl sulfoxide (DMSO) then subjected to acetone extraction and astaxanthin recovery efficiency of 67% was observed. Additionally, it was mentioned that almost the entire astaxanthin could be extracted by performing this method until the cells became colorless (Kim et al. 2022). At this point, repeating the process may affect the integrity of a product and increase total downstream cost (Mussagy et al. 2022). Hence, a balance between the efficiency and the cost should be considered carefully (Álvarez et al. 2020; Mussagy et al. 2022). On the other hand, the toxicity of solvents can affect biochemical structure of astaxanthin and increase the risk of contamination in the final product.

Furthermore, even minor losses of toxic solvents can negatively affect the success of an entire operation. Therefore, using only GRAS solvent systems seems to be the right way to avoid any doubts. For example, researchers developed a new solvent, liquid biphasic consists of food-grade ethanol and 2-propanol, and they extracted astaxanthin as 95.1% by floating the gas bubbles around the cells. Thus, the obtained astaxanthin is considered as more suitable for nutraceutical applications (Kim et al. 2022).

In addition, physical and chemical cell disruption could be carried out simultaneously (Mussagy et al. 2022). Therefore, solvent extractions can be performed at relatively higher pressure and higher temperatures to accelerate the extraction efficiency. High pressure induces the hydrolysis of the cell wall components, while high temperature accelerates the diffusion of the solvent system into the cell (Kim et al. 2022). In a study, in which the astaxanthin was recovered (>95%) in 10 min without any loss in its antioxidant activity even at 200 °C and 6 MPa is a good example for the effect of temperature and pressure changes. In another study, the effects of pressure and temperature were compared by using acetone and ethanol. The results proven that the increase in pressure (5–10 MPa) improves the extractability of astaxanthin, while the increase in temperature (50–100 °C) can decrease it (Kim et al. 2022). Physical cell disruption techniques can be listed as bead milling, high pressure homogenization, ultrasound treatment, microwave treatment, pulsed electric field treatment, and supercritical fluid extraction (Kumar et al. 2022; Miyakawa 2021).

Bead milling is a promising technique to disrupt cell integrity when the algal broth is around 200 g/L. The cells that were disrupted by bead milling before supercritical CO_2 extraction and the astaxanthin extraction yield showed an increase by 92–97%. Moreover, this method can be effectively combined with GRAS solvents. For example, the mixture of acetone and ethanol was used as solvent system and bead milling process was performed at 100 °C and 100 bar. The extraction efficiency reached up to 15 mg/g biomass and the results proven that this technique improves cell permeability. If the pressure is higher than 500 bar, it can release physical, and thermal stresses, and thus it can damage the cells. High pressure homogenization is another industrial extraction technique. During the disruption process, pressure intensity and the number of passages are the most important parameters. As a related example, in one study, high pressure microfluidization was first performed at 30,000 psi; then 80% of total astaxanthin was recovered by ethyl acetate extraction. At this point, cell aggregation and nozzle clogging should be considered as critical issues to reach highest efficiency (Sztretye et al. 2020). It was revealed that number of passages had more influence in the increase in extraction efficiency compared to pressure intensity (Mussagy et al. 2022).

Ultrasonication is another cell disruption technique and generally frequency of sound waves is preferred above 20 kHz (Kim et al. 2022; Mussagy et al. 2022). In one study, *H. pluvialis* was first pretreated by sonication at 750 W in a 2 M NaOH; then 81% of astaxanthin was recovered by extraction in methanol solution (Kim et al. 2022). In another study, natural astaxanthin in dried *H. pluvialis* was obtained via a process combined with a liquid biphasic flotation system using isopropanol. The optimum conditions were maintained at 25 °C, 200 W for 25 min. The results revealed that a high astaxanthin recovery of 95% was observed (Mussagy et al. 2022). The effectiveness of this physical method varies with biomass-solvent ratio, temperature, treatment time, ultrasound power, and frequency (Kim et al. 2022; Mussagy et al. 2022). In several studies, microwave-assisted extraction has been demonstrated as an unconventional procedure to obtain astaxanthin from *H. pluvialis* biomass (Mularczyk et al. 2020). As an example in a recent study, a high amount of astaxanthin was extracted from dried biomass by using acetone at 75 °C, 720 W for 5 min. In a different study, ethanol was used to recover astaxanthin from *H. pluvialis.* The process was performed at 30 °C, 500 W for 240 min and the extraction efficiency of 95% was achieved (Mussagy et al. 2022). Pulsed electric field technique forms pores in the cytoplasmic membrane depending on the strength of the applied forces (Kim et al. 2022).

In a study, which in the electroporation was performed as 10 pulses 5 ms at 1 kV/cm and *H. pluvialis* was subjected to aqueous incubation for 6 h. The results supported that high astaxanthin extraction was reached as 96% of total carotenoid (Mussagy et al. 2022). However, there is a lack of experimental data to understand the critical parameters in this technique (Kim et al. 2022). Besides, very short nanosecond pulse electric field treatment can induce the carotenoid expression pathways leading to enhanced astaxanthin biosynthesis in *H. pluvialis* (Radice et al. 2021; Wang et al. 2022). For example, astaxanthin accumulation increased 20–30% in *H. pluvialis* by performing this technique to improve cell mortality (up to 20%) (Kim et al. 2022). Among physical cell disruption techniques, supercritical fluid extraction is the most preferred one (Marino et al. 2020). Generally, CO_2 is used as a chemical substance because it is not expensive and toxic. In addition, it is also chemically inert, highly stable, and eco-friendly (Mussagy et al. 2022). In order to emphasize the effects of the operation parameters, astaxanthin extraction from *H. pluvialis* was performed at different conditions; extraction time (20–120 min), CO_2 flow rate (3.62–14.48 g/min), temperature (50–80 °C), and pressure (100–550 bar). The results revealed that the astaxanthin recovery yield was increased with respect to the decrease in the CO_2 flow rate, while the higher purity was reached with the high CO_2 flow rate. From another perspective, the increase in temperature had an adverse effect on extraction efficiency, while it was increased with the increase in pressure.

The maximum astaxanthin extraction efficiency of 98.6% was observed at 50 °C, 550 bars (Di Sanzo et al. 2018). Recently, using GRAS solvents to develop new co-solvent formulation is considered as a good option to improve astaxanthin extraction efficiency (Álvarez et al. 2020; Molino et al. 2018). In a recent study, astaxanthin composition of dried *H. pluvialis* was recovered as 92% by using supercritical-CO_2 with ethanol at 65 °C, 550 bar, 80 min and the CO_2 flow rate, and ethanol flow rate were 25 mL/min and 1 mL/min, respectively (Mussagy et al. 2022). In another study, *H. pluvialis* was treated via hydrothermal cell disruption; then astaxanthin was extracted with ethanol solution. A high extraction yield of 98% was observed at 55 °C and 8 MPa (Kim et al. 2022).

17.10 The New Trends in Astaxanthin Extraction from *Haematococcus* Cells

The new trend is a combination of unconventional cell disruption techniques with ionic liquids, eutectic solvents, and biosolvents (Kim et al. 2022). The main mechanism in this process depends on lots of interactions between solvent and solute such as hydrogen bonds, van der Waals, etc. (Mussagy et al. 2022). In order to recover astaxanthin from microalga, the most suitable solvent should be selected based on polarity, reuse potential, and boiling temperature of the solvent. Astaxanthin has a hydrophobic nature; therefore a non-polar solvent present higher affinity compared to polar solvents (Nasri et al. 2021; Yamashita 2021). Moreover, non-polar solvents are capable of interacting with the cell wall, thus they show remarkable effect on cell permeabilization (Sztretye et al. 2020). From an environmental point of view, non-polar solvents are considered more suitable due to ease of recycling (Mussagy et al. 2022). The efficiency of an extraction process depends on properties of the solvent system, characteristics of algal biomass, and process parameters. The solvent should be non-flammable, non-toxic, non-volatile, and effective at low temperatures (Stachowiak and Szulc 2021).

Ionic liquid (IL) treatment is preferred as an extraction process in various algal applications (Kim et al. 2022). During this process, organic cations and organic/inorganic anions are main components in organic salts (Mussagy et al. 2022). In one study, 40% 1-ethyl-3-methylimidazolium dibutylphosphate ([Emim] DBP) solution was used to extract astaxanthin from dried *H. pluvialis* at 45 °C and 77% of total astaxanthin was recovered successfully (Kim et al. 2022). In a different study, 1-ethyl-3-methylimidazolium ethylsulfate was mentioned as a potent solvent system in order to obtain astaxanthin from *H. pluvialis*. After only 1 min, it was demonstrated that this solvent system was 82% more effective compared to ethyl acetate at the same conditions (Mussagy et al. 2022). In another study, dried *H. pluvialis* was pretreated with 40% 1-butyl-3-methylimidazolium chloride ([Bmim] Cl) at 90 °C, then the biomass subjected in methanol extraction at 50 °C. The astaxanthin recovery efficiency was reported as 91%. Additionally, imidazolium-based ILs ([Bmim] Cl, [Emim] Cl, [Bmim] Bu_2PO_4, and [Emim] BF_4) presented greater extraction efficiency than pyridinium-([Bmpy] Cl) and ammonium-based ([Bmam] Cl) ILs at the same process conditions. As a conclusion, it was emphasized that IL type, IL concentration, and operating conditions are crucial parameters that affect the progress (Kim et al. 2022).

Nanomaterials have been drawing great attention in extraction processes due to their extraordinarily large surface areas. In a study, in which ultrathin alpha nanoplates (1.14 μm × 7.7 nm) were penetrated to the surface of *H. pluvialis* cell wall. Almost the entire astaxanthin content (99%) was extracted under DCM/methanol extraction. In a different study, carbon nanodots (1.0 mg/mL) were added into the *H. pluvialis* culture medium as an ingredient. Astaxanthin extraction was carried out with hexane and interestingly, the recovered astaxanthin amount was determined as 1.8 folder than control culture. As a relatively new field, nanomaterial technology can be considered a practical technique for efficiently extraction of bioactive compounds (Kim et al. 2022). Enzyme treatment is a good option to reduce the strength of the cell wall; therefore, more astaxanthin can be derived from the algal cell (Mularczyk et al. 2020). As an example in a recent study, a combination of β-1,3-glucanase and protease was used with DCM to recover astaxanthin from *H. pluvialis* at pH 4.5, 55 °C for 30 min.

A moderate astaxanthin extraction was observed at 79.3%. In another study, cellulase and pectinase were applied into ethyl acetate solvent and the extraction process was carried out with this mixture. Even optimization process was performed based on pH, temperature, and operation time, only 67% of total astaxanthin was obtained from *H. pluvialis* (Kim et al. 2022). Germination is a progress in microalgae cells which is observed when an astaxanthin-rich *H. pluvialis* cell is transferred to nutrient-rich medium. Thus, the cell wall becomes more susceptible to degradation when exposed to cell disruption techniques. In one study, the germinated *H. pluvialis* was subjected to ethyl acetate and a moderate recovery efficiency of 85% was obtained. At this point, germination period is a critical point which maintains the efficiency of the extraction procedure. In this regard, it was mentioned that astaxanthin composition can be converted to photosynthetic pigments and the amount of astaxanthin can be decreased 30% for cultivation period longer than 24 h (Kim et al. 2022).

17.11 Further Recommendations for Future Applications

17.11.1 Haematococcus *Cultivation*

Each cell has its own intracellular dynamics which control the production, accumulation, and transportation of cellular metabolites. Indeed, metabolic fluxes respond to changes in environmental conditions and nutrients during the life cycle. Therefore, cell engineering is the major strategy to manage cellular mechanisms (Yonamine et al. 2020). Genetic modifications on both mechanisms of utilization of nutrients and accumulating metabolites is a promising direction to enhance astaxanthin production efficiency (Ren et al. 2021; Villaro et al. 2021). These modifications can be presented by classical mutagenesis or chemical mutagenesis methods, or these methods can be combined with genetic pathway engineering (Stachowiak and Szulc 2021). Moreover, the obtained data can be correlated with transcriptomic analysis; thus, more solid information on astaxanthin biosynthesis can be revealed (Kumar et al. 2022; Luo et al. 2019).

New nutrient combinations could be evaluated to understand their effects on metabolite synthesis pathways (Wang et al. 2022). They may explore the culture to produce new and novel biocompounds (Ren et al. 2021). For this purpose, supercritical-CO_2 treatment can be recommended to detect various metabolites obtained from *Haematococcus* cells. During supercritical-CO_2 treatment, CO_2 acts like a lipophilic solvent and consequently extracts more non-polar compounds than polar compounds. Although astaxanthin is a non-polar compound, this technique promises an opportunity to extract other metabolites in *H. pluvialis*. At this point, GRAS solvents can be estimated as co-solvents due to their low environmental impact (Molino et al. 2018).

The metabolic effects of medium ingredients and process parameters should be examined in order to understand the production more clearly and to manipulate it more easily (Hasunuma et al. 2019). The mechanism underlying astaxanthin synthesis in stress-induced *H. pluvialis* needs to be deciphered (Nasri et al. 2021). In this regard, stable isotopes are promising probes to label a substrate and follow its changes. As microalgae use CO_2, labeling C atoms in CO_2 is a promising strategy to understand the formation of carbon sources to astaxanthin (Yonamine et al. 2020). In order to achieve maximum productivity, each parameter that affects biomass accumulation and target product synthesis needs to be considered. In more detail, the differences in cultivation modes and the effects of light intensity can be underlined (Oncel et al. 2011). On the other hand, mathematical models can be applied to determine key operation conditions (Ren et al. 2021). For example, computational fluid dynamics (CFD) method can be used to emphasize the effects of cultivation broth on the cells (Aslanbay Guler et al. 2020).

17.11.2 Commercial Natural Astaxanthin Production

Microalgae production can be detailed in each property from different perspectives. Firstly, cultivation process should be evaluated according to area, climatic conditions, the risk of contamination, and water availability. Additionally, the total production cost, automation level of the production platform, and system efficiency should be also considered to make the process more economically profitable (Mularczyk et al. 2020; Tramontin et al. 2019). In this regard, the necessary ingredients for microalgal cultivation can be obtained through a biorefinery approach. For example, CO_2 can be obtained from industrial companies. Moreover, microalgae can utilize nitrogen and phosphorous from wastewater. Also, the required water content can be applied by using seawater (Ren et al. 2021; Wang et al. 2022). In this regard, the crucial point is that *H. pluvialis* can utilize metal ions, and thus the culture can be contaminated by heavy metals (Djearamane et al. 2019). In order to avoid this issue, waste quality should be controlled and pretreatment should be involved to maintain metabolites more qualified (Mularczyk et al. 2020). Commercial production of astaxanthin derived from *H. pluvialis* has several drawbacks as high light requirements, slow growth, and the risk of cross-contamination (Naz et al. 2021). Therefore, photobioreactors can be used to maintain suitable conditions that allow growth of vegetative cells. Then, the culture can be inoculated into open ponds to accumulate astaxanthin for a short period (Saini and Keum 2019).

During cultivation, *H. pluvialis* culture actually is a collection of cells at different stages of their life cycle. Therefore, it can be aimed to separate the cells that are in the cystic phase at the harvesting stage from the cells that are still in the vegetative growth phase (Yao et al. 2020). After this stage, the production can be divided into two separate parts. In the first, astaxanthin production can be induced by cultivating the green cells under stress conditions. On the other, cells that are already in the cystic phase can be subjected to extraction processes (Radice et al. 2021). Besides that, considering the astaxanthin production process, the presence of immature cells means impurities in the product. Therefore, product stability should be controlled by managing, and processing each step carefully (Miyakawa 2021). In a study, a new methodology was mentioned to select *H. pluvialis* cysts by electrophoresis. It has been known that the biochemical composition of *H. pluvialis* cells changes during algal growth. The net negative charge of their cell walls presents an opportunity to separate the non-motile cells from the motile cells. Also, the cell diameter changes during the growth period. Therefore, loaded cells run at different speeds due to the presence of an electric field (Radice et al. 2021).

The cost of downstream operations, especially extraction process, should be advanced to promote cost reductions (Ren et al. 2021). In industrial applications, the lowest extraction time is most crucial point due to high direct/indirect costs of this process (Álvarez et al. 2020). In order to achieve high astaxanthin recovery efficiency, a synergistic combination of extraction methods, labor-efficient operations should be underlined to design a platform with cost-effective and low energy consumption (Yao et al. 2020). In addition, scale-up strategies should be considered to achieve practical applications in algal production through a biorefinery approach (Kim et al. 2022). Among different solvent systems, a preferable solvent can be evaluated by using computational models. For example, COSMO offers a great opportunity to emphasize the interactions between solvent and solute. However, this method lacks information about mass transfer mechanisms and extraction kinetics which change according to characteristics of algal biomass. Therefore, synergistic interactions should also be evaluated to understand these crucial points.

17.11.3 Utilization of Microalgal Astaxanthin as an End-Product

The complex biological and physiological features of *H. pluvialis* cells should be underlined for operating synergistic combinations of chemical, physical, and/or physicochemical treatments (Kim et al. 2022; Ren et al. 2021). When natural compounds are removed from their biological matrix, the high unsaturated structure makes them highly sensitive against the environmental conditions during isolation, manufacture, and storage. Therefore, they can lose their biological features (Stachowiak and Szulc 2021). The stability of a product is an important parameter for natural astaxanthin (Mularczyk et al. 2020; Villaro et al. 2021; Zakaria et al. 2021). Recently, encapsulation technologies of astaxanthin in liposomes (microencapsulation and nanoencapsulation) are considered as promising techniques to increase the properties of astaxanthin in terms of stability, water solubility, and bioavailability (Sztretye et al. 2020; Yamashita 2021). Today, whey protein isolate, sodium caseinate, and soluble corn fiber are accepted proteins by consumers. On the other hand, cyclic oligosaccharides are also evaluated in microencapsulation applications as hydroxypropyl-β-cyclodextrin, sulfobutyl ether β-cyclodextrin (Stachowiak and Szulc 2021).

Additionally, bioavailability of natural astaxanthin is an important property as well as its stability. After the astaxanthin-enriched extract is involved in purification/refinement process, the product should be examined to understand its properties. Then, the product can be considered as ready to be in market (Miyakawa 2021). In this regard, it was demonstrated that *H. pluvialis* biomass is required to be subjected into free-drying and vacuum-packed process, then it can be storage at 20C. However, the effects of processing and storage conditions on a bioactive compound should be evaluated to emphasize its potential effects on the human body (Villaro et al. 2021). At this point, bioavailability seem to be the key word that explains the fraction of natural compounds that enters the systemic circulation (Mularczyk et al. 2020). Therefore, it is considered as a crucial parameter in order to understand the potential effects of a natural substrate on the human body.

17.11.4 Microalgal Astaxanthin as a Diet Supplementary

There are many shortcomings in the detailed evaluation of the potential benefits and risks of a bioactive compound in the human body (Mason et al. 2020; Zhou et al. 2021). On this topic, the studies that explain the effects of micronutrients on the human body are limited (Kawamura et al. 2021; Nishida et al. 2022). Additionally, the effects of oxidative stress, antioxidant enzyme levels on the human body are unclear (Hongo et al. 2020; Mason et al. 2020). It is necessary to underline the molecular mechanisms behind the effects of a bioactive compound (Chen et al. 2021; Shokri-mashhadi et al. 2021). In this regard, there is limited evidence to explain the molecular mechanism underlying astaxanthin-induced protection (Huang et al. 2021). Therefore, there are insufficient experiments to recommend astaxanthin as an exogenous supplement to athletes (Liu et al. 2021; Mason et al. 2020). It is necessary to elucidate the effects of astaxanthin on human participants both at cellular and organ levels (Sztretye et al. 2020; Villaro et al. 2021). As a conclusion, it can be emphasized that the molecular mechanism around Nrf2 should be explored in clinical applications; thus, its safety and efficacy could be more clearly addressed (Kohandel et al. 2021). In addition, the impact of astaxanthin on fatty acid oxidations can be investigated by determining the metabolic enzymes (carnitine palmitoyltransferase 1, fatty acid transferase, AMPK) to understand the ergogenic benefit of it (Brown et al. 2021; Kato et al. 2020; Lee et al. 2020; Liu et al. 2021). On the other hand, a bioactive compound can exert different effects according to its concentration (Nasri et al. 2021). In this regard, it has been demonstrated that high doses of antioxidants can inhibit the beneficial effects of exercise (Kawamura et al. 2021). Therefore, personalized strategies for antioxidant supplementation should be discussed to evaluate the current impact of antioxidants (Mason et al. 2020). The optimal amount of astaxanthin in athlete nutrition should be determined based on pharmacokinetics to enable the success of the training. Additionally, the mechanistic factors that astaxanthin exerts on in-exercising humans should be underlined.

References

Álvarez CE, Vardanega R, Salinas-Fuentes F, Ramírez JP, Muñoz WB, Jiménez-Rondón D et al (2020) Effect of CO_2 flow rate on the extraction of astaxanthin and fatty acids from *Haematococcus pluvialis* using supercritical fluid technology. Molecules 25(24):1–17

Aslanbay Guler B, Deniz I, Demirel Z, Oncel SS, Imamoglu E (2020) Computational fluid dynamics modelling of stirred tank photobioreactor for *Haematococcus pluvialis* production: hydrodynamics and mixing conditions. Algal Res 47:1–9. https://doi.org/10.1016/j.algal.2020.101854

Bhattarai G, So HS, Kieu TTT, Kook SH, Lee JC, Jeon YM (2021) Astaxanthin inhibits diabetes-triggered periodontal destruction, ameliorates oxidative complications in stz-injected mice, and recovers nrf2-dependent antioxidant system. Nutrients 13(10):1–18

Brown DR, Warner AR, Deb SK, Gough LA, Sparks SA, McNaughton LR (2021) The effect of astaxanthin supplementation on performance and fat oxidation during a 40 km cycling time trial. J Sci Med Sport 24(1):92–97. https://doi.org/10.1016/j.jsams.2020.06.017

Chen Y, Li S, Guo Y, Yu H, Bao Y, Xin X et al (2020) Astaxanthin attenuates hypertensive vascular remodeling by protecting vascular smooth muscle cells from oxidative stress-induced mitochondrial dysfunction. Oxidative Med Cell Longev 2020:1–19

Chen Y, Yu T, Deuster P (2021) Astaxanthin protects against heat-induced mitochondrial alterations in mouse hypothalamus. Neuroscience 476:12–20. https://doi.org/10.1016/j.neuroscience.2021.09.010

Chou HY, Ma DL, Leung CH, Chiu CC, Hour TC, Wang HMD (2020) Purified astaxanthin from Haematococcus pluvialis promotes tissue regeneration by reducing oxidative stress and the secretion of collagen *in vitro* and *in vivo*. Oxidative Med Cell Longev 2020:1–13

Di Sanzo G, Mehariya S, Martino M, Larocca V, Casella P, Chianese S et al (2018) Supercritical carbon dioxide extraction of astaxanthin, lutein, and fatty acids from *Haematococcus pluvialis* microalgae. Mar Drugs 16(9):1–18

Djearamane S, Lim YM, Wong LS, Lee PF (2019) Cellular accumulation and cytotoxic effects of zinc oxide nanoparticles in microalga *Haematococcus pluvialis*. PeerJ 2019(9):1–19

García F, Lobos P, Ponce A, Cataldo K, Meza D, Farías P et al (2020) Astaxanthin counteracts excitotoxicity and reduces the ensuing increases in calcium levels and mitochondrial reactive oxygen species generation. Mar Drugs 18(6):1–15

Gürlek C, Yarkent Ç, Köse A, Oral İ, Öncel S, Elibol M (2020a) Evaluation of several microalgal extracts as bioactive metabolites as potential pharmaceutical compounds. IFMBE Proc 73:267–272

Gürlek C, Yarkent Ç, Köse A, Tuğcu B, Gebeloğlu IK, Öncel S et al (2020b) Screening of antioxidant and cytotoxic activities of several microalgal extracts with pharmaceutical potential. Health Technol 10(1):111–117

Hasunuma T, Takaki A, Matsuda M, Kato Y, Vavricka CJ, Kondo A (2019) Single-stage astaxanthin production enhances the nonmevalonate pathway and photosynthetic central metabolism in *Synechococcus* sp. PCC 7002. ACS. Synth Biol 8(12):2701–2709

Hongo N, Takamura Y, Nishimaru H, Matsumoto J, Tobe K, Saito T et al (2020) Astaxanthin ameliorated parvalbumin-positive neuron deficits and Alzheimer's disease-related pathological progression in the hippocampus of AppNL-G-F/NL-G-F mice. Front Pharmacol 11:1–15

Huang C, Wen C, Yang M, Li A, Fan C, Gan D et al (2021) Astaxanthin improved the cognitive deficits in APP/PS1 transgenic mice via selective activation of mTOR. J Neuroimmune Pharmacol 16(3):609–619

Kato T, Kasai T, Sato A, Ishiwata S, Yatsu S, Matsumoto H et al (2020) Effects of 3-month astaxanthin supplementation on cardiac function in heart failure patients with left ventricular systolic dysfunction-a pilot study. Nutrients 12(6):1–11

Kawamura A, Aoi W, Abe R, Kobayashi Y, Kuwahata M, Higashi A (2021) Astaxanthin-, β-carotene-, and resveratrol-rich foods support resistance training-induced adaptation. Antioxidants 10(113):1–11

Kim B, Youn Lee S, Lakshmi Narasimhan A, Kim S, Oh YK (2022) Cell disruption and astaxanthin extraction from Haematococcus pluvialis: recent advances. Bioresour Technol 343:1–12. https://doi.org/10.1016/j.biortech.2021.126124

Kohandel Z, Farkhondeh T, Aschner M, Samarghandian S (2021) Nrf2 a molecular therapeutic target for astaxanthin. Biomed Pharmacother 137:1–7. https://doi.org/10.1016/j.biopha.2021.111374

Krestinin R, Baburina Y, Odinokova I, Kruglov A, Fadeeva I, Zvyagina A et al (2020) Isoproterenol-induced permeability transition pore-related dysfunction of heart mitochondria is attenuated by astaxanthin. Biomedicine 8(10):1–20

Kumar S, Kumar R, Kumari A, Panwar A (2022) Astaxanthin: a super antioxidant from microalgae and its therapeutic potential. J Basic Microbiol 62(9):1064–1082

Kwak MS, Lim JW, Kim H (2021) Astaxanthin inhibits interleukin-6 expression in cerulein/resistin-stimulated pancreatic acinar cells. Mediat Inflamm 2021:1–14

Lee H, Lim JW, Kim H (2020) Effect of astaxanthin on activation of autophagy and inhibition of apoptosis in helicobacter. Nutrients 12(1750):1–18

Li F, Cai M, Lin M, Huang X, Wang J, Ke H et al (2019) Differences between motile and nonmotile cells of *Haematococcus pluvialis* in the production of astaxanthin at different light intensities. Mar Drugs 17(1):1–14

Li Y, Wang Q, Chu C, Liu S (2020) Astaxanthin protects retinal ganglion cells from acute glaucoma via the Nrf2/HO-1 pathway. J Chem Neuroanat 110:1–8. https://doi.org/10.1016/j.jchemneu.2020.101876

Liu SZ, Valencia AP, VanDoren MP, Shankland EG, Roshanravan B, Conley KE et al (2021) Astaxanthin supplementation enhances metabolic adaptation with aerobic training in the elderly. Physiol Rep 9(11):1–11

Lu W, Shi Y, Wang R, Su D, Tang M, Liu Y et al (2021) Antioxidant activity and healthy benefits of natural pigments in fruits: a review. Int J Mol Sci 22(9):1–18

Luo Q, Bian C, Tao M, Huang Y, Zheng Y, Lv Y et al (2019) Genome and transcriptome sequencing of the astaxanthin-producing green microalga, Haematococcus pluvialis. Genome Biol Evol 11(1):166–173

Marino T, Iovine A, Casella P, Martino M, Chianese S, Larocca V et al (2020) From *Haematococcus pluvialis* microalgae a powerful antioxidant for cosmetic applications. Chem Eng Trans 79:271–276

Mascia F, Girolomoni L, Alcocer MJP, Bargigia I, Perozeni F, Cazzaniga S et al (2017) Functional analysis of photosynthetic pigment binding complexes in the green alga *Haematococcus pluvialis* reveals distribution of astaxanthin in photosystems. Sci Rep 7(1):1–14

Mason SA, Trewin AJ, Parker L, Wadley GD (2020) Antioxidant supplements and endurance exercise: current evidence and mechanistic insights. Redox Biol 35:1–21. https://doi.org/10.1016/j.redox.2020.101471

Miyakawa K (2021) Commercial production of astaxanthin from the green alga *Haematococcus pluvialis*. In: Carotenoids: biosynthetic and biofunctional approaches, pp 3–11. https://link.springer.com/book/10.1007/978-981-15-7360-6#toc%0A

Molino A, Mehariya S, Iovine A, Larocca V, Di Sanzo G, Martino M et al (2018) Extraction of astaxanthin and lutein from microalga *Haematococcus pluvialis* in the red phase using CO_2 supercritical fluid extraction technology with ethanol as co-solvent. Mar Drugs 16(11):1–16

Mularczyk M, Michalak I, Marycz K (2020) Astaxanthin and other nutrients from *Haematococcus pluvialis*-multifunctional applications. Mar Drugs 18(9):1–22
Mussagy CU, Pereira JFB, Santos-Ebinuma VC, Pessoa A, Raghavan V (2022) Insights into using green and unconventional technologies to recover natural astaxanthin from microbial biomass. Crit Rev Food Sci Nutr:1–15. https://doi.org/10.1080/10408398.2022.2093326
Nasri N, Keyhanfar M, Behbahani M, Dini G (2021) Enhancement of astaxanthin production in *Haematococcus pluvialis* using zinc oxide nanoparticles. J Biotechnol 342:72–78. https://doi.org/10.1016/j.jbiotec.2021.10.004
Naz T, Yang J, Nosheen S, Sun C, Nazir Y, Mohamed H et al (2021) Genetic modification of Mucor circinelloides for Canthaxanthin production by heterologous expression of β-carotene ketolase gene. Front Nutr 8:1–11
Nishida Y, Nawaz A, Kado T, Takikawa A, Igarashi Y, Onogi Y et al (2020) Astaxanthin stimulates mitochondrial biogenesis in insulin resistant muscle *via* activation of AMPK pathway. J Cachexia Sarcopenia Muscle 11(1):241–258
Nishida Y, Nawaz A, Hecht K, Tobe K (2022) Astaxanthin as a novel mitochondrial regulator: a new aspect of carotenoids, beyond antioxidants. Nutrients 14(1):1–39
Oncel SS, Imamoglu E, Gunerken E, Sukan FV (2011) Comparison of different cultivation modes and light intensities using mono-cultures and co-cultures of *Haematococcus pluvialis* and *Chlorella zofingiensis*. J Chem Technol Biotechnol 86(3):414–420
Radice RP, Fiorentino R, De Luca M, Limongi AR, Viviano E, Bermano G et al (2021) An innovative protocol to select the best growth phase for astaxanthin biosynthesis in H. pluvialis. Biotechnol Reports 31:1–6
Ren Y, Deng J, Huang J, Wu Z, Yi L, Bi Y et al (2021) Using green alga *Haematococcus pluvialis* for astaxanthin and lipid co-production: advances and outlook. Bioresour Technol 340:125736. https://doi.org/10.1016/j.biortech.2021.125736
Saini RK, Keum YS (2019) Microbial platforms to produce commercially vital carotenoids at industrial scale: an updated review of critical issues. J Ind Microbiol Biotechnol 46(5):657–674. https://doi.org/10.1007/s10295-018-2104-7
Shokri-mashhadi N, Tahmasebi M, Mohammadi-asl J, Zakerkish M, Mohammadshahi M (2021) The antioxidant and anti-inflammatory effects of astaxanthin supplementation on the expression of miR-146a and miR-126 in patients with type 2 diabetes mellitus: a randomised, double-blind, placebo-controlled clinical trial. Int J Clin Pract 75(5):1–6
Stachowiak B, Szulc P (2021) Astaxanthin for the food industry. Molecules 26(9):1–18
Sun L, Miyaji N, Yang M, Mills EM, Taniyama S, Uchida T et al (2021) Astaxanthin prevents atrophy in slow muscle fibers by inhibiting mitochondrial reactive oxygen species *via* a mitochondria-mediated apoptosis pathway. Nutrients 13(2):1–17
Sztretye M, Singlár Z, Szabó L, Angyal Á, Balogh N, Vakilzadeh F et al (2020) Improved tetanic force and mitochondrial calcium homeostasis by astaxanthin treatment in mouse skeletal muscle. Antioxidants. 9(2):1–18
Tang X, Man Y, Hu X, Xu X, Ren L (2022) Identification of carotenoids biosynthesis pathway in *Schizochytrium* sp. and utilization in astaxanthin biosynthesis. Enzyme Microb Technol 156:1–7. https://doi.org/10.1016/j.enzmictec.2022.110018
Tramontin LRR, Kildegaard KR, Sudarsan S, Borodina I (2019) Enhancement of astaxanthin biosynthesis in oleaginous yeast Yarrowia lipolytica *via* microalgal pathway. Microorganisms 7(10):1–17
Villaro S, Ciardi M, Morillas-españa A, Sánchez-Zurano A, Acién-Fernández G, Lafarga T (2021) Microalgae derived astaxanthin: research and consumer trends and industrial use as food. Foods 10(2303):1–17
Wang X, Mou JH, Qin ZH, Bin HT, Zheng L, Buhagiar J et al (2022) Supplementation with rac-GR24 facilitates the accumulation of biomass and astaxanthin in two successive stages of *Haematococcus pluvialis* cultivation. J Agric Food Chem 70(15):4677–4689

Yamashita E (2021) Extensive bioactivity of astaxanthin from *Haematococcus pluvialis* in human. In: Carotenoids: biosynthetic and biofunctional approaches, pp 249–260. https://link.springer.com/book/10.1007/978-981-15-7360-6#toc%0A

Yao J, Kim HS, Kim JY, Choi YE, Park J (2020) Mechanical stress induced astaxanthin accumulation of: H. pluvialis on a chip. Lab Chip 20(3):647–654

Yarkent Ç, Oncel SS (2022) Recent progress in microalgal squalene production and its cosmetic application. Biotechnol Bioprocess Eng 11:1–11

Yarkent Ç, Gürlek C, Oncel SS (2020) Potential of microalgal compounds in trending natural cosmetics: a review. Sustain Chem Pharm 17:1–11

Yonamine Y, Hiramatsu K, Ideguchi T, Ito T, Fujiwara T, Miura Y, et al. Spatiotemporal monitoring of intracellular metabolic dynamics by resonance Raman microscopy with isotope labeling. RSC Adv; 2020;10(28):16679–16686

Zakaria NNA, Zamzurie NA, Harith ZT (2021) Evaluation of sunscreen cream incorporated with astaxanthin from *Haematococcus pluvialis* in different storage conditions. IOP Conf Ser Earth Environ Sci 756(1):1–7

Zhou X, Cao Q, Orfila C, Zhao J, Zhang L (2021) Systematic review and meta-analysis on the effects of astaxanthin on human skin ageing. Nutrients 13(9):1–18

Zhu HZ, Jiang S, Wu JJ, Zhou XR, Liu PY, Huang FH et al (2022) Production of high levels of 3S,3′S-Astaxanthin in *Yarrowia lipolytica via* iterative metabolic engineering. J Agric Food Chem 70(8):2673–2683

Chapter 18
Commercialization of *Haematococcus*-Based Products: Current Status and Future Forecast

Diganta Kumar Saikia, Refad Ahmed, Channakeshavaiah Chikkaputtaiah, and Natarajan Velmurugan

18.1 Introduction

Freshwater microalgae *Haematococcus pluvialis,* belong to the species of chlorophyta, is one of the potent biological sources for astaxanthin production (Sarada et al. 2002). Very limited numbers of astaxanthin producing strains are being commerically cultivated among the vast range of astaxanthin producing microorganisms. Of them, due to the potentiality to accumulate large quantities of astaxanthin, *H. pluvialis* has considered as commercially-important and promising natural source of astaxanthin (Mularczyk et al. 2020). Direct and natural production of xanthophyll carotenoid astaxanthin was reported in several bacteria and algae while the same was accumulated via food chain in some crustaceans, birds, and fishes (Li et al. 2011; Vardanega et al. 2022). Having several application of astaxanthin, the key applications are in aquaculture as feed additive, essential in

Diganta Kumar Saikia and Refad Ahmed contributed equally with all other contributors.

D. K. Saikia
Biological Sciences Division, CSIR-North East Institute of Science and Technology (CSIR-NEIST), Branch Laboratory-Itanagar, Naharlagun, Arunachal Pradesh, India

R. Ahmed · N. Velmurugan (✉)
Biological Sciences Division, CSIR-North East Institute of Science and Technology (CSIR-NEIST), Branch Laboratory-Itanagar, Naharlagun, Arunachal Pradesh, India

Academy of Scientific and Innovative Research (AcSIR), Ghaziabad, India
e-mail: natarajan@neist.res.in

C. Chikkaputtaiah
Academy of Scientific and Innovative Research (AcSIR), Ghaziabad, India

Biological Sciences and Technology Division, CSIR-North East Institute of Science and Technology (CSIR-NEIST), Jorhat, Assam, India

R. Raja et al. (eds.), *Haematococcus*, https://doi.org/10.1007/978-981-99-2901-6_18

farming of salmon and trout and dietary supplements as it has properties of anti-cancer, anti-diabetic, anti-inflammatory, anti-aging activities, and sun proofing properties, and reported to boost immune system (Li et al. 2011; Oslan et al. 2021; Vardanega et al. 2022).

At present, commerically available astaxanthin was either chemically synthesized or naturally found in different bioresources such as microalgae, yeast, and crustacean by-products (Li et al. 2011). Though synthetic astaxanthin leads the commercial market, there is a wide exploitation of biologically-synthesized astaxanthin because of high market demand for natural astaxanthin. Microalgae strains belonging to *H. pluvialis* reported to accumulate higher levels of astaxanthin content (Li et al. 2011). Astaxanthin biosynthesis naturally occur in *H. pluvialis* when they grew under stressful environments, like high salinity, high intensity light conditions, and different carbon-to-nitrogen ratio (Oslan et al. 2021). *H. pluvialis* has been considered as promising source all over the world due to a lesser production cost than currently used synthetic molecules worldwide (Colusse et al. 2021). *H. pluvialis* reported to accumulate astaxanthin about 5% of its dry weight which is considered as the best naturally-occurred, high-value carotenoid pigment astaxanthin (Shah et al. 2016). Natural astaxanthin as a supplement has no adverse effect on human health and is found safe for consumption and also as a feed additive in the animal and aquaculture feed industries. Over the last 20 years, no negative effects have been reported on the use of astaxanthin as a dietary supplement (Oslan et al. 2021).

Several multinational companies based on Israel, USA, and India have been focused on the production of *H. pluvialis* at large-scale level and produce 300 tons of *H. pluvialis* per year. Few of them are AstaReal (www.astareal.se), Cyanotech Corporation (www.cyanotech.com), and Algatech (www.algatech.com). AstaReal Group (www.astareal.com) was the first company to produce astaxanthin commercially from the year 1994. AstaReal® oleoresin, AstaReal® EL25, and AstaReal® A1010 were three different successful astaxanthin products from AstaReal group. Cosmetic products (e.g., face serum, body and face hydration creams Antiox®) incorporated with astaxanthin were originally produced by a Brazilian company Ocean Drop (www.oceandrop.com.br), and successfully leading in the global markets. Addition of *H. pluvialis*-derived astaxanthin in the bread was reported to be well preserved from degradation during baking. The bread consists of microalgal astaxanthin was originally developed by Japanese researcher (Silva et al. 2020).

Different studies reported different predictions for global market values for astaxanthin. Among them, the global market for astaxanthin, as predicted on a recent study by Vardanega et al. (2022), can grow up to a rate of 16.2%, growing to US$ 2.57 billion in 2025 and in 2027 reaching to US$ 3.4 billion. Depending on the quality and origin (synthetic or natural) of astaxanthin, the market price was found to vary between a range of US$2000 and 7500/kg (Vardanega et al. 2022). The estimated market value for nutraceutical grade astaxanthin from *H. pluvialis* was at prices ranging from US$6000 to US$7150/kg (Butler and Golan 2020; Kim et al. 2022). As of 2014, the market price of astaxanthin varied from $2500 to 7000/kg, on the other hand its future world market was valued at $447 million and anticipated at 280 metric tons of which most of the astaxanthin was derived synthetically (Panis

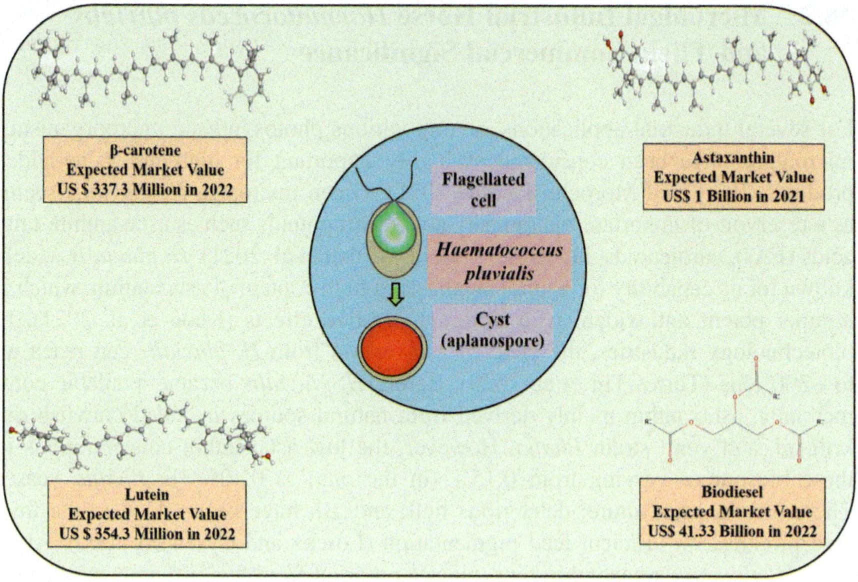

Fig. 18.1 Schematic representation of commercial products derived from *Haematococcus* in different stages from flagellated cell to aplanospore and their recent market values. Global market values of various bioproducts: (1) astaxanthin (US $ 1 billion in 2021; https://www.marketdataforecast.com/market-reports/astaxanthin-market) (2) β-carotene (US $ 337.3 million in 2022; https://www.futuremarketinsights.com/reports/beta-carotene-market) (3) lutein (US $ 354.3 million in2022; https://www.futuremarketinsights.com/reports/lutein-market) and (4) biodiesel (US $ 41.33 billion in 2022; https://www.digitaljournal.com/pr/biodiesel-market-is-expected-to-reach-value-of-us-59-bn-at-3-6-cagr-between-2022-and-2032)

and Carreon 2016). Nowadays, to stimulate green solution with less cost and natural product, the estimation for global market to exceed for astaxanthin derived from algae seemed to be acquiring potential market (Panis and Carreon 2016).

Apart from the astaxanthin synthesized by *H. pluvialis* species in high quantities under adverse growth conditions, other metabolites with commercial importance are also accumulated in the same microalgae including lipids, carbohydrates, and proteins (Fig. 18.1). The other metabolites have been used for producing biodiesel, bioethanol, biogas, biofertilizer, and animal feed (Nishshanka et al. 2022). In view of the above, the first part of this chapter briefly highlights the importance of industrially important strain *Haematococcus pluvialis* and commercial aspects of its bioactive compounds. It was further followed by the detailed discussion on commercial competitive nature of *Haematococcus*-derived bioproducts and their future perspectives.

18.2 Microalgal Industrial Horse *Haematococcus pluvialis* and Their Commercial Significance

For several industrial applications, among various photosynthetic microorganisms, microalgae have been considered as highly important for high-end value-added products (Baby and Morocho-Jácome 2021). Green microalga *H. pluvialis* seems as a reservoir of important biologically active compounds such as astaxanthin, fatty acids (FAs), carotenoids, lutein and proteins (Oslan et al. 2021). *H. pluvialis* widely known for its capability for natural synthesis of high content of astaxanthin, which is a super potent antioxidant with different curative effects (Khoo et al. 2021). In biotechnology industries, the yield of astaxanthin from *H. pluvialis* can reach up to ~300 tons (Torres-Tiji et al. 2020). Before *H. pluvialis* became available commercially, astaxanthin mainly derived from natural sources included crawfish oil, krill oil, and yeast strain *Phafia*. However, the low astaxanthin concentrations in these biosources varying from 0.15% (in the oils) to 0.40% (in *Phaffia* yeast). Therefore, in aquaculture, deleterious bulk and ash have been added to the final feed products for efficient feed pigmentation (Lorenz and Cysewski 2000). Apart from *H. pluvialis*, astaxanthin biosynthesis reported from few different microalgae strains such as *Chlorococcum* sp. and *Chlorella zofingiensis*, several yeast strains such as *Xanthophyllomyces dendrorhous*, *Phaffia rhodozyma,* and *Yarrowia lipolytica*, bacteria species includes *Brevibacterium* sp., *Paracoccus* sp., and *Agrobacterium* sp. and thraustochytrids such as *Aurantiochytrium*. However, the astaxanthin content reported was between 0.04 and 2.7% of dry weight in most of these microorganisms, whereas the largest natural source of astaxanthin content was found in *H. pluvialis* having a content of up to 5% (Mota et al. 2022). Other microorganisms producing astaxanthin shown various characteristics such as variable yields of astaxanthin and growth rates. Therefore, the feasibility of these microorganisms for industrial use depends upon these characteristics, and very sceptical.

Astaxanthin is a reddish carotenoid pigment naturally synthesized under stress condition by *H. pluvialis* microalgae and which make appearances in the pink-red color in salmon, lobster, shrimp, and other seafoods (Davies 2008; Li et al. 2011; Vardanega et al. 2022). Despite astaxanthin utilization in aquaculture, value of astaxanthin-based market products for human consumption was constantly growing and was estimated at US$35–60 million in 2008 (Le-Feuvre et al. 2020). Global production of astaxanthin was raised between 43,279 and 73,717 kg with a rate of 11.24% from 2011 to 2016. Natural astaxanthin market price is mainly determined by its purity and concentration. Based on the astaxanthin concentration, market price for dry-form of *H. pluvialis* may vary from US$2500 to 7000 kg^{-1} (Le-Feuvre et al. 2020). At the same time, a high market values have been proposed for the astaxanthin oleoresin prepared through supercritical CO_2 extraction method. The value of these oleoresin usually cost around US$8000 to 15,000 kg^{-1}. The astaxanthin content in these oleores differs between 5, 7, 10, or 15% based on individual commercial product (Astax Chile SPA) (Le-Feuvre et al. 2020).

European, North American countries, and China along with few other Asian countries were globally dominating the market of these oleoresins (Le-Feuvre et al. 2020). In the course of life cycle, the *H. pluvialis* cells changes from green to red condition where ultra-structural changes occur in the cells (Oslan et al. 2021). *H. pluvialis* can synthesize astaxanthin naturally via a one-stage process or two-stage process generally used as a dietary supplement owing to its easy availability, coupled with cardiovascular benefits (Kumar et al. 2022; Vardanega et al. 2022). 3,3′-dihydroxy-β-carotene-4,4′-dione, a brilliant blood-red color secondary carotenoid can be obtained from *H. pluvialis* by directly applying stress condition (Oslan et al. 2021).

18.3 Commercial Products Derived from *Haematococcus* and Their Market Value

18.3.1 Astaxanthin

The biotechnological, scientific, and commercial sectors have been showing considerably great interest in red colored carotenoid astaxanthin due to its great importance in human nutrition and health because of its apparent bio-functional characteristics. Currently, the synthetic route is mainly used for astaxanthin production for commercial purposes. However, there is a rapidly rising demand for sustainable and naturally derived source of astaxanthin from *H. pluvialis* for the production of nutraceutical and cosmetic products (Miki 1991). Extensive review has been made on the astaxanthin potential to protect human against a wide range of diseases including neurodegenerative diseases, UV light oxidative damage to the skin and eyes, liver diseases, cancer, inflammation, and gastric lesions. Astaxanthin is approved in countries such as Japan, USA, and some European countries to be used as coloring essence in salmon feed and as a nutraceutical material for human uses (Fábryová et al. 2020). Astaxanthin diesters (25%) and astaxanthin monoesters (70%) were the main form of esters present in astaxanthin derived from *H. pluvialis.*

H. pluvialis-based astaxanthin has much higher antioxidant (about 20 times) properties then that of the synthetically produced astaxanthin. Though *H. pluvialis-serived astaxanthin* possesses several biologically-important properties, high cost of *H. pluvialis*-based astaxanthin production hampers its share in the global market (Panis and Carreon 2016). The production cost of synthetic astaxanthin is about US$1000/kg which has market value over US$2000/kg. The Badische Anilin and Soda-Fabrik (BASF) and Dutch State Mines (DSM) were the leading global producers of astaxanthin (Jannel et al. 2020). Even if it is more expensive, for direct human consumption, consumers now prefer natural forms of astaxanthin due to the increase concerns of environment and health. *H. pluvialis* has been cultivated at industrial scale-level for extraction of astaxanthin for human uses (Jannel et al. 2020). The use of *H. pluvialis* as a human nutritional supplement is approved by

US FDA and EFSA. However, *H. pluvialis* contributes to about only 1% of the total global astaxanthin production (Jannel et al. 2020) .There is a growing demand of astaxanthin derived from *H. pluvialis* in the world market because of the rising trend of consumers to use nutraceutical and cosmetic product containing natural ingredients (Fábryová et al. 2020).

H. pluvialis-based astaxanthin market price was reported about US$2500–7000/kg, and in some cases, it goes up to US$15,000/kg. In 2014, its market potentiality was 280 metric tons, US$447 million; and was increased to 670 metric tons, US$1.1 billion, in 2020 (Shah 2019). A market research reported that the international market of the natural and synthetic derived astaxanthin was priced at US$555.4 million in 2016 as indicated in Algae Industry Magazine. Astaxanthin global market transaction can reach over US$250 million per year (Zhao et al. 2020).In 2018, astaxanthin market value reached US$600 million and by 2026 the market value is estimated to reach US$880 million (Silva et al. 2020). The price of *H. pluvialis* produced astaxanthin has estimated to reach US$7000 kg^{-1} due to its high cost of production, and in an adequate location, it is estimated to be almost US$1800 kg^{-1}. In 2019, this market reached approximately about US$123 billion and some economical analysts projected that the compound annual growth rate (CAGR) would expand at a rate of 8.2% growing the market to US$230 billion in 2027 (Hu 2019; Jannel et al. 2020). This dynamism could have been followed considering the growing evidence of the nutritional values and acquire a representative place in the world dietary supplement market in the forthcoming years.

Algatechnologies Ltd. in Israel, a subsidiary of Fuji Chemical Industry in Toyama, Japan, AstaReal Inc. in Hawaii, USA, BGG (Beijing Ginko Group) in Yunnan Province, China, Cyanotech Corporation in Hawaii, USA, and Parry's Pharmaceuticals in India are major *H. pluvialis*-based astaxanthin producers worldwide (Hu 2019; Jannel et al. 2020). For astaxanthin produced by *H. pluvialis,* functional food industries can come up as another dynamic market. Due to astaxanthin properties as a colorant and antioxidant, it may have various technological functionalities as a food ingredient and can provide protection during processing and storage of functional foods (Martínez-Álvarez et al. 2020). In 2018, the world market for functionalized foods was calculated at US$161 billion and by 2025 it is estimated to grow to a CAGR of 7.9% (GVR 2019). There is a huge economical market opportunity for astaxanthin derived from *H. pluvialis* if the regulatory hurdles to its production removed (Jannel et al. 2020).

Salmon production facilities widely utilizing astaxanthin derived from microalgae as additive material for their feeds. *H. pluvialis*-derived astaxanthin gives salmonids the distinctive salmon pigmentation, which is required by the customers. Alga technologies (Israel), cyanotech (Israel), Jingzhou Natural Astaxanthin Inc. (China), Algae tech International (Malaysia), and Parry Nutraceuticals (India) are the main astaxanthin-based feed producers in the market at present. In 2009, US$300 million and US$30 million were estimated for the astaxanthin market for animal feed and nutraceuticals, respectively. It was predicted that the market price of astaxanthin-based feed may reach about US$350 million by 2027 with natural astaxanthin market CAGR of 6% from 2021 to 2027 (www.

gminsights.com/industry-analysis/astaxanthin-market). Grand View Research, Inc. reported that the global market of astaxanthin would reach US$2.57 billion till 2025.

H. pluvialis-derived natural astaxanthin is predicted to lead the market and by 2024 rise to US$770 million with 7.7% CAGR. In nutraceutical industry, in 2017, natural astaxanthin derived from *H. pluvialis* had occupied 54.8% which considered as notably the largest contribution, and by 2024, the production level is estimated to reach 190 metric tons (Mu et al. 2019). Though various commercial companies (Cyanotech Corporation, Mera Pharmaceuticals Inc., Algatechnologies, Fuji Chemical Industry Co. Ltd., etc.) were involved in huge scale production of *H. pluvialis,* astaxanthin production capacity from *H. pluvialis* is far beyond the world demand of natural astaxanthin. In pharmaceutical and nutraceutical markets, the naturally produced astaxanthin is almost 3–4 folds more important than the synthetic astaxanthin (Shah et al. 2016). As market demand for naturally produced astaxanthin is increased gradually due to certain commercial applications, it is likely to replace the synthetic astaxanthin, there is a huge potential and appealing business opportunity for industrial level mass cultivation of *H. pluvialis* (Shah et al. 2016). It is anticipated that in future, the production cost of *H. pluvialis*-derived natural astaxanthin can be highly competitive owing to promising progress in the development of different production strategies. Even since mid-1990s, several major companies accomplished industrial-level production and global marketing of *H. pluvialis*-derived natural astaxanthin (Table 18.1). The space for astaxanthin in the nutraceutical market is increasing day to day. A very attractive market has been predicted for *Haematococcus* astaxanthin though the price of these products is considerably high (Fig. 18.2). To meet the global demand of consumers, the *Haematococcus* producers must focus more in increasing the astaxanthin yield without compromising quality. It is worth noting that in recent years, the cultivation ability of several manufacturers has already doubled. Besides, the increases in existing player's production ability, production giants like BGG with major production capacities have entered the market (Shah et al. 2016).

18.3.2 β-Carotene

Commercially, the first high-value material produced from microalgae is β-carotene (Borowitzka 2013). Aside from being an antioxidant, it is also a precursor for vitamin A synthesis. β-carotene is successful in controlling the levels of cholesterol and minimizing the risk of cardiovascular diseases and are effective in prevention and inhibition of different types of tumors in human body (Rammuni et al. 2019). Ionic radiation causes the decline of white blood cells and platelets which is prevented by β-carotene thus improving the immune system and shielding against exposures to undesired radiation (Rammuni et al. 2019).

From an economical perspective, in 2010, the universal demands led to a market value of US$261 million (Zarandi-Miandoab et al. 2019). In 2015, the world market price for β-carotene was valued at US$432.2 million (Joel 2016). In 2014, the world

Table 18.1 List of *Haematococcus*-based commercial products available in the global market

Microbial source	Astaxanthin content	Product
Haematococcus pluvialis	NAP[a]	Name: Night shift
		Company:Cipher
		Ingredients:Acai fruit oil, prebiotic green microalgae, glycyrrhetinic acid, panthenol, beta glucan, potassium azeloyl diglycinate, niacinamide, triple-enzyme DNA repair complex, alpha arbutin, glabridin, hemp seed oil, madecassoside, palmitoyl tripeptide-38
		Price: US$ 42–US$ 132
H. pluvialis	NAP[a]	Name: Ctrl z
		Company:Cipher
		Ingredients: prebiotic green microalgae, panthenol, palmitoyl tripeptide-38, resurfacing yeast oligopeptide, diglucosyl gallic acid, pineapple enzyme, prickly pear seed oil, madecassoside, undecylenoyl phenylalanine, salicylic acid, polyhydeoxy acid, alpha hydroxyl acids.
		Price: US$ 45–US$ 158
H. pluvialis	10%	Name: Astaxanthin Extract Powder
		Country: India
		Company: Maple life sciences
		Price: US$ 151.53
H. pluvialis	>10%	Name: Haematococcus pluvialis Extract capsule
		Company: YunDao Biotech production
		Price: US$ 76.19
H. pluvialis	NAP[a]	Name: BioAstin® Hawaiian Astaxanthin®
		Country: United States
		Company: Nutrex Hawaii
H. pluvialis	NAP[a]	Name: Astaxanthin 60
		Country: Germany
		Company: Dr. Peter Hartig
H. pluvialis	5%, 10% 20% Astaxanthin Oleoresin	Name: AstaPure®
		Country: United States
		Company: Algatech
H. pluvialis	3% Astaxanthin whole algae powder	Name: AstaPure® ARAVA
		Country: United States
		Company: Algatech
H. pluvialis	2.5% Astaxanthin CWD powder	Name: Astrapure®
		Country: United States
		Company: Algatech

(continued)

Table 18.1 (continued)

Microbial source	Astaxanthin content	Product
H. pluvialis	2.5% Astxanthin Beadlets and 2% Astxanthin vegetarian Beadlets	Name: AstaPure®
		Country: United States
		Company: Algatech
H. pluvialis	NAP[a]	Name: NOVASTA® Astaxanthin
		Country: Sweden
		Company: AstaReal
H. pluvialis	6%–8%	Name: AstaZine®
		Country: United States
		Company: Algaehealth
H. pluvialis	5%–10% Astaxanthin oleoresins	Name: Astalif 5 and Astalif 10
		Country: Iceland
		Company: Algalif
		Ingredient: 4 mg or 12 mg astaxanthin, formulated with vitamin E (D-α-tocopherol), sunflower oil, safflower oil, and rosemary extract
H. pluvialis	5%	Name: ALGAMO ORGANIC DIRED BIOMASS
		Country: Czech Republic
		Company: ALGAMO
H. pluvialis	5%–10%	Name: ALGAMO ORGANIC OLEORESIN
		Country: Czech Republic
		Company: ALGAMO
H. pluvialis	1%, 5% or 2%	Name: ALGAMO CWD
		Country: Czech Republic
		Company: ALGAMO
H. pluvialis	NAP[a]	Name: Algamo Astaxanthin
		Country: Czech Republic
		Company: ALGAMO
H. pluvialis	NAP[a]	Name: Algamo OED Forte
		Country: Czech Republic
		Company: ALGAMO
H. pluvialis	NAP[a]	Name: Algamo Bio Astaxanthin
		Country: Czech Republic
		Company: ALGAMO
H. pluvialis	NAP[a]	Name: Algamo Astaxanthin + CBD 10 mg
		Country: Czech Republic
		Company: ALGAMO
H. pluvialis	NAP[a]	Name: Algastin BL
		Country: Czech Republic
		Company: ALGAMO

(continued)

Table 18.1 (continued)

Microbial source	Astaxanthin content	Product
H. pluvialis	NAP[a]	Name: Algastin SHN
		Country: Czech Republic
		Company: ALGAMO
H. pluvialis	NAP[a]	Name: Algastin CV
		Country: Czech Republic
		Company: ALGAMO
H. pluvialis	NAP[a]	Name: Algastin
		Country: Czech Republic
		Company: ALGAMO
H. pluvialis	NAP[a]	Name: Algastin MC
		Country: Czech Republic
		Company: ALGAMO
H. pluvialis	NAP[a]	Name: SuperPure® Astaxanthin
		Country: United States
		Company: PURE SYNERGY
		Ingredient: *H. pluvialis* algae, Sunflower Lecithin, whole red wine grape, vegan (Pullulan) capsule
H. pluvialis	NAP[a]	Name: Super-Light Wrinkle Defense SPF 30
		Country: United States
		Company: Paula's choice skincare
H. pluvialis	NAP[a]	Name: Super Antioxidant Concentrate Serum with Retinol
		Country: United States
		Company: Paula's choice skincare
H. pluvialis	NAP[a]	Name: Super Antioxidant Concentrate Serum with Retinol
		Country: United States
		Company: Paula's choice skincare

[a]*NAP* not available for public

market value for β-carotene was US$200 million and reached to US$247 million by 2019 (Gervasi et al. 2017). According to the Market Watch report of 2019, the market value of β-carotene is expected to reach a level of US$618.94 million globally by 2026. From 2018 to 2026, it is anticipated to rise at a CAGR of 3.8% (Silva et al. 2020). In 2020, the commercial global market value of β-carotene was expected to be US$314.14 billion and with a yearly growth rate of 3.9% from 2020 to 2025, which is further expected to reach US$380.37 billion in 2025. Nowadays the synthesis of β-carotene is mostly via chemical route to fulfill the demand in different sectors. The development of the natural carotenoids sector has been triggered with an increase in preference of consumers towards clean and natural

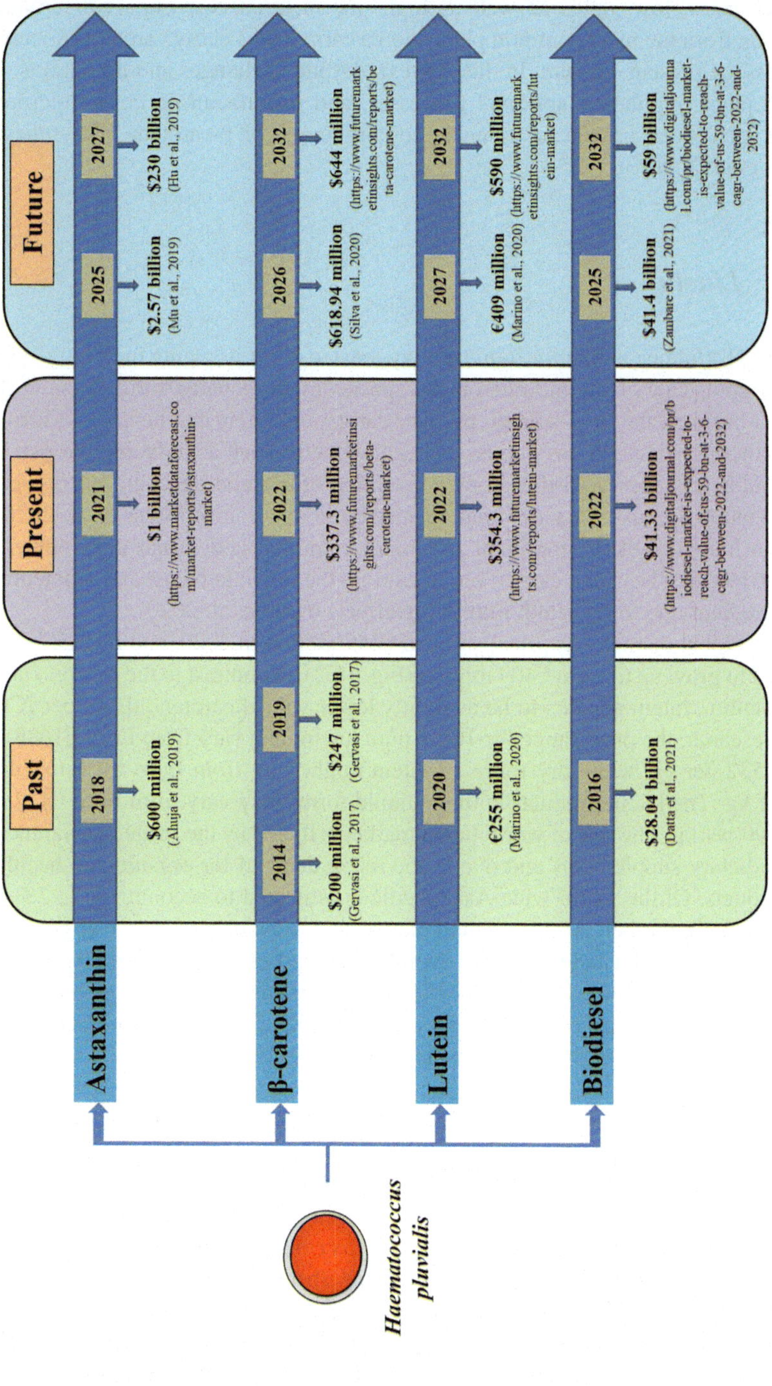

Fig. 18.2 Timeline representing the overall market values of *Haematococcus*-based products. Global market values of astaxanthin, β-carotene, lutein, and biodiesel were mentioned in past, present, and future perspective

label products (Papadaki and Mantzouridou 2021). Due to seasonal and geographical differences, low yields as well as high raw material cost impact β-carotene production from the most common plant source carrots thus carrots are not expected to address the present demand. In this context, despite challenges and limitations on scale-up plans, regulatory approval processes, and manufacturing costs, microorganisms have been proven to be an important source of β-carotene biosynthesis (Marino et al. 2020).

18.3.3 Lutein

Xanthophyll (lutein) is mainly found in oxygenated form in plants including spinach, marigold, kale and some microalgae species in oxygenated form (Saha et al. 2020). In humans, the lipid-soluble primary carotenoid lutein can be accessed from diet. Furthermore, Lutein has numerous health benefits such as reducing the risk of stroke and heart attack, assisting in the prevention of macular degenerative disease, and mitigation against other debilitating metabolic syndromes (Saha et al. 2020). Lutein in human milk is important for infant cognitive and visual development. Intensive research has been carried out to explore the lutein to be used as a potential food ingredient due to its health improving effects (Saha et al. 2020).

Lutein global market was priced at Euro €255 million in 2020, and by 2027, it is estimated to grow up to Euro €409 million (Fig. 18.2). In contrast to the market value of astaxanthin, lutein appears to be a slightly lower valued carotenoid. As per ICIS market research, the price range for 100% pure lutein may vary from Euro €1688 to Euro €2532 per kg, while dry forms of lutein might vary from Euro €126 to Euro €253 per kg. The lutein products with the liquid forms may vary from Euro €422 to Euro €590 per kg. The rise of world lutein market is fueled by the surge in awareness towards dietary supplements and due to the rising demand for organic and healthy food products. Of the world wide Asia pacific is estimated to account for 23.2% of the world lutein market. A high growth is expected to be observed in fast growing countries such as India and China. An overall 29% market share was held by dietary supplements in 2019 and is expected to keep their dominance over the global lutein market (https://www.prnewswire.com/news-releases/lutein-market-to-reach-usd-454-8-million-by-2026%2D%2Dreports-and-data-300941985.html).

18.3.4 Lipid (Biodiesel/Ethanol)

As per recent report, specific strains of *H. pluvialis* consist of 36–40% carbohydrates and 32–37% lipid on dry weight basis during the red phase of the microalga (Nishshanka et al. 2022). A study by Damiani et al. (2010) reported that the triacylglycerols (TAG) content of *H. pluvialis* can share 53.5% in their total lipid content. *H. pluvialis* can be utilized as a feedstock to produce biodiesel, animal feed, bioethanol for which integrated bio-refinery approaches would ensure the maximum use of input resources while lowering the amount of waste generated. In general, microalgae under stress conditions can accumulate high-value compounds along with biofuels. The upregulation of astaxanthin in *H. pluvialis* was always accompanied with the accumulation of non-polar TAGs and FAs, which were excellent precursors for biodiesel, demonstrated that cells exposed to nitrogen starvation or continuous highlight(HL) resulted in the accumulation of 32.99% and 34.85% dry weight, respectively (Ren et al. 2021).

The international biodiesel market size in 2016 was enumerated at US$28.04 billion (Datta et al. 2021). Eco-friendly fuels that reduce GHG emissions have seen increase in demand which has increased the interest towards biodiesel market. The international biodiesel market size in 2016 was enumerated at US$28.04 billion and would possibly reach US$41.2 billion by 2021 at a 3.8% CAGR. However, the world biodiesel market in the foreseeable future is anticipated to gain 5.255 CAGR from 2021 to 2028. The biodiesel demand is estimated to exceed US$41.4 billion liters by 2025 in which *Haematococcus* could potentially make significant contribution (Zambare et al. 2021; Datta et al. 2021). Though *H. pluvialis* has high biotechnological potential, the cost associated with production process for commercialization is challenging by considering the target as a single compound. Therefore, to overcome the cost associated with the production, researchers are aiming at producing several other compounds from same feedstock (Nishshanka et al. 2022). Accordingly, intracellular lipids have also been reported to accumulate in *H. pluvialis* under specific stress conditions. The lipid bodies in *H. pluvialis* composed of important fatty acids, and those can be used for various application after appropriate refinery downstream processes (Ren et al. 2021). Docosahexaenoic acid (DHA) and eicosapentaenoic acid (EPA) have been considered as high-value polyunsaturated fatty acids (PUFAs) generally obtained from microalgae. These have been found to be abundant in many species of microalgae those are important in health issue of neuron (Ren et al. 2021).

18.4 Competitive Potential of *Haematococcus* in Global Astaxanthin Market

The approach for commercial production of microbial carotenoids is more environmentally safe than the carotenoids manufactured synthetically by chemical route (Bhosale 2004). Despite the availability of different synthetic and other bioresources-derived carotenoids, there is a growing interest towards microbial derived carotenoids (Bhosale 2004). The green alga *H. pluvialis* and the heterobasidiomycetous yeast *Xanthophyllomyces dendrorhous* were currently well known for the astaxanthin to be produced biologically at the industrial scale. To be more competitive in the market, quantitative and qualitative improvement in microbial production of astaxanthin is a key requirement. Under continuous aeration and illumination fermentation condition, *X. dendrorhous* had produced a maximum of

4.7 mg/g astaxanthin in dry cell weight (Rodríguez-Sáiz et al. 2010). On the other hand, under continuous illumination without aeration, the maximum astaxanthin accumulated was reported as 9.2 mg/g in wet cell weight (up to 4%–5% by dry weight). At present the astaxanthin market is dominated by the synthetic manufactures, however, as an alternative both of these biotechnological processes can be considered.

Several recent industrial scale-up studies related to biotechnological processes involving *X. dendrorhous* and *H. pluvialis* were providing cost-effective procedures for the commercial utilization of astaxanthin (Rodríguez-Sáiz et al. 2010; Ding et al. 2020). At present, we are obtaining astaxanthin through chemical process or fermenting natural resource material like crustacean products (Barredo et al. 2017). Due to the lower production price (around US$1000/kg), 95% of global market was dominated by synthetic astaxanthin than naturally produced astaxanthin. Nowadays, NHU (China), DSM (The Netherlands), and BASF (France) were reported for large-scale production of synthetic astaxanthin. However, the final utilization of synthetically produced astaxanthin is limited due to its petrochemical origin, which in fact increasing the interest on naturally produced astaxanthin sources. Thus, astaxanthin obtained from natural sources such as *H. pluvialis*, *X. dendrorhous*, and bacteria such as *Paracoccus carotinifaciens* are a great alternative to synthetic astaxanthin (Barredo et al. 2017).

18.5 Forecast for *Haematococcus*-Based Commercially-Important Compounds

As eco-friendly products are obtaining great attentions these days, for which the market for astaxanthin production by biotechnological process is expected to rise to US$148.1 million (Bauer and Minceva 2021). In nutraceutical industry, the world market for astaxanthin from *H. pluvialis* is estimated to grow to 190 metric tons by 2024 with a price of US$770 million (Ding et al. 2020). In 2021, the total world astaxanthin market size was worth US$1.0 billion, and would reach US$3.4 billion with an annual compound rate of 16.2% increase between 2022 and 2027 (https://www.marketdataforecast.com/market-reports/astaxanthin-market). The global market valuation of β-carotene was reached US$337.3 million in 2022. The growth of the overall sales of β-carotene products is anticipated to increase at a rate of CAGR of 6.7% between 2022 and 2032 amounting up to US$644 million by the end of 2032 (https://www.futuremarketinsights.com/reports/beta-carotene-market). On 2022, the value of the world lutein market is estimated at US$354.3 million and with a growth at a CAGR of 5.2%, the market is estimated to be valued around US $590 million in 2032 (https://www.futuremarketinsights.com/reports/lutein-market). By the end of 2022, the biodiesel world market is anticipated to grow to a value of US$59 billion. It is projected that the biodiesel market will increase at 3.6% CAGR between 2022 and 2032, and is anticipated to grow a valuation of US

$59 billion in the year 2023 (https://www.digitaljournal.com/pr/biodiesel-market-is-expected-to-reach-value-of-us-59-bn-at-3-6-cagr-between-2022-and-2032).

18.6 Conclusions

H. pluvialis has its application in different industries such as cosmetics, food, and pharmaceutical industries. Expansion of nutraceutical, animal feed, and cosmetic markets demand high quantity of astaxanthin. Furthermore, the demand for natural sourced biomaterials as aquaculture feed and dietary supplements is increasing day to day. *H. pluvialis* could potentially fill these gaps in the markets. However, astaxanthin produced from natural sources such as *H. pluvialis* must compete with the synthetic form of astaxanthin which still dominates the industrial market. Although the natural astaxanthin production from *Haematococcus* requires advance technology and high production cost, these can be lowered by utilization and development of effective bioprocessing and cultivation approaches. As high growth and demand has been seen in the world market for bioactive compounds, *H. pluvialis*-derived products can potentially dominate the market in future.

Acknowledgments The authors wish to thank the Director, CSIR-NEIST, Jorhat, Assam for his kind encouragement in carrying out this work. This chapter was supported by research funds of DST-SYST program (SP/YO/2019/1591) funded by the Department of Science and Technology (DST), Government of India.

References

Baby AR, Morocho-Jácome AL (2021) Dermocosmetic applications of microalgal pigments. Adv Appl Microbiol 117:63–93

Barredo J, García-Estrada C, Kosalkova K, Barreiro C (2017) Biosynthesis of astaxanthin as a main carotenoid in the Heterobasidiomycetous yeast *Xanthophyllomyces dendrorhous*. J Fungi 3(3): 44

Bauer A, Minceva M (2021) Techno-economic analysis of a new downstream process for the production of astaxanthin from the microalgae *Haematococcus pluvialis*. Bioresour Bioprocess 8:111

Bhosale P (2004) Environmental and cultural stimulants in the production of carotenoids from microorganisms. Appl Microbiol Biotechnol 63:351–361

Borowitzka MA (2013) High-value products from microalgae—their development and commercialisation. J Appl Phycol 25(3):743–756

Butler T, Golan Y (2020) Astaxanthin production from microalgae. In: Alam M, Xu JL, Wang Z (eds) Microalgae biotechnology for food, health and high value products. Springer, Singapore, pp 175–242. https://doi.org/10.1007/978-981-15-0169-2_6

Colusse GA, Duarte MER, de Carvalho JC, Noseda MD (2021) Production of astaxanthin by *Haematococcus pluvialis*: lab processes to scale up including the cost considerations. In: Ravishankar GA, Rao AR (eds) Global perspectives on astaxanthin. Academic Press, pp 121–130. ISBN 97801282330147. https://doi.org/10.1016/B978-0-12-823304-7.00019-2

Damiani MC, Popovich CA, Constenla D, Leonardi PI (2010) Lipid analysis in Haematococcus pluvialis to assess its potential use as a biodiesel feedstock. Bioresour Technol 101(11): 3801–3807

Datta I, Ghosh A, Acharjee A, Rakshit A, Saha B (2021) Overview on biodiesel market. Vietnam J Chem 59(3):271–284

Davies SJ (2008) Colouration and flesh quality in farmed salmon and trout. In: Lie Q (ed) Improving farmed fish quality and safety improving farmed fish quality and safety. Woodhead Publishing, pp 446–470. ISBN 9781845692995. https://doi.org/10.1533/9781845694920.2.446

Ding Y, Zhang A, Wen X, Wang Z, Wang K, Geng Y, Li Y (2020) Application of surfactants for controlling destructive fungus contamination in mass cultivation of *Haematococcus pluvialis*. Bioresour Technol 317:124025

Fábryová T, Tůmová L, da Silva DC, Pereira DM, Andrade PB, Valentão P, Cheel J (2020) Isolation of astaxanthin monoesters from the microalgae *Haematococcus pluvialis* by high performance counter current chromatography (HPCCC) combined with high performance liquid chromatography (HPLC). Algal Res 49:101947

Gervasi T, Pellizzeri V, Benameur Q, Gervasi C, Santini A, Cicero N, Dugo G (2017) Valorization of raw materials from agricultural industry for astaxanthin and β-carotene production by *Xanthophyllomyces dendrorhous*. Nat Prod Res 32(13):1554–1561

Hu IC (2019) Production of potential coproducts from microalgae. In: Paney A, Chang J, Soccol CR, LEe D, Chisti Y (eds) Biofuels from algae, 2nd edn, pp 345–358. ISBN 9780444641922. https://doi.org/10.1016/B978-0-444-64192-2.00014-7

Jannel S, Caro Y, Bermudes M, Petit T (2020) Novel insights into the biotechnological production of *Haematococcus pluvialis*-derived astaxanthin: advances and key challenges to allow its industrial use as novel food ingredient. J Mar Sci Eng 8(10):789

Joel J (2016) Carotenoids market by type (astaxanthin, beta-carotene, lutein, lycopene, canthaxanthin, zeaxanthin, and others) for feed, food, supplements, cosmetics, and pharmaceuticals - global industry perspective, comprehensive analysis, size, share, growth, segment, trends and forecast, 2015 – 2021. Zion Market Research, United States, p 110

Khoo KS, Ooi CW, Chew KW, Foo SC, Lim JW, Tao Y, Jiang N, Ho SH, Show PL (2021) Permeabilization of *Haematococcus pluvialis* and solid-liquid extraction of astaxanthin by CO2-based alkyl carbamate ionic liquids. Chem Eng J 411:128510

Kim B, Lee SY, Narasimhan AL, Kim S, Oh YK (2022) Cell disruption and astaxanthin extraction from Haematococcus pluvialis: recent advances. Bioresour Technol 343:126124

Kumar S, Kumar R, Diksha K, A., & Panwar, A. (2022) Astaxanthin: A super antioxidant from microalgae and its therapeutic potential. J Basic Microbiol 62(9):1064–1082

Le-Feuvre R, Moraga-Suazo P, Gonzalez J (2020) Biotechnology applied to *Haematococcus pluvialis* fotow: challenges and prospects for the enhancement of astaxanthin accumulation. J Appl Phycol 32:3831–3852

Li J, Zhu D, Niu J, Shen S, Wang G (2011) An economic assessment of astaxanthin production by large scale cultivation of *Haematococcus pluvialis*. Biotechnol Adv 29(6):568–574

Lorenz RT, Cysewski GR (2000) Commercial potential for *Haematococcus* microalgae as a natural source of astaxanthin. Trends Biotechnol 18(4):160–167

Marino T, Casella P, Sangiorgio P, Verardi A, Ferraro A, Hristoforou E, Molino A, Musmarra D (2020) Natural beta-carotene: a microalgae derivate for nutraceutical applications. Chem Eng Trans 79:103–108

Martínez-Álvarez Ó, Calvo MM, Gómez-Estaca J (2020) Recent advances in astaxanthin micro/nanoencapsulation to improve its stability and functionality as a food ingredient. Mar Drugs 18: 406

Miki W (1991) Biological functions and activities of animal carotenoids. Pure Appl Chem 63:141–146

Mota GCP, Moraes LBSD, Oliveira CYB, Oliveira DWS, Abreu JLD, Dantas DMM, Gálvez AO (2022) Astaxanthin from *Haematococcus pluvialis*: processes, applications, and market. Prep Biochem Biotechnol 52(5):598–609

Mu N, Mehar JG, Mudliar SN, Shekh AY (2019) Recent advances in microalgal bioactives for food, feed, and healthcare products: commercial potential, market space, and sustainability. Compr Rev Food Sci Food Saf 18(6):1882–1897

Mularczyk M, Michalak I, Marycz K (2020) Astaxanthin and other nutrients from *Haematococcus pluvialis*—multifunctional applications. Mar Drugs 18(9):459

Nishshanka GKSH, Liyanaarachchi VC, Nimarshana PHV, Ariyadasa TU, Chang JS (2022) *Haematococcus pluvialis*: a potential feedstock for multiple-product biorefining. J Clean Prod 344:131103

Oslan SNH, Tan JS, Oslan SN, Matanjun P, Mokhtar RAM, Shapawi R, Huda N (2021) *Haematococcus pluvialis* as a potential source of astaxanthin with diverse applications in industrial sectors: current research and future directions. Molecules 26(21):6470

Panis G, Carreon JR (2016) Commercial astaxanthin production derived by green alga *Haematococcus pluvialis*: a microalgae process model and a techno-economic assessment all through production line. Algal Res 18:175–190

Papadaki E, Mantzouridou FT (2021) Natural *β*-carotene production by *Blakeslea trispora* cultivated in Spanish-style green olive processing wastewaters. Foods 10(2):327

Rammuni MN, Ariyadasa TU, Nimarshana PHV, Attalage RA (2019) Comparative assessment on the extraction of carotenoids from microalgal sources: astaxanthin from *H. pluvialis* and β-carotene from *D. salina*. Food Chem 277:128–134

Ren Y, Deng J, Huang J, Wu Z, Yi L, Bi Y, Chen F (2021) Using green alga *Haematococcuspluvialis* for astaxanthin and lipid co-production: advances and outlook. Bioresour Technol 340:125736

Rodríguez-Sáiz M, de la Fuente JL, Barredo JL (2010) Xanthophyllomyces dendrorhous for the industrial production of astaxanthin. Appl Microbiol Biotechnol 88(3):645–658

Saha SK, Ermis H, Murray P (2020) Marine microalgae for potential lutein production. Appl Sci 10(18):6457

Sarada R, Tripathi U, Ravishankar G (2002) Influence of stress on astaxanthin production in *Haematococcus pluvialis* grown under different culture conditions. Process Biochem 37(6): 623–627

Shah MMR (2019) Astaxanthin production by microalgae *Haematococcus pluvialis* through wastewater treatment: waste to resource. In: Gupta S, Bux F (eds) Application of microalgae in wastewater treatment. Springer, Cham, pp 17–39. ISBN 978-3-030-13908-7. https://doi.org/10.1007/978-3-030-13909-4_2

Shah MMR, Liang Y, Cheng JJ, Daroch M (2016) Astaxanthin-producing green microalga *Haematococcus pluvialis*: from single cell to high value commercial products. Front Plant Sci 7:531

Silva SC, Ferreira ICFR, Dias MM, Barreiro MF (2020) Microalgae-derived pigments: a 10-year bibliometric review and industry and market trend analysis. Molecules 25(15):3406

Torres-Tiji Y, Fields FJ, Mayfield SP (2020) Microalgae as a future food source. Biotechnol Adv 41:107536

Vardanega R, Cerezal-Mezquita P, Veggi PC (2022) Supercritical fluid extraction of astaxanthin-rich extracts from *Haematococcus pluvialis*: economic assessment. Bioresour Technol 361: 127706

Zambare V, Patankar R, Bhusare B, Christopher L (2021) Recent advances in feedstock and lipase research and development towards commercialization of enzymatic biodiesel. PRO 9(10):1743

Zarandi-Miandoab L, Hejazi MA, Bagherieh-Najjar MB, Chaparzadeh N (2019) Optimization of the four most effective factors on β-carotene production by *Dunaliella salina* using response surface methodology. Iran J Pharm Res 18(3):1566–1579

Zhao Y, Hou Y, Chai W, Liu Z, Wang X, He C, Hu Z, Chen S, Wang W, Chen F (2020) Transcriptome analysis of Haematococcus pluvialis of multiple defensive systems against nitrogen starvation. Enzym Microb Technol 134:109487

Chapter 19
Bioenergy Applications of *Haematococcus*

Nilay Kumar Sarker, Tirthankar Mukherjee, and Prasad Kaparaju

19.1 Introduction

Microalgae are microscopic organisms classified as eukaryotic micro-organisms or prokaryotic cyanobacteria (Barbera et al. 2016). Till date more than 25,000 species have been isolated and identified (Mata et al. 2010). Similar to plants, microalgae perform photosynthesis using CO_2 and other nutrients. Growth mechanism of microalgae may have similarity and they can be referred as plant-like organisms, but they have a lot of dissimilarity to terrestrial plants. Similarly, microalgae are unicellular and have simple reproductive structure, they do not have roots, stems, leaves, and vascular tissues, they grow and then multiply (Randrianarison and Ashraf 2017). Microalgae can grow in autotrophic, heterotrophic, and mixotrophic ways (Chen et al. 2011; Adesanya et al. 2014). During the autotrophic growth, microalgae grow and consume CO_2 as carbon source and light as energy and produce organic matters and energy contents. While in autotrophic growth, microalgae use organic compounds as source of energy and carbon. The mixotrophic pathway consists of two-stage growth, heterotrophic as the first stage when organic carbon matters are high, followed by autotrophic growth with the induction of photosynthesis.

Evidence of microalgal use for human consumption shows that people are familiar with many benefits of this microorganism and accepted as a food in their lifelong time ago. For example, Spirulina was cultivated as an ingredient of cake named "Tecuitlatl" in ancient Aztecs civilization in Mexico (Abalde et al. 1995). Natives to Chad used *Spirulina* to prepare "Dihe." In China for more than 2000 years another microalgae, *Nostoc* flagella for me was used to prepare a type of food called

N. K. Sarker · T. Mukherjee · P. Kaparaju (✉)
School of Engineering and Built Environment, Griffith University (Nathan Campus), Nathan, QLD, Australia
e-mail: p.kaparaju@griffith.edu.au

R. Raja et al. (eds.), *Haematococcus*, https://doi.org/10.1007/978-981-99-2901-6_19

"facai" (Gao 1998). Research on microalgal to explore its applications started around 80 years ago, such as lipid production and wastewater treatment (Harder and von Witsch 1942; Golueke et al. 1957). The oil crisis of 1973 motivated new research projects to find alternative sources of energy and some of them were focused on microalgae-based gaseous and liquid biofuels. After conducting research for more than two decades the "US Aquatic Species Program" at the "National Renewable Energy Laboratory," formerly known as the "Solar Energy Research Institute" concluded in the mid-2000s that microalgal biofuel production technologies were not commercially viable (Sheehan et al. 1998). Till now, microalgae are commercially produced mainly to produce food and feed. Microalgae can be grown commercially in any aqueous medium, where nutrients and light are available. Microalgae can be produced in large scale in open ponds and closed photobioreactors (PBRs). Open ponds require less construction and maintenance cost which made it popular way of commercial algal cultivation. But open ponds have disadvantages too. Contaminations, insufficient light distribution and dilutions by rain are the most important ones. Closed PBRs offer the opportunities to address these problems. But the operation and maintenance cost of PBRs are high. Because of the cost related issues, only a handful of companies have built PBR based microalgae production facilities.

Different types of algal strains are popular for different types of bioenergy applications. As the urge for renewable biofuel is increasing, we need to explore new biomass resources. Keeping that in mind, different types of algal strains, which were not considered before for bioenergy applications should be investigated to expand the range of options. In this chapter we will discuss about the bioenergy applications of *Haematococcus* species. They are green microalgae, belong to Chlorophycean genus. *Haematococcus* species can be feedstocks of high value products such as nutraceuticals, pharmaceuticals, and cosmetics along with aquaculture feeds and other food products (Guerin et al. 2003; Ambati et al. 2014). Specially, *Haematococcus pluvialis* is considered a promising feedstock of β-carotene, lutein, canthaxanthin, and astaxanthin (Amaya and Nickell 2015). But bioenergy applications of *Haematococcus* have not explored enough. Very few studies were carried out on *Haematococcus* to explore their energy potentials. This book chapter highlights the energy applications of *Haematococcus*. In this chapter we discussed all those available articles. It should be noted that, almost all of articles on energy applications on *Haematococcus* are recent studies. Through this book chapter we highlighted what has been done on this matter. For the future research studies interested on energy applications of *Haematococcus*, this book chapter will help them to know what has not been done and where need to focus to expand the knowledge on energy applications of *Haematococcus*.

19.2 Bioenergy Potential of *Haematococcus*

Several factors determine the bioenergy potentials of biomass, such as high higher heating value (HHV), high volatile contents, lower moisture and low ash content are preferable features of biomass for energy applications. These parameters not only represent the stored energy in biomass but also act as indicator of other factors for commercial uses, such as handling and processing cost of the bioenergy application. *Haematococcus* showed satisfactory HHV comparable to *Chlorella* sp. a popular algal strain for bioenergy research, high volatile matter, and low ash content but more than desirable moisture content (Haque et al. 2017). This makes this strain not ideal but potent for bioenergy. As of till now, main application of *Haematococcus* is astaxanthin production, a lot of interest is found for residual *Haematococcus*. It was found that ideally because of higher moisture content, residual *Haematococcus* is less competent than whole *Haematococcus*, additionally whole *Haematococcus* also have higher HHV (Haque et al. 2017), but laboratory experiments showed that energy conversion of residual biomass of *Haematococcus* is more convenient.

Any type of bioenergy conversion possesses concern of environmental pollution, especially NOx and SOx emission. Because Sox can cause corrosion to the process equipment. Ultimate analysis showed that *Haematococcus* have satisfactory limit of N and S, which will cause the formation of environmentally favorable NO_x and SO_x (Haque et al. 2017), which is comparable to rice husk, a popular biomass used to power generation commercially. The ratio of hydrogen and carbon of *Haematococcus* is similar to Chlorella, *Nannochloropsis* and *Spirulina* species (Chow et al. 2013; Phukan et al. 2011; Zhu et al. 2013), these species are being widely investigated for energy applications. *Haematococcus* is well known for algal research on food and medicine ingredients. But the similarity of stored energy profile shows it has also adequate potential for energy as well.

19.3 Biomass Conversion of Haematococcus for Energy Applications

The history of biomass utilization dates back to the history of inhabitation of ancient people on earth. Accessing bioenergy was easy for the ancient people because biomass is a source of combustible carbon, which was burnt directly to create light and heat, and to cook food. For thousands of years combustion was the only way to use biomass. Other forms of conversion techniques such as thermochemical and biochemical conversions for industrial uses were being adopted from around the nineteenth century. Converting biomass into energy through thermochemical processes can be categorized into three major routes. All these processes require high operating temperatures. Due to the cost efficiency and ease to operate such processes, all these thermochemical routes have been commercialized and are used to generate energy from biomass (Kumari and Karmee 2022). The three major thermochemical routes are: pyrolysis, gasification, and liquefaction. Biomass also can be converted to energy through biochemical conversion which is considered environment friendly approaches. Two most popular biochemical conversion routes used for biofuel production are fermentation and anaerobic digestion (AD).

19.4 Pyrolysis

Fundamentals of Pyrolysis: Around 5000 years ago, pyrolysis mechanism was first developed in some European countries and the Middle East to produce charcoal. These countries utilized the generated biochar from pyrolysis for waterproofing their boats. Normally biomass is comprised of lignin, hemicellulose, and cellulose. Biomass pyrolysis takes place in an inert environment via heat treatment process (Jahirul et al. 2012). The characteristics of the pyrolysis products are governed by several parameters such as type of biomass, temperature of the system, and also heating rate. The principal pyrolysis products are pyro-gas, biochar, and bio-oil which is formed from the condensable gases (Mukherjee et al. 2019). The pyro-gas generated constitutes of non-condensable gases such as CH_4, CO_2, CO, and H_2. Gas-solid heterogeneous and gas-phase homogeneous reactions occur during thermal decomposition of biomass. Predominantly in gas-phase reactions the condensable vapors dissociate into tiny molecules of non-condensable gases like carbon dioxide and carbon monoxide. Pyrolysis oil formed from the thermal decomposition of solid biomass has exhibited capability as an alternative source of energy in power plants and other various industrial applications for internal combustion engines or steam boilers (Debnath et al. 2021).

One disadvantage of pyrolysis is that the process is cost intensive. This indicates that this process is only feasible for establishments which have huge quantities of biomass waste so that they can convert it into biofuel and other pyro-products. The cost of this process increases because it requires a continuous supply of fuel to run in full capacity. The conversion of biomass to bio-oil via pyrolysis still needs to overcome a few challenges. Better understanding of this technology is essential which can effectively use the biomass for large scale commercial production. Biomass pyrolysis was investigated thoroughly because it offers better bio-oil conversion, flexible adjustment of desired output such as pyro-gas, oil or bio char, comparatively lower operation cost and lower nitrogen and sulfur in bio-oil (Azizi et al. 2018).

At present, the key usage of the pyrolysis technology is for electricity generation and carbon dioxide reduction with socio-economic advantages. Although several pyrolysis plants have been installed, these are vastly reliant on biomass cost and availability, labor costs, fuel price, etc. Generally, the pyrolysis technology is operated in pilot scale for research use while only a few are used in industrial practices for production of steam, electricity, and heat (Mitra et al. 2021). Moreover, scientists are focusing more on maximizing the overall pyro-oil yield rather than considering upgrading bio-oil. Ensyn and DynaMotive are two of the few pyrolysis companies that have been leading in North America since the 1990s. Both the companies consume renewable biomass like agriculture wastes and forest residues as the feedstock. Another two renowned pyrolysis companies are Pyrovac and Renewable Oil International that fundamentally use vacuum pyrolysis technology (Kumari and Karmee 2022).

19.4.1 Pyrolysis of Haematococcus

Gong et al. (2020) conducted pyrolysis on fresh and residual biomass of *Haematococcus pluvialis* and obtained oil, gas, and char from both of them (Gong et al. 2020). Although both types of biomasses showed potential to be feedstock of bioenergy, reaction parameters and results of pyrolysis were not same. For example, average activation energy was found 204.72 and 178.51 kJ/mol for fresh and residual *Haematococcus,* respectively. The produced gas from both biomasses had CO_2, CO, H_2, and CHs, CHs were relatively lower in pyro-gas of residual biomass but CH_4 contents were higher. Pyrolyzing residual biomass can be an effective way to add value to the so-called waste by-product. In their experiments, they conducted pyrolysis of raw biomass and residual biomass at 400, 500, 600, 700, and 800 °C. For the raw biomass, they found char yield decreased to one-third for increasing temperature from 400 to 500 °C. When temperature increased from 500 to 800 °C, char yield remained almost the same. This happened because the lipid decomposed at 400 °C. Temperature increase from 400 to 700 °C increased oil yield 2.5 times. But beyond that temperature oil production rate decreased because high temperature caused liquid content to gasify. The gas yield showed a comparatively complex trend. Temperature increases from 400 to 600 °C caused increase of gas yield to 5% and then decreased by 7% at 700 °c. Then again gas yield increased 9% at 800 °C. From 500 to 800 °C, among all the products the gaseous products were found highest followed by oil and char products. When whole biomass was pyrolyzed, above 400 ° C, the recyclable part was more than 80% suggesting a great recovery value. The product distribution pattern of pyrolysis of residual biomass showed similar trend. Although products of residual biomass showed higher recovery value with a certain amount of pyrolysis gas and oil yield. The liquid hydrocarbons from pyrolysis which is known as pyrolysis oil has versatile uses for energy such as can be refined and upgraded to gasoline, diesel, and heavy oil (Gong et al. 2017). The produced pyro-oil from both whole and residue biomass showed no significant differences at their boiling point curves from simulated distillation tests. For example, when residual biomass was pyrolyzed at a range of 180–350 °C, mass recovery initially decreased before gradually coinciding above 400 °C. Different types of commercial oil products can be identified based on the ranges of boiling products. The ranges of boiling point curves of the liquid products of this experiment showed the found pyro-oil mainly consisted of gasoline and diesel (whole biomass, >90%; residual biomass, >79%). The content of diesel and heavy oil increased when pyrolysis temperature was increased but gasoline content decreased slightly. The explanation is that the higher temperature accelerated the decomposition of macromolecules and the gasification of short-chain hydrocarbons.

Ashokkumar et al. (2021) produced biochar from *Haematococcus* residue through pyrolysis. They synthesized mini pellets using this biochar and hematite (Fe_2O_3). The biomass residue found after the extraction of astaxanthin was pyrolyzed to produce biochar. This biochar was utilized for the production of metallic iron successfully. Then they studied hematite iron ore (Fe_2O_3) reduction with *Haematococcus* biochar. They found high temperature (≥975 °C) supported the conversion. This was a cost-effective process where natural astaxanthin, a high value product can be produced as well as biochar, a solid biofuel can be produced which can replace coke to produce iron. In this way algal biochar can reduce energy utilization, increase cost-effectiveness, and enhance CO_2 mitigation in ironmaking industry (Ashokkumar et al. 2021).

While producing biochar from *Haematococcus* using pyrolysis, the heating rate during pyrolysis determines the biochar yield. A positive relation was found between the pyrolysis heating rate and biochar yield. A faster heating process is also responsible for less residence time. Complete degradation (or higher degradation) requires higher residence time. Therefore, faster heating rate is responsible for the incomplete degradation at the end of the conversion process. Another study observed pyrolysis and biochar yield from several microalgae species (Vasudev et al. 2021). They found *Haematococcus pluvialis* yielded slightly less biochar compared to Chlorella vulgaris when heating rate was increased. But compared to many other terrestrial biomass, *Haematococcus pluvialis* showed no observable differences among the three pyrolysis heating rates investigated (Vasudev et al. 2021).

Recently, biochar received attention from researcher communities because of its versatile potential applications, for example, as a replacement solid fuel, soil improvement (in agricultural field), CO_2 sequestration (in environment field) (Anto et al. 2021). Studies argued that biochar production can offer a potential biorefinery approach which is also capable of carbon sequestration (Hagemann et al. 2017). Yu et al. (2017) claimed microalgae as a sustainable and renewable biomass which can produce sustainable biochar (Yu et al. 2017). A pyrolysis process was carried out without the presence of any catalyst. During the pyrolysis, heat was provided at a rate of 30 °C/min at 450 °C. The result showed highest biochar yield of 43.2 wt%. Woolf et al. (2010) suggested biochar as a promising tool to mitigate global climate change. They suggested that biochar has the larger potential than the combustion of sustainable biomass for bioenergy to mitigate climate change. Earlier reports show that the approach of using algal biomass pyrolysis with reduced emissions to produce biochar appears to be a promising pathway to sustainability (Smith et al. 2013). Many researchers also worked co-pyrolysis of microalgae with other substances. Technologically, several attempts found successful. Such as co-pyrolysis of microalgae with coal (Chen et al. 2012) and textile dyeing sludge (Peng et al. 2015) was found suitable to produce different forms of energy. It was also found that algal ratio of more than 80% during co-pyrolysis guarantee a lower average activation energy and desired pyro-gas products.

19.5 Gasification

19.5.1 Primary Concepts of Gasification

Biomass gasification takes place in reduced air environment where the carbon rich materials are decomposed into syngas occurs. This process is a mixture of diverse categories of thermochemical conversion technology. It includes pyrolysis, combustion, and followed by reduction. The key product of gasification is syngas, which has multiple uses in process industries uses. The subsequent sections discuss the importance of the different reactions occurring in each zone during the gasification process (Neubauer 2013).

19.5.1.1 Pyrolysis Zone

This zone is endothermic in nature and the moisture from the biomass is removed at 200 °C. Gases such as CO_2, H_2O, and acetic acid are driven off between 200 and 300 °C. The main devolatilization occurs between 300 and 500 °C. In this stage, huge amounts of tar comprising of methyl alcohol and CO_2 rich gases are produced. Moreover, H_2 is produced between 500 and 700 °C temperature.

19.5.1.2 Oxidation Zone

The burning of carbon rich material in oxygenated environment is the combustion process. This stage is exothermic in nature and the oxidation temperature is about 1400 °C. The main reactions in this stage are as follows:

$$C + O_2 \rightarrow CO_2 \; (\Delta H_0 = -391.6 \text{ KJ/mol}) \tag{19.1}$$

$$2C + O_2 \rightarrow 2CO \; (\Delta H_0 = -286 \text{ KJ/mol}) \tag{19.2}$$

19.5.1.3 Reduction Zone

The products obtained from the oxidation zone are passed over a red-hot dried bed of charcoal. The reduction reactions take place in this zone and the chief products depends upon the gasifying medium (Neubauer 2013). The main reactions in this stage are as follows:

$$C + H_2O \rightarrow CO + H_2 \; (\Delta H_0 = +131.79 \text{ KJ/mol}) \tag{19.3}$$

$$C + CO_2 \rightarrow 2CO \ (\Delta H_0 = +179.3 \ KJ/mol) \qquad (19.4)$$

$$CO + H_2O \rightarrow CO_2 + H_2 \ (\Delta H_0 = -47.49 \ KJ/mol) \qquad (19.5)$$

$$C + 2H_2 \rightarrow CH_4 \ (\Delta H_0 = -22 \ KJ/mol) \qquad (19.6)$$

Reactions 3 and 4 are the major endothermic reactions taking place in this zone. Thus, in this zone, the temperature lies between 800 and 1000 °C. The temperature of the reduction zone directly influences the calorific value of the produced gases.

Thus, the gaseous products obtained throughout the gasification process is a combination of gases like H_2, CO, CO_2, CH_4, and C_2H_4, along with several additional gases such as nitrogen, sulfur, alkali compounds, and tars. This is an endothermic route, and thus it involves heat energy for additional reaction. Gasification can be carried out via several carrier gases such as air, O_2, steam, CO_2, or even a combination of these gases in different ratios. The gasifying medium is a significant parameter in its product composition. More H_2 is formed when steam is utilized as a gasifying medium because of water gas shift and methane reforming reactions (Neubauer 2013). Gasification products have various usages involving power or combined heat and power, generation of heat, or it is additionally processed for the synthesis of liquid fuel and other organic/inorganic chemicals.

19.6 Gasification of *Haematococcus*

Thomassen et al. (2019) worked on a complete process flow of commercial gasification plant of *Haematococcus*. Microalgae were cultivated in multi-stage cultivation units. Then biomass was separated through centrifuge and dried via spray drying. After that it was pre-treated in a bead mill and then a component was extracted. After that the residual biomass was dried by evaporation and the dried residual biomass went through gasification process which produced gasoline, syngas, biochar, and diesel. The authors conducted several processes, such as extracting astaxanthin in one process and B-carotene in another process. They found that both of the processes showed adequate technical feasibility, but commercial feasibility was still far from reality. Main reason was high overall energy requirement. Their study clearly showed that *Haematococcus* is capable for energy applications through gasification but its commercialization requires technological innovation and effective energy management (Thomassen et al. 2019).

19.7 Thermochemical Liquefaction

19.7.1 Basics of Thermochemical Liquefaction

The process of converting gas or solid materials into liquid is called liquefaction. This process functions at moderate temperatures of 280–370 °C and at very high pressures at about 1025 MPa, which helps the formation of biocrude. The greatest advantage of this process is that it can convert any biomass containing any type of moisture content into biocrude. Liquefaction is regarded as a clean technology because it does not produce harmful by-products such as NH_3, NO_x, and SO_x (Garcia-Garcia et al. 2022). This process possesses a high conversion rate and pure product formation, which reforms the organic fraction present in the biomass feedstocks. This process can be operated in the presence of H_2, catalyst and also at higher pressure to enhance the reaction rate and the quality of the produced fuel. At extremely high temperature, H_2O in supercritical form, behaves as a reactant, which reforms the organic matter via the addition of H_2 ions into the hydrocarbons. High temperature is not suitable for hydrothermal liquefaction. Because high temperature may cause to Bourdard reaction which will be responsible for production of more gaseous contents than of liquid contents (Osman et al. 2021). The products from liquefaction, i.e., bio-oil is generally used as fuels in engines, boilers, and also turbines for heat and power generation.

19.7.2 Thermochemical Liquefaction of **Haematococcus**

Liquefaction of microalgae offers several benefits. One benefit is it produces high quality biocrude and other one is it can be done with wet biomass. As microalgae are cultivated in aqueous medium, one major energy consuming process for algal processing is drying. However, hydrothermal liquefaction (HTL) saves this energy. But till date, it is not a mature technology. Research is going on about the use of suitable catalysts. In laboratories, effective catalysts are available. However, almost all of them are not suitable for practical use because of their cost and availability. Hong et al. (2021) conducted experiments on HTL of *Haematococcus pluvialis* using metal-supported montmorillonite catalysts. These catalysts are of low cost and can be prepared conveniently. They experimented different types of metals. Among them iron showed best performance. At low temperature (200 °C) the combination of Fe/montmorillonite catalyst and methanol solvent provided maximal biocrude yield of around 70% and the highest selectivity was found more than 99%. This catalyst increased the viscosity of the biocrude by changing the composition of bio crude. The catalyst was responsible for enhancing the cracking reactions of non-volatile tarry residue and this incident caused changed composition of biocrude. Due to this catalyst, esters were the major compounds of the biocrude, and fatty acids contents and Nitrogen containing compounds in biocrude were decreased significantly. A co-solvent made of methanol and ethanol showed a good synergistic effect and helped for liquefaction of microalgae. Although selectivity and yield found for co-solvent was slightly less compared with single solvent, but the reaction time was reduced significantly.

19.8 Fermentation

19.8.1 Fundamentals of Fermentation

Fermentation is a biological process of biomass conversion. Two types of fermentation process available: (1) fermentation in presence of air known as aerobic fermentation and (2) without the presence of air known as anaerobic fermentation. Conversion of biomass to biofuel includes a series of steps: pre-treatment, hydrolysis (acid or enzymatic) and followed by fermentation. The pre-treatment stage targets to damage the cell wall of the biomass material and also expose cellulose and hemicellulose for hydrolysis. It can be grouped into four main classes; (1) physical or mechanical treatment, i.e., crushing; (2) physicochemical, i.e., wet-air oxidation, (3) chemical, i.e., oxidizing/reducing agents or organic solvents; and (4) biological treatment using enzymes or mixture of them. Secondly, pre-treated biomass is disintegrated into its monomers by acid/enzymatic hydrolysis (Garcia-Garcia et al. 2022). Finally, the transitional monomers are transformed into alcohols using yeast or bacteria. The last step can be in the presence or absence of air depending on the type of bacteria used. Fermentation is typically utilized for the preservation of sour foods such as pickled cucumbers, kombucha, and yogurt because it produces lactic acid, and it is also used to produce alcoholic beverages like beer and wine. However, ethanol can be produced via fermentation and it can be used as a fuel to operate generators for electricity.

19.8.2 Fermentation of Haematococcus

Haematococcus pluvialis have to go through green and red stage in their life cycle. When a stage changes, the chemical composition of biomass also changes. *H. pluvialis* in the green stage have gelatinous extracellular matrix at their Macrozooids during the green stage. On the other hand, carbohydrate content in their thick sporopollenin cell wall at the red stage becomes more than half to almost three quarter (Haque et al. 2017). At the red stage, carbohydrate content can reach up to more than twice that of green stage (Grewe and Griehl 2012). It was found that monosaccharide conversion of microalgal carbohydrates was higher compared to lignocellulosic feedstock. Moreover, microalgae biomass does not require pretreatments to hydrolyze cellulosic matter (Kafarov and González Delgado 2011). These characteristics made *H. pluvialis* a suitable candidate for bioethanol production.

Residual algal biomass has higher moisture content (Haque et al. 2017). Therefore, converting this biomass into energy requires drying and commercial fast drying is an energy intensive process. But fermentation does not require drying. Additionally, biomass having more carbohydrate than lipid shows better performance. Residual *Haematococcus* with high carbohydrate fits the criteria. Haque et al. (2017) studied an integrated approach of production of *H. pluvialis* using wastewater and utilization of residual biomass utilization after it was used to obtain high value products (Haque et al. 2017). In their proposed system, *H. pluvialis* was cultivated using wastewater produced from a bioethanol plant which uses fermentation technology. After the extraction of astaxanthin, the residual algal biomass will be fed to the saccharification unit along with other feedstocks. As there is an established bioethanol plant already available, it will reduce the upfront cost around 24%, such as cost for infrastructure, expenditure on land procurement, and wastewater supply station (wastewater for culture), along with operation and maintenance costs such as procurement and sterilization of media and cost of carbon dioxide. Because of using wastewater cost will be saved on KNO_3, Na_2HPO_4, $NaHCO_3$, CO_2, water, and wastewater treatment. The approximate cost save would be around 22%. The ethanol plant employed ozonation to treat their wastewater, so no additional cost is expected by the integration. The integrated system will reduce the cost of astaxanthin production significantly.

The lower market price of bioethanol is not sufficient enough to recover the processing cost with current available technology. Additionally, less than 50% of sugar was released by raw *H. pluvialis* compared with residual biomass. This suggests cell disruption is suggesting the requirement of cell disruption necessary before hydrolysis (Hosseini et al. 2020). This is an additional energy extensive process. The residual biomass obtained from astaxanthin extraction already goes through cell disruption. Therefore, residual *H. pluvialis* seems more economical for bioethanol production compared to raw whole biomass. All the results given on biorefinery are here based on laboratory experiments. Laboratory experiments conducted on bench scale may show technical feasibility, but commercial potential may need more. A more practical solution can be investigating biorefinery which involves processing of residual *Haematococcus* biomass into biofuels through pilot scale studies to better understand its techno-economic feasibility. Scaling up the pilot scale data to commercial application is more convenient to understand the potential outcome.

19.9 Biogas

19.9.1 Primary Concept on Biogas

One of the most used biochemical conversion routes for the generation of biofuels from various biomasses is AD. This process is efficient in converting a varied class of biomass materials into biogas which is a combination of CH_4 and CO_2. AD involves four stages: hydrolysis, acidogenesis, acetogenesis, and methanogenesis (Kaparaju and Felby 2010). As the four stages can take place in a single reactor, however, two-stage AD reactors can be used to produce both H_2 and CH_4. Pre-treatment of biomass takes place before the process and upgrading of biogas takes place after the process in the commercial plants. AD processes can be both wet and dry, depending on the dry matter content of the feed. During the digestion process, the biomass carbohydrates, proteins, and lipids of the feedstock are converted into biogas via a variety of intermediate complex pathways carried by a mixture of bacteria. Hydrolysis aids the transformation of large macromolecules into its monomers or smaller components that is used by the acidogenic bacteria, with the help of the enzymes secreted by the hydrolytic bacteria. In the second stage the acidogenic bacteria absorbs the hydrolysis products and produce volatile fatty acids like acetates besides larger organic acids like butyrate and propionate are also produced. During the acetogenesis stage, the larger acids are transformed to acetate, while lipids are also changed via their own acetogenic process. H_2 is generated in third stage also. In the final stage, i.e., methanogenesis, CH_4 is generated by the methanogenic species via acetolactic route or the hydrogenotrophic route (Moya et al. 2017; Kaparaju et al. 2009). The biogas produced can be directly utilized as a fuel, in combined heat and power gas engines or it can be upgraded to natural gas-quality biomethane.

19.9.2 Biogas from Haematococcus

Biogas production through anaerobic digestion is an effective method to convert wet biomass to energy without drying. This process utilizes the whole biomass without the necessity of extracting any components. Hosseini et al. (2020) produced biogas from raw *H. pluvialis*. Its biochemical composition was 40.7% carbohydrates, 26.1% proteins, and 23.2% lipids (Hosseini et al. 2020). Their experiment yielded a biomethane of 91.9 mL/g volatile solid which was lower than other algal strains popular for biogas such as *Chlorella reinhardtii* and *Scenedesmus obliquus*. The lower yield happened because the cell wall of *Haematococcus* is comparatively thicker which slowed down the breakdown of the cell during conversion process. One potential solution can be cell disruption which will pretreat the biomass and make it easier for anaerobic digestion. Other methods also can be used. Aydin (2016) used bioaugmentation with *Clostridium thermocellum* when they processed *H. pluvialis* in an anaerobic digester (Aydin 2016). They found higher methane yield because this process made the cells easier for biodegradation. In their experiment, 5, 10, and 15% bioaugmentation increased yield of volatile solids 26%, 44%, and 56% (Nishshanka et al. 2022). The growth of *Clostridium thermocellum* enhanced the degradation of thick wall of *H. pluvialis*. The residual biomass of *Haematococcus* is easier to process for biogas. For example, if astaxanthin is extracted, this extraction process causes cell disruption and damages cell wall.

Therefore, further cell breakdown becomes easier (Nishshanka et al. 2022). Biomethane yield of residual biomass was 3 times than whole *H. pluvialis*.

19.10 Esterification

19.10.1 Basics of Esterification

To enhance the production of biofuel from biomass, a typical chemical route has been used in recent times known as esterification. Esterification is the process of merging an organic acid with an alcohol to form an ester and water. Industrial applications of esterification are seen in the field of pharmaceuticals, food and beverage, and renewable biodiesel production. Long chain free fatty acids (FFAs) undergo esterification with the presence of homogenous acid catalyst. Using this approach, the high free fatty acid content of fats, oils, and greases obtained from a wastewater feedstock is transformed into biodiesel. Long before, transesterification was carried out using homogeneous base catalysts such as KOH and NaOH. But now these catalysts are considered out-of-date to treat these kinds of feedstocks (Ahmed and Huddersman 2022). Biodiesel can be produced by three technologies as follows:

1. "Alkaline catalyzed transesterification (suitable for feedstock with low FFA content)."
2. "Acid catalyzed transesterification/esterification (good for feedstock with high FFA content)."
3. "Transesterification double step process (good for feedstock with high FFA content)."

Biodiesel produced via esterification can be blended with existing petroleum diesel and used in engines for transportation or heat and power generators.

19.10.2 Esterification of Haematococcus

The intermediate chain length fatty acids (C16–18) were found comparatively higher in *H. pluvialis* which is similar to conventional energy crops used as feedstocks for biodiesel (Nishshanka et al. 2022). This biochemical structure suggests that *H. pluviaris* is also a potential candidate for biodiesel which biodiesel can be used in existing engines potentially. If the cultivation condition can be manipulated to produce *H. pluvialis* with high lipid content, it will serve as feedstock for biodiesel production. Similar to other forms of fuels, algal biodiesel is still not commercially feasible because of its high processing cost. A feasible energy option would be to integrate the production of astaxanthin along with biodiesel which will outweigh the associated costs.

It was already found that methyl esters from jatropha (Prueksakorn et al. 2010), palm oil (Pleanjai and Gheewala 2009), rapeseed oil (Yee et al. 2009), and used cooking oil (Yee et al. 2009) produced using existing technology showed favorable energy balances. The results do not preclude the possibility that either of these systems would be desirable from a greenhouse gas-reduction viewpoint. A financially viable process may be possible if the energy products can be viewed as mere by-products as with the case of H. pluvialis, or if biodiesel production takes place within a multifunctional biorefinery system, wherein these systems may be coupled with carbon capture and storage to potentially achieve negative life cycle CO_2 emissions (Fatih Demirbas 2009).

19.11 Algal Biofuel for Food Security

19.11.1 Impact of Traditional Biofuel on Food Security

Food security has become a rising concern in many parts of the world. Food security of a country or region can be closely connected to other countries or regions. Food security means year around access to sufficient and nutritious food (Barrett 2010). Food security depends on several factors. Local production, availability, and price of agro-chemicals and fuel price, natural disasters and political stability are the major ones. For climate change and energy security, worldwide many countries are leaning toward biofuel. Current trend of commercial biofuels is mainly coming from first generation biofuels, which are produced from energy crops. It is to be noted that, fertile cultivable agricultural lands are limited. Many of the energy crops are also consumed by human. Therefore, the production of energy crops is responsible for land use change for food production. This has a direct impact on food accessibility because of less food production and higher price. Increased demand of biofuel raises the concern of food security as biofuel competes for land, water, labor, and other resources (Renzaho et al. 2017; Schut and Florin 2015). Specially, in 2007–2008 there was a hike of global food price which was thought instigated by rising demand of bioenergy feedstocks (Zilberman et al. 2013; Hochman et al. 2018).

Increased food price reduces food access, especially in the food importing countries and low-income urban households (Hao et al. 2017; Tomei and Helliwell 2016). It was reported that in Africa farmers leaned to produce bioenergy feedstocks instead of food crops which reduced the availability and access to food (Schut and Florin 2015; Thondhlana 2015). Furthermore, the households dependent on crop production for livelihood, extreme weather events create an instability of food market and make the families vulnerable (Reardon et al. 1988). Reports also suggested the cases of land-grabbing, where international companies bought a large quantity of land to cultivate bioenergy feedstocks which threatened food security by displacing farmers and their crop cultivation (Thondhlana 2015; Rulli et al. 2016). As with land, in water scarce areas, competition for irrigation water can reduce food security. Furthermore, the competition for water can potentially reduce the possibilities for cleaning and cooking food, which negatively impacts food security. The effects depend on the local conditions, as well as on how crops for bioenergy are produced and how the land is managed.

One case study from China will make it easy to understand the concern of food security due to biofuel. Chinese government started biofuel program in 2002, motivated by energy security (Weng et al. 2019). They expected that it would not be a danger for food security. The government banned normal corn to be used for bioethanol production, only inferior corns were allowed to be used from reserved stocks after a period of 2 or 3 years. But inferior corn production was also decreasing. Therefore, government changed policy and allowed normal corn. All the bioethanol facilities in Heilongjiang and Jilin, which were the heart of bioethanol production in China, used normal corn as a feedstock to produce bioethanol, because there was shortage of inferior corn supply. Eventually, normal corn was used more to produce bioethanol after that. There was a clear competition for supply of corn for bioethanol and supply of corn for feed, food, and processed food.

The Chinese government tried to diversify bioethanol production by using cassava and sweet potato, sweet sorghum, and other crops to make corn secure as food. However, this approach did not ensure food security as well. Sweet potato for fuel competed with food use and agricultural resources. Sweet sorghum was planted in ecologically fragile marginal lands. This plantation had negative environmental impacts associated with the expansion of non-grain biofuel feedstock on this land will be inevitable. Sweet sorghum also competed with feed use and agricultural resources. Because these places were water shortage areas. Growing sweet sorghum not only reduces water availability for agricultural use, but also for other uses in the region.

The need to understand the linkage between food security and bioenergy is emphasized by the 2nd Sustainable Development Goals "Zero hunger" and 7th "Clean and affordable energy" (United Nations 2015). At one side there is increasing demand of biofuel in the developing countries and on another side the concern of food security is going to make a conflict of interest for overall sustainability of many countries. We cannot deny the necessity of any of them as both of them hold great importance. We must go forward with both of them and to do so, we need to find technological solutions through scientific innovations.

19.11.2 Role of Algal Biofuel to Address the Concern

One unique feature of microalgae is that, irrespective of strains, they can grow anywhere. Microalgae can be cultivated commercially in open ponds or closed photobioreactors, in indoor or over the roof (Sarker and Salam 2019). Microalgae can be grown in any place where minimal nutrients are available in any form. Microalgae also can be grown in wastewater (Sarker 2021). For example, if *Haematococcus* are cultivated commercially with wastewater and bioenergy is produced from there, it will not compete with agricultural land or fresh water or any sort of food, as well as it will sequestrate CO_2 (Bhadra et al. 2020; Sarker and Salam 2020). This CO_2 sequestrating feature showed biogas upgrading potential by microalgae, which will be upgrading a gaseous renewable energy with a renewable feedstock for biofuel. In traditional biogas upgrading system, CO_2 is absorbed by chemicals, similar to upgrading natural gas (Sarker 2016, 2020). The traditional process produces upgraded fuel gas only, but microalgae-based biogas upgrading system is capable of upgrading gas as well as producing algal biomass in the same cultivation unit. Integrating different processes may increase overall cost but reduce per unit cost (Sarker and Sarkar 2018). Biofuel from *Haematococcus* will not be a threat for food security in any form. Bioenergy from *Haematococcus is* capable of addressing many concerns of food security. We cannot deny the importance of biofuel for climate change and future planet. But we must ensure the availability of biofuel without threatening food security. Algal biofuel has the potential of that. That is the reason, it is called third generation biofuel.

19.12 Prospects and Challenges

Countries, around the world including major oil producing countries are changing their policies to increase the share of renewable energy in their domestic consumption. Transport sector is one of the major sectors which contribute to global warming. Therefore, many countries such as UAE and USA have set specific targets and goals for the transport sector. The renewable transport fuel obligation statistics from the UK department of transport, recommends that every transport fuel supplier must make sure proportions of renewable source (biofuel) are present their products (Adeniyi et al. 2018). Transport sector is mostly dominant by liquid fuel. Therefore, liquid bioenergy will be the major tool to meet these targets. Specially, legislations and policies regarding CO_2 emission reduction or eliminations from fossil fuels has accelerated the global biofuel market. Globally positive attitude toward renewable energy and its commercial expansion is also supported by political entities.

Aviation sector is going to push the biofuel market further in coming years. A lot of research activities are going on to find and implement renewable alternatives to fossil fuels in the aviation industries. Microalgae-based additives can play a vital role here (Nair and Paulose 2014). Several studies have shown potentiality of economic and ecological sustainability of bio jet fuel (Adeniyi et al. 2018). Currently, "drop in" replacements of traditional bio jet fuels are being produced in significant quantities. They are algae-based products combined with petroleum-based fuel as blends or emulsion (Adeniyi et al. 2018). In September 2008 Solazyme, a California-based renewable oil producer produced the first ever algal-based jet fuel that passed all tested specification including the most challenging ASTM D1655 specification for aviation turbine fuel (Hudson et al. 2016). This development led to the two successful algae-based biojet powered flights the following year on 7th and 30th January, with the hope of more to come (Nair and Paulose 2014). This could as well benefit the UK aviation sector in the reduction of up to 24% CO_2 emission by 2050, but there must be a step change adjustment in the policies and investment framework for sustainable aviation fuel (Vijayaraghavan and Hemanathan 2009).

One major issue of liquid bioenergy (included algal biofuel) is that current engines are not designed for them. Because of the different physicochemical properties of biofuel, the engines will experience operational challenges (Acharya et al. 2017). Therefore, complete replacement of biofuel will require a major overhaul of transport engine mechanisms which is not realistic. Another solution is producing biofuels with certain characteristics to satisfy the engine requirements. But that will need long term plan. Short- and medium-term solution can be fuel blend. Fuel blend is a mixture of different fuels which offers better overall fuel property to achieve overall engine performance (Neto et al. 2013). Algal biodiesel can be blended with petroleum diesel and bioethanol or in some cases used as an additive which can be used in existing engines (Barabás et al. 2010). One study reported 20% algal biodiesel blend (Adeniyi et al. 2018) provided overall satisfactory result in terms of emission and engine performance.

A positive market force has been created in favor of biofuel. But we have discussed before that if we cannot meet our demand of biofuel with the feedstocks those do not possess a threat to food security, this tool to combat climate change will appear as danger. It will cause land use change, create economic burden over the middle class, lower middle class, and poor people. We do not want a solution for climate change that will create a class division in many regions. Effective solutions are already in our hands. Algal bioenergy is one of them. As a comparatively new research field, a long distance to be covered to commercialize it. Already, algal bioenergy is in use in aviation sector. To expand periphery of algal bioenergy, we need to explore unknown and non-investigated microalgal strains. Additionally, algal species like *Haematococcus* which are used for different purposes should be investigated for bioenergy purposes. The more knowledge we acquire, the more we will march forward by overcoming the obstacles to commercialize algae-based bioenergy.

19.13 Conclusion

We have discussed the technological potential of *Haematococcus* biomass to convert to bioenergy. Although *Haematococcus* have adequate potential to be an energy producing feedstock, but residual *Haematococcus* showed better conversion potential than whole raw biomass. It offers another benefit, *Haematococcus* biomass can be used for astaxanthin production as well as energy. Energy and other chemical production will not contradict, and this will also make sure zero waste approach of *Haematococcus* use. But it needs a lot more research. Research on energy applications on *Haematococcus* just began in recent years and a long way to go. For example, the use of catalysts during pyrolysis, liquefaction, gasification or esterification or innovative solvent for esterification or enhancer for biogas production. As algal bioenergy does not threaten food security and in future we will need diverse feedstock for algal biofuel, more research should be carried on potential feed stocks like *Haematococcus*.

Bibliography

Abalde J, Cid A, Fidalgo Paredes P, Torres E, Herrero C (1995) Microalgas: cultivo y aplicaciones. Universidade, Servizo de Publicacións, A Coruña. https://doi.org/10.17979/spudc.9788497497695. ISBN: 978-84-97497-69-5

Acharya N, Nanda P, Panda S, Acharya S (2017) Analysis of properties and estimation of optimum blending ratio of blended mahua biodiesel. Eng Sci Technol 20:511–517

Adeniyi OM, Azimov U, Burluka A (2018) Algae biofuel: current status and future applications. Renew Sust Energ Rev 90:316–335

Adesanya VO, Davey MP, Scott SA, Smith AG (2014) Kinetic modelling of growth and storage molecule production in microalgae under mixotrophic and autotrophic conditions. Bioresour Technol 157:293–304

Ahmed R, Huddersman K (2022) Review of biodiesel production by the esterification of wastewater containing fats oils and grease (FOGs). J Ind Eng Chem 110:1–14

Amaya E, Nickell D (2015) 11 - Using feed to enhance the color quality of fish and crustaceans. In: Davis DA (ed) Feed and feeding practices in aquaculture. Woodhead Publishing, Oxford, pp 269–298

Ambati RR, Siew Moi P, Ravi S, Aswathanarayana RG (2014) Astaxanthin: sources, extraction, stability, biological activities and its commercial applications—a review. Mar Drugs 12:128–152

Anto S, Sudhakar MP, Shan Ahamed T, Samuel MS, Mathimani T, Brindhadevi K, Pugazhendhi A (2021) Activation strategies for biochar to use as an efficient catalyst in various applications. Fuel 285:119205

Ashokkumar V, Chen W-H, Kumar G, Satjarak A, Chanthapatchot W, Ngamcharussrivichai C (2021) A biorefinery approach for high value-added bioproduct (astaxanthin) from alga Haematococcus sp. and residue pyrolysis for biochar synthesis and metallic iron production from hematite (Fe2O3). Fuel 304:121150

Aydin S (2016) Enhancement of microbial diversity and methane yield by bacterial bioaugmentation through the anaerobic digestion of Haematococcus pluvialis. Appl Microbiol Biotechnol 100:5631–5637

Azizi K, Keshavarz Moraveji M, Abedini Najafabadi H (2018) A review on bio-fuel production from microalgal biomass by using pyrolysis method. Renew Sust Energ Rev 82:3046–3059

Barabás I, Todoruţ A, Băldean D (2010) Performance and emission characteristics of an CI engine fueled with diesel–biodiesel–bioethanol blends. Fuel 89:3827–3832

Barbera E, Sforza E, Kumar S, Morosinotto T, Bertucco A (2016) Cultivation of Scenedesmus obliquus in liquid hydrolysate from flash hydrolysis for nutrient recycling. Bioresour Technol 207:59–66

Barrett CB (2010) Measuring food insecurity. Science 327:825–828

Bhadra S, Salam PA, Sarker NK (2020) Microalgae-based biodiesel production in open raceway ponds using coal thermal flue gas: a case of West Bengal, India. Environ Qual Manag 29:27–36

Chen C-Y, Yeh K-L, Aisyah R, Lee D-J, Chang J-S (2011) Cultivation, photobioreactor design and harvesting of microalgae for biodiesel production: a critical review. Bioresour Technol 102:71–81

Chen C, Ma X, He Y (2012) Co-pyrolysis characteristics of microalgae Chlorella vulgaris and coal through TGA. Bioresour Technol 117:264–273

Chow MC, Jackson WR, Chaffee AL, Marshall M (2013) Thermal treatment of algae for production of biofuel. Energy Fuel 27:1926–1950

Debnath B, Saha I, Mukherjee T, Mitra S, Das A, Das A (2021) Chapter 12 - Sorbents from waste materials: a circular economic approach. In: Núñez-Delgado A (ed) Sorbents materials for controlling environmental pollution. Elsevier, Amsterdam, pp 285–322

Fatih Demirbas M (2009) Biorefineries for biofuel upgrading: a critical review. Appl Energy 86: S151–S161

Gao K (1998) Chinese studies on the edible blue-green alga, Nostoc flagelliforme: a review. J Appl Phycol 10:37–49

Garcia-Garcia G, McCord S, Styring P (2022) Chapter 6 - Environmental assessment of biomass to biofuels: biochemical conversion routes. In: Fokaides PA, Kylili A, Georgali P-Z (eds) Environmental assessment of renewable energy conversion technologies. Elsevier, Amsterdam, pp 129–156

Golueke CG, Oswald WJ, Gotaas HB (1957) Anaerobic digestion of algae. Appl Microbiol 5:47–55

Gong Z, Du A, Wang Z, Fang P, Li X (2017) Experimental study on pyrolysis characteristics of oil sludge with a tube furnace reactor. Energy Fuel 31:8102–8108

Gong Z, Fang P, Wang Z, Li X, Wang Z, Meng F (2020) Pyrolysis characteristics and products distribution of Haematococcus pluvialis microalgae and its extraction residue. Renew Energy 146:2134–2141

Grewe CB, Griehl C (2012) 8: The carotenoid astaxanthin from Haematococcus pluvialis. In: Clemens P, Christian W (eds) Microalgal biotechnology: integration and economy. De Gruyter, Berlin, pp 129–144

Guerin M, Huntley ME, Olaizola M (2003) Haematococcus astaxanthin: applications for human health and nutrition. Trends Biotechnol 21:210–216

Hagemann N, Joseph S, Schmidt H-P, Kammann CI, Harter J, Borch T, Young RB, Varga K, Taherymoosavi S, Elliott KW, McKenna A, Albu M, Mayrhofer C, Obst M, Conte P, Dieguez-Alonso A, Orsetti S, Subdiaga E, Behrens S, Kappler A (2017) Organic coating on biochar explains its nutrient retention and stimulation of soil fertility. Nat Commun 8:1089

Hao N, Pedroni P, Colson G, Wetzstein M (2017) The linkage between the U.S. ethanol market and developing countries' maize prices: a panel SVAR analysis. Agric Econ 48:629–638

Haque F, Dutta A, Thimmanagari M, Chiang YW (2017) Integrated Haematococcus pluvialis biomass production and nutrient removal using bioethanol plant waste effluent. Process Saf Environ Prot 111:128–137

Harder R, von Witsch H (1942) Ueber Massenkultur von Diatomeen. Berichte der Deutschen Botanischen Gesellschaft 60:14–153

Hochman G, Zilberman D, Ethanol C, U.S. (2018) Biofuel policy 10 years later: a quantitative assessment. Am J Agric Econ 100:570–584

Hong W, Chen J, Ding Q, Gao Y, Ye L, Yin Y, Tu S (2021) Efficient thermochemical liquefaction of microalgae Haematococcus pluvialis for production of high quality biocrude with high selectivity over Fe/montmorillonite catalyst. J Energy Inst 97:73–79

Hosseini A, Jazini M, Mahdieh M, Karimi K (2020) Efficient superantioxidant and biofuel production from microalga Haematococcus pluvialis via a biorefinery approach. Bioresour Technol 306:123100

Hudson L, Jefferson A, Bauen A, Nattrass L (2016) Chapter 14 - sustainable aviation: a UK roadmap for sustainable aviation fuels. In: Chuck CJ (ed) Biofuels for aviation. Academic Press, London, pp 315–337

Jahirul MI, Rasul MG, Chowdhury AA, Ashwath N (2012) Biofuels Production through biomass pyrolysis—a technological review. Energies 5:4952–5001. https://doi.org/10.3390/en5124952

Kafarov V, González Delgado ÁD (2011) Microalgae based biorefinery: issues to consider. CT&F 4:5–21

Kaparaju P, Felby C (2010) Characterization of lignin during oxidative and hydrothermal pre-treatment processes of wheat straw and corn Stover. Bioresour Technol 101:3175–3181

Kaparaju P, Serrano M, Thomsen AB, Kongjan P, Angelidaki I (2009) Bioethanol, biohydrogen and biogas production from wheat straw in a biorefinery concept. Bioresour Technol 100:2562–2568

Kumari G, Karmee SK (2022) Chapter 22 - thermochemical routes applying biomass: a critical assessment. In: Sahay S (ed) Handbook of biofuels. Academic Press, London, pp 435–451
Mata TM, Martins AA, Caetano NS (2010) Microalgae for biodiesel production and other applications: a review. Renew Sust Energ Rev 14:217–232
Mitra S, Mukherjee T, Kaparaju P (2021) Prediction of methyl orange removal by iron decorated activated carbon using an artificial neural network. Environ Technol 42:3288–3303
Moya D, Aldás C, López G, Kaparaju P (2017) Municipal solid waste as a valuable renewable energy resource: a worldwide opportunity of energy recovery by using waste-to-energy technologies. Energy Procedia 134:286–295
Mukherjee T, Rahaman M, Ghosh A, Bose S (2019) Optimization of adsorbent derived from non-biodegradable waste employing response surface methodology toward the removal of dye solutions. Int J Environ Sci Technol 16:8671–8678
Nair S, Paulose H (2014) Emergence of green business models: the case of algae biofuel for aviation. Energy Policy 65:175–184
Neto AAD, Fernandes MR, Neto ELB, Dantas TNC, Moura MCPA (2013) Effect of biodiesel/diesel-based microemulsions on the exhaust emissions of a diesel engine. Braz J Petrol Gas 7
Neubauer Y (2013) 6 - Biomass gasification. In: Rosendahl L (ed) Biomass combustion science, technology and engineering. Woodhead Publishing, Sawston, pp 106–129
Nishshanka GKSH, Liyanaarachchi VC, Nimarshana PHV, Ariyadasa TU, Chang J-S (2022) Haematococcus pluvialis: a potential feedstock for multiple-product biorefining. J Clean Prod 344:131103
Osman AI, Mehta N, Elgarahy AM, Al-Hinai A, Al-Muhtaseb AH, Rooney DW (2021) Conversion of biomass to biofuels and life cycle assessment: a review. Environ Chem Lett 19:4075–4118
Peng X, Ma X, Lin Y, Guo Z, Hu S, Ning X, Cao Y, Zhang Y (2015) Co-pyrolysis between microalgae and textile dyeing sludge by TG–FTIR: kinetics and products. Energy Convers Manag 100:391–402
Phukan MM, Chutia RS, Konwar BK, Kataki R (2011) Microalgae chlorella as a potential bio-energy feedstock. Appl Energy 88:3307–3312
Pleanjai S, Gheewala SH (2009) Full chain energy analysis of biodiesel production from palm oil in Thailand. Appl Energy 86:S209–S214
Prueksakorn K, Gheewala SH, Malakul P, Bonnet S (2010) Energy analysis of Jatropha plantation systems for biodiesel production in Thailand. Energy Sustain Dev 14:1–5
Randrianarison G, Ashraf MA (2017) Microalgae: a potential plant for energy production. Geol Ecol Landsc 1:104–120
Reardon T, Matlon P, Delgado C (1988) Coping with household-level food insecurity in drought-affected areas of Burkina Faso. World Dev 16:1065–1074
Renzaho AMN, Kamara JK, Toole M (2017) Biofuel production and its impact on food security in low and middle income countries: implications for the post-2015 sustainable development goals. Renew Sust Energ Rev 78:503–516
Rulli MC, Bellomi D, Cazzoli A, De Carolis G, D'Odorico P (2016) The water-land-food nexus of first-generation biofuels. Sci Rep 6:22521
Sarker NK (2016) Theoretical effect of concentration, circulation rate, stages, pressure and temperature of single amine and amine mixture solvents on gas sweetening performance. Egypt J Pet 25:343–354
Sarker NK (2020) Effect of concentration, circulation rate, stages, pressure and temperature of pure glycols on natural gas dehydration performance and sales gas characteristics. Pet Coal 62:477–502
Sarker NK (2021) Exploring the potential of wastewater reclamation by means of outdoor cultivation of microalgae in photobioreactors, energy. Ecol Environ 7:473
Sarker NK, Salam PA (2019) Indoor and outdoor cultivation of Chlorella vulgaris and its application in wastewater treatment in a tropical city—Bangkok, Thailand. SN Appl Sci 1:1645
Sarker NK, Salam PA (2020) Design of batch algal cultivation systems and ranking of the design parameters, energy. Ecol Environ 5:196–210

Sarker NK, Sarkar S (2018) A comparative study on cost analysis, efficiency, and process mechanism of effluent treatment plants in Bangladesh. Environ Qual Manag 27:127–133

Schut M, Florin MJ (2015) The policy and practice of sustainable biofuels: between global frameworks and local heterogeneity. The case of food security in Mozambique. Biomass Bioenergy 72:123–135

Sheehan J, Dunahay T, Benemann J, Roessler P (1998) Look back at the US department of energy's aquatic species program: biodiesel from algae; close-out report. National Renewable Energy Lab, Golden

Smith CR, Buzan EM, Lee JW (2013) Potential impact of biochar water-extractable substances on environmental sustainability. ACS Sustain Chem Eng 1:118–126

Thomassen G, Van Dael M, You F, Van Passel S (2019) A multi-objective optimization-extended techno-economic assessment: exploring the optimal microalgal-based value chain. Green Chem 21:5945–5959

Thondhlana G (2015) Land acquisition for and local livelihood implications of biofuel development in Zimbabwe. Land Use Policy 49:11–19

Tomei J, Helliwell R (2016) Food versus fuel? Going beyond biofuels. Land Use Policy 56:320–326

United Nations (2015) Sustainable development goals, United Nations, S-1018, New York

Vasudev V, Ku X, Lin J (2021) Combustion behavior of algal biochars obtained at different pyrolysis heating rates. ACS Omega 6:19144–19152

Vijayaraghavan K, Hemanathan K (2009) Biodiesel production from freshwater algae. Energy Fuel 23:5448–5453

Weng Y, Chang S, Cai W, Wang C (2019) Exploring the impacts of biofuel expansion on land use change and food security based on a land explicit CGE model: a case study of China. Appl Energy 236:514–525

Woolf D, Amonette JE, Street-Perrott FA, Lehmann J, Joseph S (2010) Sustainable biochar to mitigate global climate change. Nat Commun. https://doi.org/10.1038/ncomms1053

Yee KF, Tan KT, Abdullah AZ, Lee KT (2009) Life cycle assessment of palm biodiesel: revealing facts and benefits for sustainability. Appl Energy 86:S189–S196

Yu KL, Show PL, Ong HC, Ling TC, Chi-Wei Lan J, Chen W-H, Chang J-S (2017) Microalgae from wastewater treatment to biochar – feedstock preparation and conversion technologies. Energy Convers Manag 150:1–13

Zhu Y, Albrecht KO, Elliott DC, Hallen RT, Jones SB (2013) Development of hydrothermal liquefaction and upgrading technologies for lipid-extracted algae conversion to liquid fuels. Algal Res 2:455–464

Zilberman D, Hochman G, Rajagopal D, Sexton S, Timilsina G (2013) The impact of biofuels on commodity food prices: assessment of findings. Am J Agric Econ 95:275–281

The manufacturer's authorised representative in the EU is Springer Nature Customer Service Centre GmbH, Europaplatz 3, 69115 Heidelberg, Germany. If you have any concerns regarding our products, please contact ProductSafety@springernature.com

Printed and bound by CPI Group (UK) Ltd, Croydon, CR0 4YY
01/10/2026
02225866-0013